Barron's Regents Exams and Answers

Algebra 2/ Trigonometry

Meg Clemens
Mathematics Department Chair and Instructor
Hugh C. Williams High School
Canton, New York

Glenn Clemens
Mathematics Instructor
Norwood-Norfolk High School
Norwood, New York

Barron's Educational Series, Inc.

All inquiries should be addressed to:
Barron's Educational Series, Inc.
250 Wireless Boulevard
Hauppauge, NY 11788
www.barronseduc.com

ISBN: 978-0-7641-4512-4
Library of Congress Control Number: 2010921952

PRINTED IN THE UNITED STATES OF AMERICA
9 8 7 6 5 4 3 2 1

Contents

Preface

This book is designed to prepare you for the Algebra 2/ Trigonometry Regents examination while strengthening your understanding and mastery of the material on which this test is based. In addition to providing questions from previous mathematics Regents examinations, this book offers these special features:

- ***Step-by-Step Solutions to All Regents Questions.*** Careful study of the solutions and answer explanations will improve your mastery of the subject. Each explanation is designed to show you how to apply the facts and concepts you have learned in class. Since the explanation for each solution has been written with emphasis on the reasoning behind each step, its value goes well beyond the application to that particular question.

- ***Unique System of Self-Analysis Charts.*** Each set of solutions for a particular Algebra 2/Trigonometry Regents examination ends with a Self-Analysis Chart and a classification of exam questions by topic. These charts will help you identify weaknesses and direct your study efforts where needed. In addition, the charts classify the questions on each exam into an organized set of topic groups. This feature will also help you to locate other questions on the same topic in other Algebra 2/Trigonometry exams.

- ***General Test-Taking Tips.*** Tips are given that will help to raise your grade on the actual Algebra 2/Trigonometry Regents exam.

- ***Graphing Calculator Skills.*** The main features of the Texas Instruments TI-83/TI-84 graphing calculator are reviewed.

- ***Algebra 2/Trigonometry Refresher.*** A brief review of key mathematics facts and skills that are tested on the Algebra 2/ Trigonometry Regents exam is included for easy reference and quick study.

- ***Glossary.*** Definitions of important terms related to the Algebra 2/ Trigonometry Regents examination are conveniently organized in a glossary.

Frequency of Topics—Algebra 2/Trigonometry

Questions on the Algebra 2/Trigonometry Regents exams fall into one of 17 topic categories. The "Key Algebra 2/Trigonometry Facts and Skills" that appears later in this book covers these concepts. The **Frequency Chart** that follows shows the number of questions that are asked in each topic category on each exam.

To help guide you further in your exam preparation, a **Self-Analysis Chart** is included after each exam. This chart identifies exactly which questions on an exam fall within each topic category. When you score a completed exam, this chart will help you identify any potential areas of weakness and then will easily allow you to find additional practice questions on those same topics on the other exams or in the "Key Algebra 2/Trigonometry Facts and Skills" section.

Frequency of Topics—Algebra 2/Trigonometry

	Number of Questions				
	Sample Test 1	**Sample Test 2**	**Official Test Sampler**	**June 2010**	**August 2010**
1. Exponents and Radicals: operations, equivalent expressions, simplifying, rationalizing	3	2	3	5	2
2. Complex Numbers: operations, powers of i, rationalizing	1	2	1	2	3
3. Quadratics and Higher-Order Polynomials: operations, factoring, binomial expansion, formula, quadratic equations and inequalities, nature of the roots, sum and product of the roots, quadratic-linear systems, completing the square, higher-order polynomials	5	3	5	6	6
4. Rationals: operations, equations	2	2	2	1	2
5. Absolute Value: equations and inequalities	1	1	1	0	0
6. Direct and Inverse Variation	1	1	0	0	0
7. Circles: center-radius equation, completing the square	1	1	1	1	1
8. Functions: relations, functions, domain, range, one-to-one, onto, inverses, compositions, transformations of functions	4	5	7	1	4
9. Exponential and Logarithmic Functions: equations, common bases, logarithm rules, e, ln	3	3	3	4	2
10. Trigonometric Functions: radian measure, arc length, cofunctions, unit circle, inverses	3	4	3	3	4
11. Trigonometric Graphs: graphs of basic functions and inverse functions, domain, range, restricted domain, transformations of graphs, amplitude, frequency, period, phase shift	3	1	1	4	1
12. Trigonometric Identities, Formulas, and Equations	2	2	2	2	2

	Number of Questions				
	Sample Test 1	Sample Test 2	Official Test Sampler	June 2010	August 2010
13. Trigonometry Laws and Applications: area of a triangle, sine and cosine laws, ambiguous cases	2	2	2	2	3
14. Sequences and Series: sigma notation, arithmetic, geometric	3	3	2	3	3
15. Statistics: studies, central tendency, dispersion including standard deviation, normal distribution	2	3	3	1	2
16. Regressions: linear, exponential, logarithmic, power, correlation coefficient	1	2	1	1	1
17. Probability: permutation, combination, geometric, binomial	2	2	2	3	3

How to Use This Book

This section explains how you can make the best use of your study time. As you work your way through this book, you will be following a carefully designed five-step study plan that will improve your understanding of the topics that the Algebra 2/Trigonometry Regents examination tests while raising your exam grade.

Step 1. Know What to Expect on Test Day. Before the day of the test, you should be thoroughly familiar with the format, the scoring, and the directions for the Algebra 2/Trigonometry Regents exam. This knowledge will help you build confidence and prevent errors that may arise from misunderstanding the directions. The next section in this book, "Getting Acquainted with the Algebra 2/Trigonometry Regents Exam," provides this important information.

Step 2. Become Testwise. The section titled "Ten Test-Taking Tips" will alert you to easy things you can do to become better prepared and to be more confident when you take the actual test.

Step 3. Know How to Use Your Graphing Calculator. Become proficient in using your graphing calculator. Some questions on the Algebra 2/Trigonometry Regents exam can be answered only by using a graphing calculator, while others become easier to answer when solved with a graphing calculator. Many can be checked with the use of a graphing calculator.

Step 4. Review Algebra 2/Trigonometry Topics. Since the Algebra 2/Trigonometry Regents exam will test you on topics that you may not have studied recently, the section entitled "Key Algebra 2/Trigonometry Facts and Skills" provides a quick refresher of the major topics tested on the examination. This section also includes illustrative practice exercises with worked-out solutions.

Step 5. Take Practice Exams Under Exam Conditions. This book contains sample Algebra 2/Trigonometry Regents examinations and two actual Regents exams with carefully worked-out solutions for all the questions. When you reach this part of the book, you should do these things:

- After you complete an exam, check the answer key for the entire test. Circle any omitted questions or questions you answered incorrectly. Study the explained solutions for these questions.
- On the Self-Analysis Chart, find the topic under which each question is classified, and enter the number of points you earned if you answered that question correctly.
- Figure out your percentage for each topic by dividing your earned points by the number of test points on that topic, carrying the division to two decimal places. If you are not satisfied with your percentage on any topic, reread the solutions for the questions you missed. Then locate related questions in other Regents examinations by using the Self-Analysis Charts to see which questions are listed for the troublesome topic. Attempting to solve these questions and then studying their solutions will provide you with additional help. You may also find it helpful to review the appropriate sections in "Key Algebra 2/Trigonometry Facts and Skills." If you need more detailed explanations and additional practice problems with answers, you may want to get Barron's companion book, *Let's Review: Algebra 2/Trigonometry*.

Tips for Practicing Effectively and Efficiently

- When taking a practice test, do not spend too much time on any one question. If you cannot come up with a method to use or if you cannot complete the solution, circle the question number. When you have completed as many questions as you can, return to the unanswered questions and try them again.
- After finishing a practice test, compare each of your solutions with the solutions that are given. Read the explanations provided even if you have answered the question correctly. Each solution has been carefully designed to provide additional insight into the topic that may be valuable when answering a different question on the same topic.
- In the weeks before the actual test, devote at least one-half hour each day to preparation. It is better to spread out your time in this way than to cram by preparing for, say, four hours during the day before the exam. As the test day gets closer, take at least one complete exam under actual test conditions.

Getting Acquainted with the Algebra 2/ Trigonometry Regents Exam

This section explains things about the Algebra 2/Trigonometry Regents examination that you may not know, such as how the exam is organized, how your exam will be scored, and where you can find a complete listing of the topics tested by this exam.

WHEN DO I TAKE THE ALGEBRA 2/TRIGONOMETRY REGENTS EXAM?

The Algebra 2/Trigonometry Regents exam is administered in January, June, and August of every school year. Most students will take this exam after completing a one-year high school level Algebra 2/Trigonometry course.

HOW IS THE ALGEBRA 2/TRIGONOMETRY REGENTS EXAM SET UP?

The Algebra 2/Trigonometry Regents exam is divided into four parts with a total of 39 questions. All of the questions in each of the four parts must be answered. You will be allowed a maximum of 3 hours in which to complete the test.

Part I consists of 27 standard multiple-choice questions with four possible answers labeled (1), (2), (3), and (4).

Parts II, III, and IV contain eight, three, and one question(s), respectively. The answers and the accompanying work for the questions in these three parts must be written directly in the question booklet. You must show or explain how you arrived at your answer by indicating the necessary steps involved, including appropriate formula substitutions, diagrams, graphs, and charts. If you use a guess-and-check strategy to arrive at a numerical answer for a problem, you must label and organize your checks and show the work for at least three guesses.

Since scrap paper is not permitted for any part of the exam, you may use the blank spaces in the question booklet as scrap paper. An extra sheet of graph paper is at the back of the question booklet. It will not be scored but may be used as scrap paper for Part I. If you need to draw a graph, graph paper will be provided in the question booklet. All work should be done in pen, except graphs and diagrams, which may be drawn in pencil.

The accompanying table summarizes the breakdown of the Algebra 2/Trigonometry Regents exam.

Question Type	Number of Questions	Credit Value
Part I: Multiple choice	27	$27 \times 2 = 54$
Part II: 2-credit open ended	8	$8 \times 2 = 16$
Part III: 4-credit open ended	3	$8 \times 4 = 12$
Part IV: 6-credit open ended	1	$1 \times 6 = 6$
	Test = 39 questions	Test = 88 points

WHAT TYPE OF CALCULATOR DO I NEED?

Graphing calculators are *required* for the Algebra 2/Trigonometry Regents examination. During the administration of the Regents exam, schools are required to make a graphing calculator available for the exclusive use of each student. You will need to use your calculator to work with trigonometric functions of angles, evaluate roots and logarithms, and perform routine calculations. Knowing how to use a graphing calculator gives you an advantage when deciding how to solve a problem. Rather than solving a problem algebraically with pen and paper, it may be easier to solve the same problem using a graph or table created by a graphing calculator. A graphical or numerical solution found using a calculator can also be used to help confirm an answer obtained using standard algebraic methods. You can find additional information about how to use a graphing calculator in the "Graphing Calculator Skills" section in this book.

WHAT GETS COLLECTED AT THE END OF THE EXAMINATION?

At the end of examination, you must return:

- Any tool provided to you by your school, such as a graphing calculator.
- The question booklet. Check that you have printed your name and the name of your school in the appropriate boxes near the top of the first page of the question booklet.
- The Part I answer sheet as well as any other special answer form that is provided by your school. You must sign the statement at the bottom of the Part I answer sheet indicating that you have not received any unlawful assistance in answering any of the questions. Your answer paper will not be accepted until you sign this declaration.

HOW IS THE EXAM SCORED?

Your answers to the 27 multiple-choice questions in Part I are scored as either correct or incorrect. Each correct answer receives 2 points. The eight questions in Part II are worth 2 points each, the three questions in Part II are worth 4 points each, and the question in Part IV is worth 6 points. Solutions to the questions in Parts II, III, and IV that are not completely correct receive partial credit according to a special scoring guide provided by the New York State Education Department.

HOW IS YOUR FINAL SCORE DETERMINED?

The maximum total raw score for the Algebra 2/Trigonometry Regents exam is 88 points. A conversion table provided by the New York State Education Department is used to convert your raw score to a final test score that falls within the usual 0 to 100 scale.

ARE ANY FORMULAS PROVIDED?

The Algebra 2/Trigonometry Regents Examination test booklet will include a reference sheet containing the formulas shown below. This formula sheet, however, does not necessarily include all of the formulas that you may need to know.

Area of a Triangle

$$K = \frac{1}{2}ab \sin C$$

Law of Cosines

$$a^2 = b^2 + c^2 - 2bc \cos A$$

Functions of the Sum of Two Angles

$$\sin (A + B) = \sin A \cos B + \cos A \sin B$$

$$\cos (A + B) = \cos A \cos B - \sin A \sin B$$

$$\tan (A + B) = \frac{\tan A + \tan B}{1 - \tan A \tan B}$$

Functions of the Double Angle

$$\sin 2A = 2 \sin A \cos A$$

$$\cos 2A = \cos^2 A - \sin^2 A$$

$$\cos 2A = 2 \cos^2 A - 1$$

$$\cos 2A = 1 - 2 \sin^2 A$$

$$\tan 2A = \frac{2 \tan A}{1 - \tan^2 A}$$

Functions of the Difference of Two Angles

$$\sin(A - B) = \sin A \cos B - \cos A \sin B$$

$$\cos(A - B) = \cos A \cos B + \sin A \sin B$$

$$\tan(A - B) = \frac{\tan A - \tan B}{1 + \tan A \tan B}$$

Functions of the Half Angle

$$\sin \frac{1}{2} A = \pm\sqrt{\frac{1 - \cos A}{2}}$$

$$\cos \frac{1}{2} A = \pm\sqrt{\frac{1 + \cos A}{2}}$$

$$\tan \frac{1}{2} A = \pm\sqrt{\frac{1 - \cos A}{1 + \cos A}}$$

Law of Sines

$$\frac{a}{\sin A} = \frac{b}{\sin B} = \frac{c}{\sin C}$$

Sum of a Finite Arithmetic Sequence

$$S_n = \frac{n(a_1 + a_n)}{2}$$

Sum of a Finite Geometric Sequence

$$S_n = \frac{a_1(1 - r^n)}{1 - r}$$

Binomial Theorem

$$(a + b)^n = {}_nC_0 a^n b^0 + {}_nC_1 a^{n-1} b^1 + {}_nC_2 a^{n-2} b^2 + \cdots + {}_nC_n a^0 b^n$$

$$(a + b)^n = \sum_{r=0}^{n} {}_nC_r a^{n-r} b^r$$

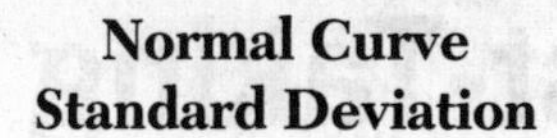

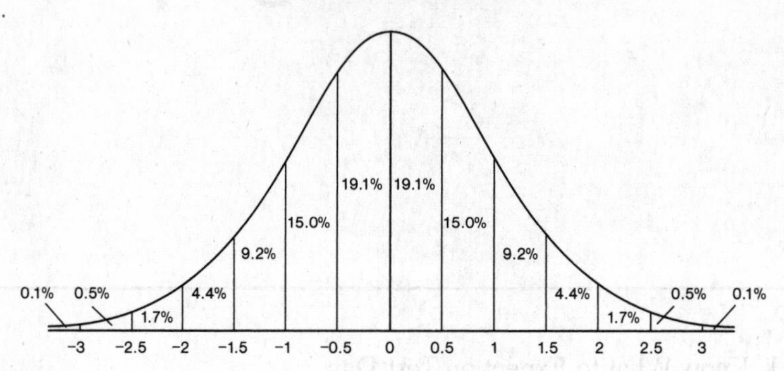

WHAT IS THE CORE CURRICULUM?

The *Core Curriculum* is the official publication of the New York State Education Department that describes the topics and skills required by the Algebra 2/Trigonometry Regents examination. This examination can test you on a wide range of topics, which include:

- Operations with real and complex numbers; algebraic operations with fractions and radicals; factoring.
- Solving quadratic equations, including those with irrational and complex roots; solving systems of equations.
- Linear, quadratic, logarithmic, exponential, and trigonometric functions and their graphs.
- Transformations and functions.
- Statistics, including normal curve; fitting a line or curve to data using least squares regression; scatter plots; correlation coefficient.
- Probability, including counting methods and probability in two-outcome experiments.
- Trigonometric equations and laws.
- Series and sequences.

If you have Internet access, you can view the *Core Curriculum* on the New York State Education Department's web site at *http://www.emsc.nysed.gov/3-8/MathCore.pdf*

Ten Test-Taking Tips

1. Know What to Expect on Test Day
2. Avoid Last-Minute Studying
3. Be Well Rested and Come Prepared on Test Day
4. Know How to Use Your Calculator
5. Know When to Use Your Calculator
6. Have a Plan for Budgeting Your Time
7. Make Your Answers Easy to Read
8. Answer the Question That Is Asked
9. Take Advantage of Multiple-Choice Questions
10. Don't Omit Any Questions

These ten practical tips can help you raise your grade on the Algebra 2/Trigonometry Regents examination.

TIP 1

Know What to Expect on Test Day

SUGGESTIONS

- Become familiar with the format and directions for the Algebra 2/ Trigonometry Regents exam.

- Know where you should write your answers for the different parts of the exam.
- Ask your teacher to show you an actual test booklet for a previously given mathematics Regents examination.

TIP 2

Avoid Last-Minute Studying

SUGGESTIONS

- Start your Regents exam preparation early by making a regular practice of:
  1. Taking detailed notes in class and then reviewing them when you get home;
  2. Completing all written homework assignments in a neat and organized way;
  3. Writing down any questions you may have about your homework so that you can ask your teacher about them;
  4. Saving your classroom tests so you can use them as an additional source of practice questions.
- Get a review book early in your preparation so that additional practice examples and explanations, if needed, will be at your fingertips. The recommended review book is Barron's *Let's Review: Algebra 2/Trigonometry*. This easy-to-follow book has been designed for fast and effective learning. It includes numerous demonstration and practice examples with solutions as well as graphing calculator approaches.
- Build skill and confidence by completing all the exams in this book and studying the accompanying solutions before the day of the Algebra 2/Trigonometry Regents exam. Because each exam takes up to 3 hours to complete, you should begin this process no later than several weeks before the exam is scheduled to be given.

- As the day of the actual exam gets closer, take the exams in this book under the timed examination conditions that you will encounter during the actual test. Then compare each of your answers with the explained answers contained in the book.
- Use the Self-Analysis Chart at the end of each exam to help pinpoint any weaknesses.
- If you do not feel confident in a particular area, study the corresponding topic in Barron's *Let's Review: Algebra 2/ Trigonometry*.
- As you work your way through the exams in this book, make a list of any formulas or rules that you need to know and learn them before the day of the exam.

TIP 3

Be Well Rested and Come Prepared on Test Day

SUGGESTIONS

- On the night before exam day, lay out all the things you must take with you. Check the items against the following list:
  1. Your exam room number as well as any personal identification that your school may require;
  2. Two blue or black ink pens;
  3. Two sharpened pencils with erasers;
  4. A ruler (your school may provide this);
  5. A graphing calculator (with fresh batteries); your school is required to make one available to you during the examination;
  6. A watch.
- Eat wisely and go to bed early so you will be alert and well rested when you take the exam.
- Be certain you know when your exam begins. Set an alarm clock to give yourself plenty of time to eat a meal and travel to school.

- Tell your parents what time you will need to leave the house in order to get to school on time.
- Arrive at the exam room on time and with confidence that you are well prepared.

TIP 4

Know How to Use Your Calculator

SUGGESTIONS

- Bring to the Algebra 2/Trigonometry Regents exam room the same calculator that you used when you completed the practice exams at home. Keep in mind that a graphing calculator is required for the Algebra 2/Trigonometry Regents exam.
- If you are required to use a graphing calculator provided by your school, make sure you practice with it in advance because not all graphing calculators work the same way.
- Before you begin the test, know if your calculator is in degree or radian mode. Set DiagnosticOn.
- Be prepared to use your calculator to:

1. Find powers and roots of numbers;
2. Evaluate permutations (${}_nP_r$), combinations (${}_nC_r$), logarithms, and trigonometric functions;
3. Graph linear, quadratic, exponential, logarithmic, and trigonometric functions in an appropriate viewing window so their key features can be seen;
4. Solve equations, inequalities, and systems of equations graphically by finding x-intercepts and points of intersection;
5. Calculate the mean and standard deviation of a set of data values;
6. Determine the line or curve that best fits paired data and then use the regression equation for interpolation and extrapolation;
7. Create a scatter plot and determine the correlation coefficient between paired data.

- When you use your graphing calculator to draw a graph as part of an answer, you are expected to show each of the following:
  1. an accurate sketch of the viewing window;
  2. scales indicated on the coordinate axes; and
  3. clearly labeled x- and y-intercepts and coordinates of points of intersection, if needed for the solution.
- When answering statistics-related questions with a graphing calculator, you will need to show the key elements of your solution.
  1. For *standard deviation* questions, show the number of scores, the mean, and the correct standard deviation (population or sample).
  2. For *regression questions*, write the regression equation with coefficients rounded in accordance with the instructions in the question. If needed, show any required substitutions for x in order to obtain the predicted values for y.
  3. For *correlation coefficient questions*, show the regression equation and indicate how the correlation coefficient was used in your solution.

TIP 5
Know When to Use Your Calculator

SUGGESTIONS

- Do not expect to have to use your calculator for each question.
- Avoid careless errors by using your calculator to perform routine arithmetic calculations that are not easily performed mentally.
- When a problem does not specify the solution method, be alert for opportunities to use your graphing calculator to solve a problem by creating a table or a graph.
- Whenever appropriate, check an answer obtained using standard algebraic methods by using your graphing calculator to create a suitable table or graph.

TIP 6

Have a Plan for Budgeting Your Time

SUGGESTIONS

- During the **first ninety minutes** of the 3-hour exam, complete the 27 multiple-choice questions in Part I. When answering troublesome multiple-choice questions, first rule out any choices that are impossible. If the answer choices are numbers, you may be able to identify the correct answer by plugging these numbers back into the original question to see if they work. If the answer choices contain variables, you may be able to substitute easy numbers for the variables in both the test question and in each choice. Then try to match the numerical result produced in the question to the answer choice that evaluates to the same number. This approach is explained more fully in Tip 9.
- During the **next fifty minutes** of the exam, complete the eight Part II questions and the three Part III questions. In order to maximize your credit for each question, clearly write down the steps you followed to arrive at each answer. Include any equations, formula substitutions, diagrams, tables, graphs, and explanations.
- During the **next twenty minutes** of the exam, complete the one Part IV question. Again, be sure to show how you arrived at each answer.
- During the **last twenty minutes**, review your entire test paper for neatness, accuracy, and completeness.
  1. Check that all answers (except graphs and diagrams) are written in ink. Make sure you have answered all the questions in each part of the exam and that all your Part I answers have been recorded accurately on the separate Part I answer sheet.
  2. Before you submit your test materials to the proctor, check that you have written your name in the reserved spaces on the front page of the question booklet and on the Part I answer sheet. Also, do not forget to sign the declaration that appears at the bottom of the Part I answer sheet.

TIP 7

Make Your Answers Easy to Read

SUGGESTIONS

- Make sure your solutions and answers are clear, neat, and logically organized. When solving problems algebraically, define what the variables stand for, as in "Let $x =$ . . . ."
- Use a pencil to draw graphs and diagrams so that you can erase neatly if necessary. Use a ruler to draw straight lines and coordinate axes.
- Label the coordinate axes. Write y at the top of the vertical axis and x to the right of the horizontal axis. Label each graph with its equation.
- When answering a question in Parts II, III, or IV, write out your approach as well as your final answers. Provide enough details to enable someone who does not know how you think to understand why and how you moved from one step of the solution to the next. If the teacher who is grading your paper finds it difficult to figure out what you wrote, he or she may mark your work as incorrect and give you little, if any, partial credit.
- Draw a box around your final answer to each Part II, III, and IV question.

TIP 8

Answer the Question That Is Asked

SUGGESTIONS

- Make sure each of your answers is in the form required by the question. For example, if a question asks for an approximation, round off your answer to the required decimal place value. If a question asks that you write an answer in lowest terms, make sure that the numerator and denominator of a fractional answer do not have any common factors other than 1 or –1. If the question calls for the answer in simplest radical form, make sure you simplify a square root radical so that the radicand does not contain any perfect square factors greater than 1.

- If the question asks for the x-coordinate (or y-coordinate) of a point, do not give both the x- and the y-coordinates.
- After solving a word problem, check the original question to make sure your *final* answer is the quantity that the question asks you to find.
- If units of measurement are given, as in area problems, check that your answer is expressed in the correct units.
- If the question requires a positive root of a quadratic equation, as in geometric problems in which the variable usually represents a physical dimension, make sure you reject the negative root.

TIP 9

Take Advantage of Multiple-Choice Questions

SUGGESTIONS

- When the answer choices for a multiple-choice question contain only numbers, try plugging each answer choice into the original question until you find the one that works.

EXAMPLE 1

What is the solution set of the equation $\sqrt{x+5}+1=x$?

(1) {–2, 3}   (3) {3}
(2) {–1, 4}   (4) {4}

Solution: The four choices include four different possible solutions: –2, –1, 3, and 4. Check each one in the original equation:

- Choice (1): Check $x = -2$: $\sqrt{-2+5} + 1 = -2$

  $2.732 \stackrel{?}{=} -2$ No

- Choice (2): Check $x = -1$: $\sqrt{-1+5} + 1 = -1$

  $3 \stackrel{?}{=} -1$ No

- Choice (3): Check $x = 3$: $\sqrt{3+5} + 1 = 3$

  $3.828 \stackrel{?}{=} 3$ No

- Choice (4): Check $x = 4$: $\sqrt{4+5} + 1 = 4$

  $4 \stackrel{?}{=} 4$ Yes

Among the given solutions, only $x = 4$ checks. The correct choice is **(4)**.

- **If the answer choices for a multiple-choice question contain variables, replace the variables with easy numbers in both the question and in each choice. Then work out the problem using these numbers. Using –1, 0, or 1 as choices can lead to surprising results, so choose numbers such as 3, 5, or 7 instead.**

EXAMPLE 2

Expressed as a single fraction, what is $\frac{1}{x+1} + \frac{1}{x}$, $x \neq 0, -1$?

(1) $\frac{2x+3}{x^2+x}$

(2) $\frac{2x+1}{x^2+x}$

(3) $\frac{2}{2x+1}$

(4) $\frac{3}{x^2}$

Solution: Suppose $x = 3$. Then the original expression evaluates to $\frac{1}{3+1}+\frac{1}{3}=0.58333=\frac{7}{12}$. Now try $x = 3$ in each of the choices:

- Choice (1): $\frac{2(3)+3}{3^2+3}=\frac{9}{12}$ Not the same
- Choice (2): $\frac{2(3)+1}{3^2+3}=\frac{7}{12}$ Agrees with the original
- Choice (3): $\frac{2}{2(3)+1}=\frac{2}{7}$ Not the same
- Choice (4): $\frac{3}{3^2}=\frac{3}{9}$ Not the same

The correct choice is **(2)**.

Even after finding a choice that agrees with the original expression, continue to check the remaining choices anyway. If two different choices both match with the original expression, choose a different x-value and check the two choices.

TIP 10

Don't Omit Any Questions

SUGGESTIONS

- Keep in mind that on each of the four parts of the test, you must answer all the questions.
- If you get stuck on a multiple-choice question in Part I, try one of the problem-solving strategies discussed in Tip 9. If you still cannot figure out the answer, try to eliminate any impossible answers. Then guess from the remaining answer choices.

- If you get stuck on a question from Parts II, III, or IV, try to maximize your partial credit by writing down any formula, diagram, or mathematics facts you think might apply. If appropriate, organize and analyze the given information by making a table or a diagram. You can then try to arrive at the correct answer by guessing, checking your guess, and then revising your guess as needed. If you use a trial-and-error procedure, be sure to write down the work for at least three guesses with checks.

A GENERAL PROBLEM-SOLVING APPROACH

1. ***Read each test question through*** the first time to get a general idea of the type of mathematics knowledge the question requires. For example, one question may ask you to solve an algebraic equation, while another question may ask you to apply some geometric principle. Then read the problem through a second time to pick out specific facts. Identify what is given and what you need to find.
2. ***Decide how you will solve the problem.*** You may need to use one of the special problem-solving strategies discussed in this section. If you decide to solve a word problem using algebraic methods, you may first need to translate the conditions of the problem into an equation or an inequality.
3. ***Carry out your plan.*** Show the steps you followed in arriving at your answer.
4. ***Verify that your answer is correct*** by making sure it works in the original question.

Graphing Calculator Skills

Because of its popularity and availability, the Texas Instruments TI-83 Plus/TI-84 Plus graphing calculator will be used as the "reference" calculator. If you are using a different graphing calculator, you may need to consult the manual that came with your calculator.

1. PERFORMING CALCULATIONS

Routine arithmetic calculations are performed in the "home screen." Enter the home screen by pressing 2nd MODE. After you enter an arithmetic expression, press ENTER to see the answer. The result is displayed at the end of the next line, as shown in the accompanying screen shot. Notice that the calculator key for an exponent is ^.

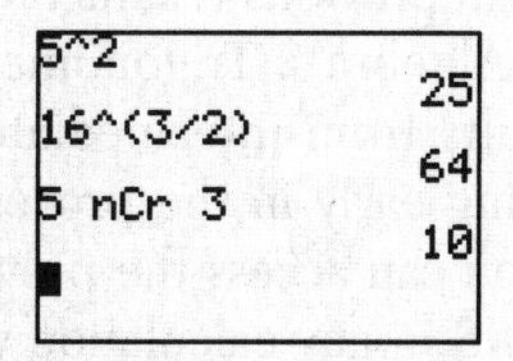

Unlike a scientific calculator, the TI-83/TI-84 does not have special keys for evaluating combinations and permutations. Instead, you have to access the calculator's math library of special functions. To evaluate ${}_5C_3$ (or ${}_5P_3$) using a TI-83/TI-84 calculator:

- Enter 5.
- Press MATH ▶ ▶ ▶ 3 to select ${}_nC_r$ (or 2 to select ${}_nP_r$) from the MATH PRB menu.
- Enter 3 and press ENTER. The result is 10.

Know how to use the calculator to change decimals to fractions. After a calculation yields a decimal result, select MATH **1:Frac** to convert the decimal to a fraction. For example, ${}_4C_3\left(\frac{7}{8}\right)^3\left(\frac{1}{8}\right)^1$ represents the probability that an event with probability of occurrence equal to 7/8 will happen exactly 3 times out of 4 attempts. Answers in fraction form are exact (no points are lost for rounding errors), and multiple-choice answers are frequently given as reduced fractions.

4 nCr 3(7/8)^3(1
/8)^1
.3349609375
Ans▸Frac
343/1024

Your calculator will not be able to convert all decimal values to fractions. Both rational numbers with large denominators and irrational numbers will remain in decimal form even after you select MATH **1:Frac**.

Store values and recall previous results on your calculator. Some of the calculations in Algebra 2/Trigonometry can result in long decimal answers. Results from intermediate steps should not be rounded at all. Rounding early in the process may cause the final answer to be wrong. You can access the previous answer by selecting 2nd (–). Use ANS in any calculation where you want to use the previous result without retyping all the numbers.

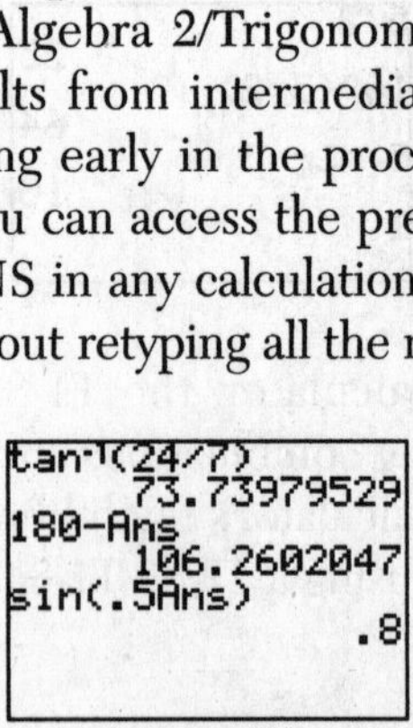
tan⁻¹(24/7)
73.73979529
180-Ans
106.2602047
sin(.5Ans)
.8

Use the store feature if you need to recall several values. After your expression, press STO▸ and then a letter (ALPHA MATH gives the letter A). Anytime you use that letter, it will represent the assigned value until you store a new value in the same letter or reset your calculator.

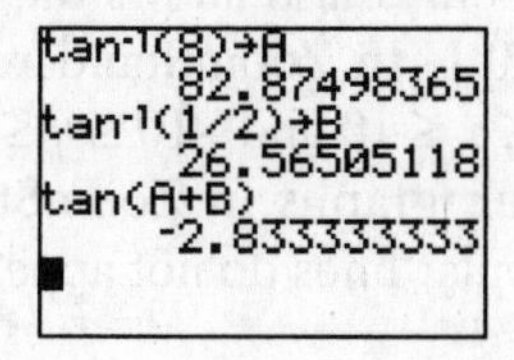

Storing values can make evaluating complicated formulas simpler. For example, use the Law of Cosines to find a when $b = 20$, $c = 15$, and $m\angle A = 50°$. Store your values as different letters, and then enter your expression. Storing values can reduce errors because of reduced keying mistakes and correct order of operations.

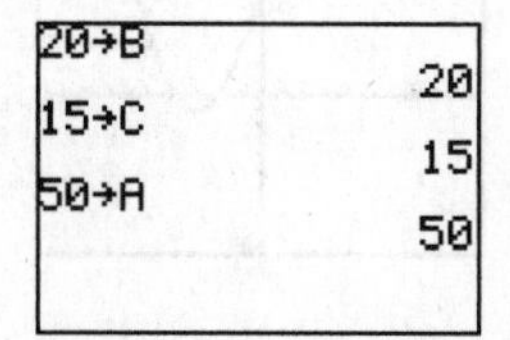

2. GRAPHING FUNCTIONS AND SETTING WINDOWS

To graph the function $y = x^2 - 6x + 8$, you must enter it in the Y= editor. Variable x is entered by pressing X,T,θ,n. Notice that it is possible to enter more than one equation.

Set a viewing window by pressing ZOOM and then selecting option 4 for a decimal window or option 6 for a standard window. In ZoomDecimal, the axes are scaled so that $-4.7 \le x \le 4.7$ and $-3.1 \le y \le 3.1$. This makes one unit in the x-direction the same length as one unit in the y-direction so graphs will have the proper proportions on the screen. It also allows the screen cursor to move in consistent steps of 0.1. In ZoomStandard, the coordinate axes are scaled so that $-10 \le x \le 10$ and $-10 \le y \le 10$. This gives a larger viewing area. However, graphs in ZoomStandard are vertically compressed; perpendicular lines do not appear perpendicular.

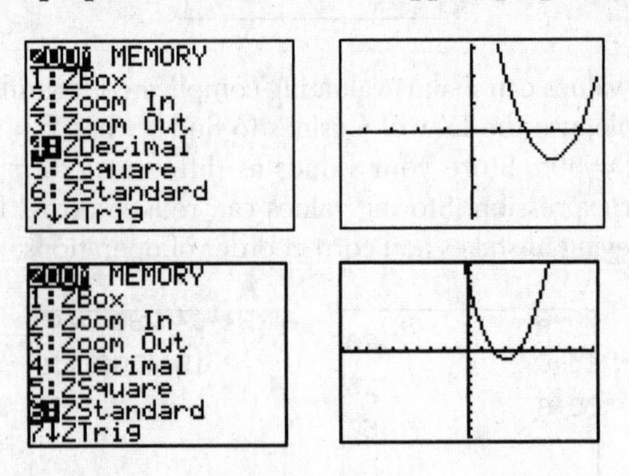

The ZOOM **7:ZTrig** (ZoomTrig) window is useful for graphing trigonometric functions. However, inverse trigonometric functions, such as $y = \sin^{-1} x$ display better in ZoomDecimal. When graphing trigonometric functions, be sure the calculator is in the correct mode. Typically, trigonometric functions are graphed in radian mode. The exception is if you are solving a trigonometric equation graphically and want the answers in degrees, then the calculator should be in degree mode.

If a graph does not fit well in the decimal, standard or trig windows, you can do several things:

- Press WINDOW and modify the values of Xmin, Xmax, Ymin, and/or Ymax. To show more graph to the right, make Xmax larger. To show more graph to the left, make Xmin smaller (or more negative).

- Read the problem. It may give information to help set the window dimensions. For example, if the problem says the domain of a function is $0 \le x \le 12$, set Xmin = 0 and Xmax = 12. If a graph is given, set the WINDOW values to match the given scale.
- Use ZoomFit. If you know appropriate values for Xmin and Xmax, set those and then choose **0:ZoomFit** in the ZOOM menu. The calculator will set an appropriate y-scale.

3. CREATING TABLES

If you want to graph $y = x^2 - 6x + 8$ using graph paper, you will need a table of values. After the equation of the parabola has been entered in the Y= editor, press 2nd WINDOW, which is TBLSET. If you know the x-value where you want the table to start, enter it in TblStart. If not, consider starting at 0. Most of the time, it is convenient to have the x-values increase in steps of 1, so set ΔTbl = 1. If you want different increments, for example, x to increase by 0.5, set ΔTbl accordingly. Then press 2nd GRAPH to display the table. To see additional values not in the starting window, use the up and down arrow keys.

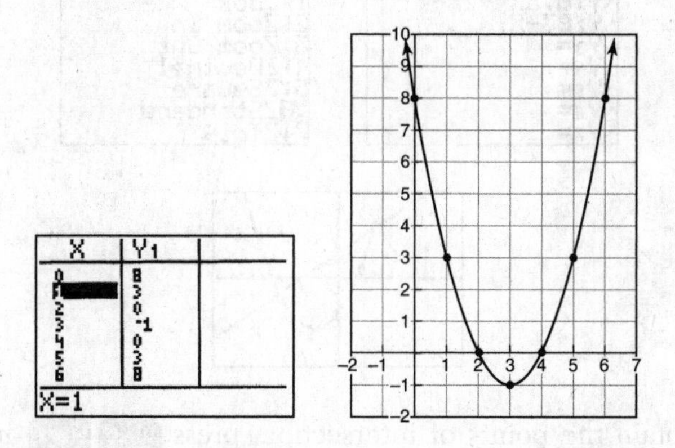

To evaluate functions at specific values, such as when checking multiple-choice answers, use the Ask feature in table setup. For example, to check if $x = 2$, $x = 4$, or $x = 6$ is a root to $y = x^2 - 6x + 8$, go to TBLSET (press 2nd WINDOW). Go to the Indpnt line, and select Ask. Then go to the TABLE (press 2nd GRAPH), and key in the values you want to evaluate. As you can see below, both $x = 2$ and $x = 4$ are roots, but $x = 6$ is not. Make sure you know how to change the table back to automatic.

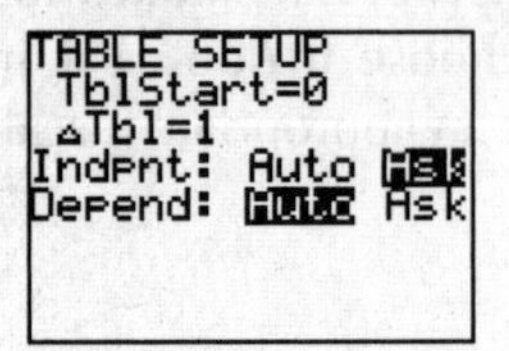

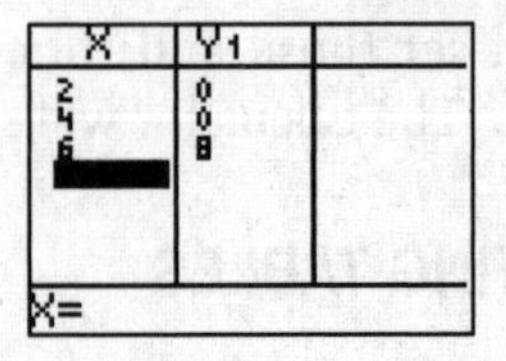

4. FINDING INTERSECTION POINTS

When two graphs intersect, you can find the coordinates of their point(s) of intersection using the intersect feature of your graphing calculator. For example, to solve the system $y = 0.5x^2 - 2x - 4$ and $y = 3 - x$, enter both equations in the Y= editor. Then graph these equations in a window that shows the point(s) of intersection.

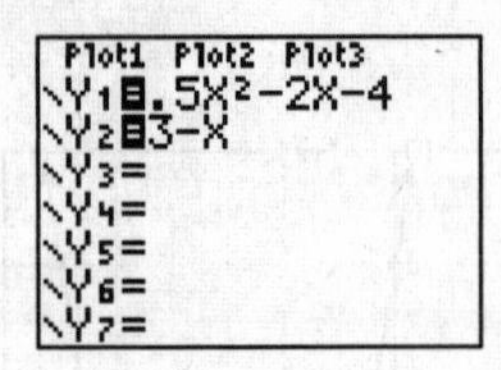

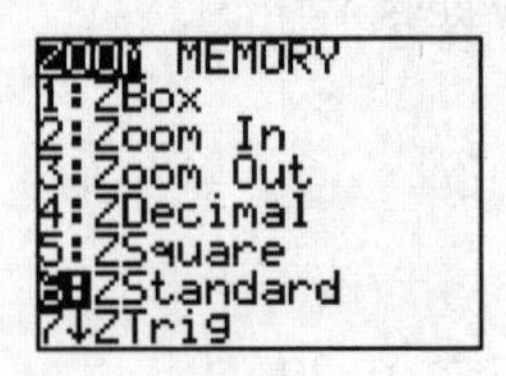

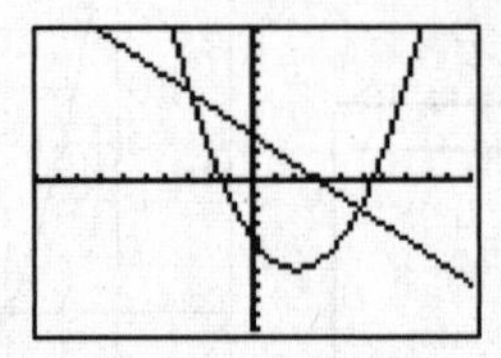

To obtain the points of intersection, press 2nd TRACE and then select **5:intersect** from the CALCULATE menu. Press ENTER ENTER, and then move the cursor to a point that is close to the point of intersection and press ENTER. The coordinates of the point of intersection will appear at the bottom of the screen.

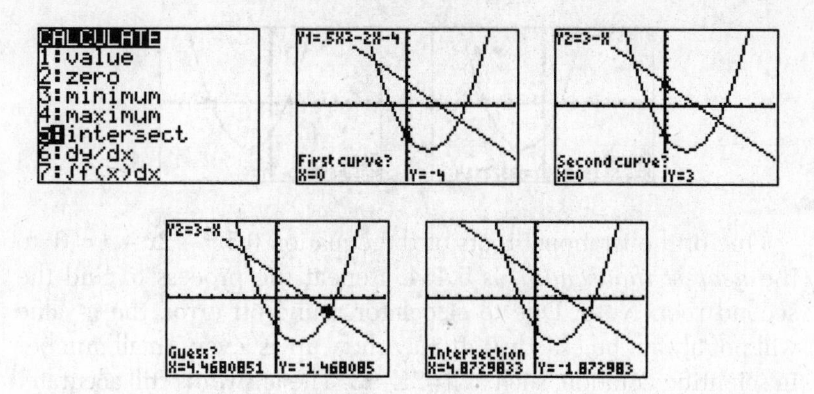

In this example, you must repeat the process to get the second intersection point on the left.

This same method can be used to solve a single equation, such as $e^{0.5x} = 12$. Graph each side of the equation separately and proceed as above.

5. FINDING ROOTS

To find a root (x-intercept or zero) of the graph of $y = 0.5x^2 - 2x - 4$, select **2:zero** from the CALC (2nd TRACE) menu. Move the cursor to a point on the graph that is slightly to the left of the x-intercept. Press ENTER. Then move the cursor to a point on the graph that is slightly to the right of the same x-intercept. Press ENTER ENTER to display the coordinates of that x-intercept. You can also key in a value on the x-axis to the left of the root and then to the right of the root instead of moving the cursor.

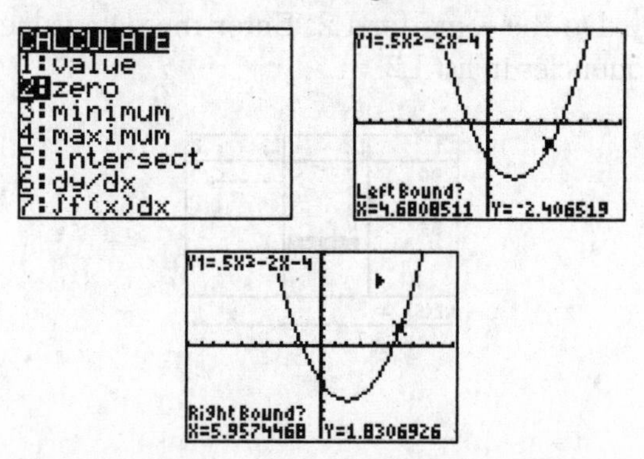

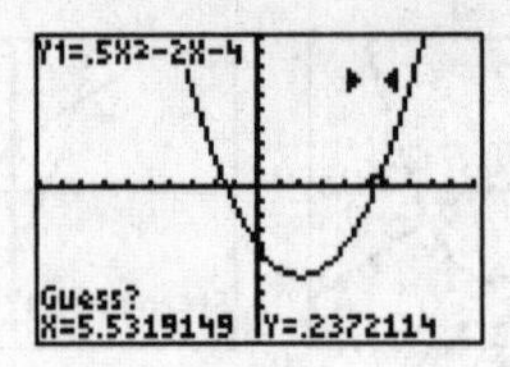

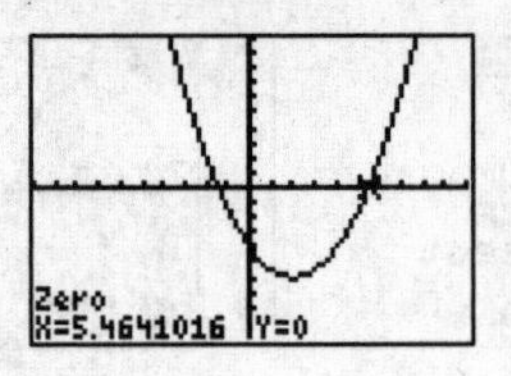

One of the irrational roots of the equation $0.5x^2 - 2x - 4 = 0$, to the *nearest thousandth*, is 5.464. Repeat the process to find the second root. Note: Due to calculator round-off error, the y-value will not always be exactly 0. It may show up as a very small number in scientific notation, such as Y=2E–13. The answer is still accurate.

Another way to find a root is to set Y2 = 0 and use the intersect feature described in the previous section.

6. CALCULATING MEAN AND STANDARD DEVIATION

Test Score	Frequency
80	5
85	7
90	9
95	4

To calculate the mean and standard deviation of the data in the accompanying table, press STAT ENTER. If the lists already contain numbers, use the arrow keys to highlight L1. Then press CLEAR ENTER. Do the same for L2. Enter the data values in list L1 and the frequencies in list L2.

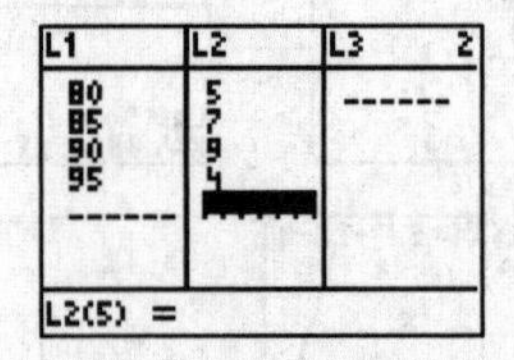

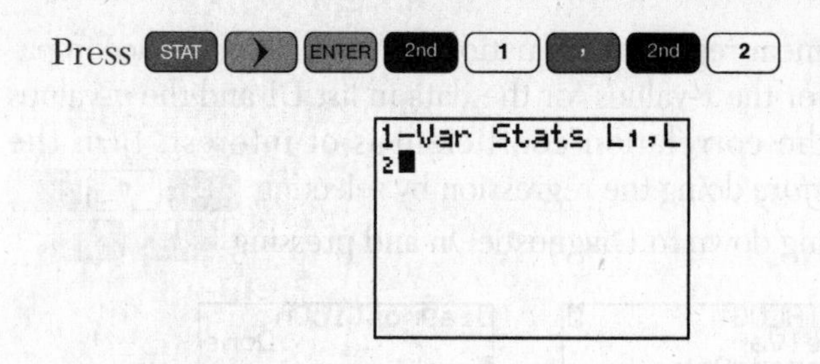

Press ENTER to get the mean, $\bar{x} = 87.4$, and the population standard deviation, $\sigma_x \approx 4.92$.

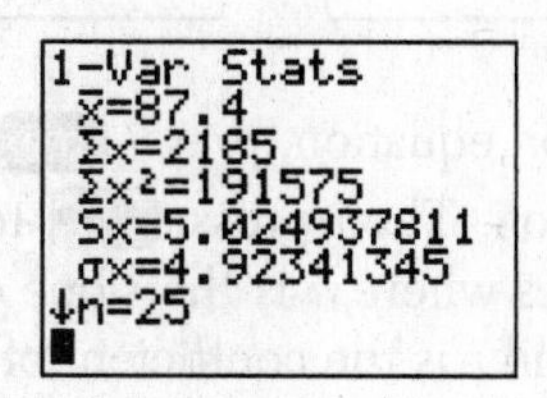

When it is not possible to collect data for an entire population, data from a smaller subset of that population are collected. If in the example, the set of 25 test scores were a sample from a larger population of test scores, then the sample standard deviation, denoted by s_x, would be used. In this example, $s_x \approx 5.02$. Problems on the Algebra 2/Trigonometry Regents exam will specify which standard deviation to use.

7. FITTING A REGRESSION LINE TO DATA

The line or curve (exponential, power, or logarithmic) that best fits a paired set of data values can be obtained using the regression feature on a graphing calculator.

Minutes Studied (x)	15	40	45	60	70	75	90
Test Grade (y)	50	67	75	75	73	89	93

To find the linear regression equation for the data in the accompanying table, enter the x-values for the data in list L1 and the y-values in list L2. If the correlation coefficient is of interest, turn the diagnostic on *before* doing the regression by selecting 2nd 0 x^{-1} and then scrolling down to DiagnosticOn and pressing ENTER ENTER.

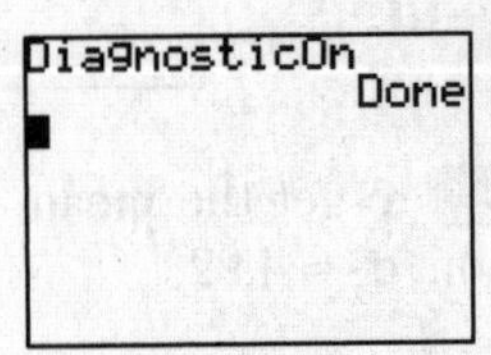

To get the regression equation, press STAT ❯ and choose the **4:LinReg(ax+b)** option. Then press ENTER to show a summary of the regression statistics where a is the slope of the regression line, b is the y-intercept, and r is the coefficient of linear correlation. An equation of the line of best fit with the regression coefficients rounded off to the *nearest hundredth* is $y = 0.53x + 44.70$.

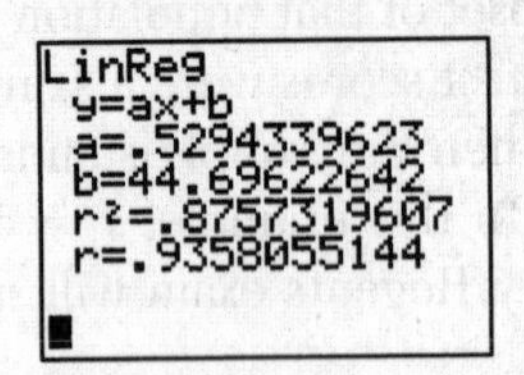

- To predict the value of y when $x = 80$ minutes, substitute 80 for x in the regression equation with the coefficients rounded according to the directions:

$$y = 0.53x + 44.70 = 0.53(80) + 44.7 = 87.1$$

- To find the value of x, to the *nearest hundredth*, that would predict $y = 98$, substitute 98 for y in the regression equation and solve for x:

$$98 = 0.53x + 44.70$$
$$x = 100.57$$

8. GRAPHING CALCULATOR TIPS

- Unless otherwise specified, use the full power/display of the calculator. Intermediate values obtained in a long calculation should be stored in the calculator memory and *not* rounded. Rounding should be done only after the final answer is reached. Unless otherwise specified, the π key on a calculator should be used in computations involving this number.
- For statistics questions involving standard deviation, indicate on your answer paper the number of scores, the mean, and the appropriate standard deviation (population or sample).
- The calculator should *not* be used to obtain normal curve probabilities. The normal curve provided on the formula sheet in the test booklet contains the only values that should be used. If your calculator has this capability, you can use it to check the answer obtained after using the curve provided in the test booklet.
- When a test question requires the calculation of a regression equation, you will be directed to use a linear, exponential, power, or logarithmic regression. After calculating the regression equation coefficients, write the regression equation in standard form. If further calculations are done with the equation, show the substitution made for each calculation.

In this book, you will see many Calculator Check examples to show you how the graphing calculator can be used to help you check a solution. Sometimes the calculator displays values or graphs in surprising ways. You will find Calculator Caution notes throughout the solutions to alert you to these discrepancies and help you enter data and formulas accurately in your calculator. Calculator Tips will help you use your calculator to solve problems in ways you may not have learned before.

Formulas

The following formulas are provided during the exam on the Algebra 2/Trigonometry reference sheet at the back of the exam booklet.

Area of a Triangle

$$K = \frac{1}{2}ab \sin C$$

Law of Cosines

$$a^2 = b^2 + c^2 - 2bc \cos A$$

Functions of the Sum of Two Angles

$$\sin(A + B) = \sin A \cos B + \cos A \sin B$$

$$\cos(A + B) = \cos A \cos B - \sin A \sin B$$

$$\tan(A + B) = \frac{\tan A + \tan B}{1 - \tan A \tan B}$$

Functions of the Double Angle

$\sin 2A = 2 \sin A \cos A$

$\cos 2A = \cos^2 A - \sin^2 A$

$\cos 2A = 2 \cos^2 A - 1$

$\cos 2A = 1 - 2 \sin^2 A$

$$\tan 2A = \frac{2 \tan A}{1 - \tan^2 A}$$

Functions of the Difference of Two Angles

$\sin(A - B) = \sin A \cos B - \cos A \sin B$

$\cos(A - B) = \cos A \cos B + \sin A \sin B$

$$\tan (A - B) = \frac{\tan A - \tan B}{1 + \tan A \tan B}$$

Functions of the Half Angle

$$\sin \frac{1}{2} A = \pm \sqrt{\frac{1 - \cos A}{2}}$$

$$\cos \frac{1}{2} A = \pm \sqrt{\frac{1 + \cos A}{2}}$$

$$\tan \frac{1}{2} A = \pm \sqrt{\frac{1 - \cos A}{1 + \cos A}}$$

Law of Sines

$$\frac{a}{\sin A} = \frac{b}{\sin B} = \frac{c}{\sin C}$$

Sum of a Finite Arithmetic Sequence

$$S_n = \frac{n(a_1 + a_n)}{2}$$

Sum of a Finite Geometric Sequence

$$S_n = \frac{a_1(1 - r^n)}{1 - r}$$

Binomial Theorem

$$(a+b)^n = {}_nC_0 a^n b^0 + {}_nC_1 a^{n-1} b^1 + {}_nC_2 a^{n-2} b^2 + \cdots + {}_nC_n a^0 b^n$$

$$(a+b)^n = \sum_{r=0}^{n} {}_nC_r a^{n-r} b^r$$

Normal Curve
Standard Deviation

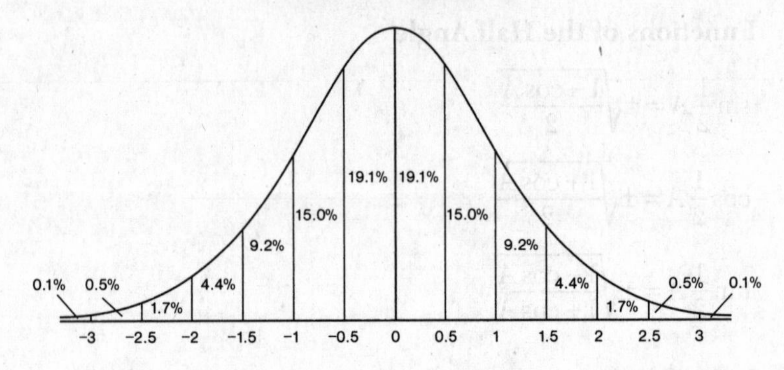

SOME ADDITIONAL FORMULAS YOU SHOULD KNOW

1. CIRCUMFERENCE, ARC LENGTH, AND AREA

Quantity	Formula
Circumference	$C = 2\pi \times \text{Radius}$
Arc length, where θ is the intercepted angle measured in radians	$s = \theta \times \text{Radius}$
Area of a circle	$A = \pi \times (\text{Radius})^2$

2. COORDINATE FORMULAS

Description	Formula
Length of line segment from (x_A, y_A) to (x_B, y_B)	$d = \sqrt{(x_A - x_B)^2 + (y_A - y_B)^2}$
Midpoint of line segment from (x_A, y_A) to (x_B, y_B)	$\left(\frac{x_A + x_B}{2}, \frac{y_A + y_B}{2}\right)$
Center-radius equation of a circle with center at (h, k) and radius $= r$	$(x - h)^2 + (y - k)^2 = r^2$
Vertex form of the equation of a parabola with vertex at (h, k)	$y = a(x - h)^2 + k$

3. QUADRATIC EQUATIONS

Feature of $ax^2 + bx + c = 0$ $(a \neq 0)$	Formula
Two roots	$x = \frac{-b \pm \sqrt{b^2 - 4ac}}{2a}$
Sum of roots	$-\frac{b}{a}$
Product of roots	$\frac{c}{a}$

4. EXPONENT LAWS

Name	Law
Product	$x^a x^b = x^{a+b}$
Quotient	$\frac{x^a}{x^b} = x^{a-b} = \frac{1}{x^{b-a}}$
Power	$(x^a)^b = x^{ab}$
Logarithm equivalence	$y = \log_b x \Leftrightarrow x = b^y$

5. LOGARITHM LAWS

Name	Law
Product	$\log(xy) = \log x + \log y$
Quotient	$\log\left(\frac{x}{y}\right) = \log x - \log y$
Power	$\log(x^y) = y \log x$
Change of base	$y = \log_b x = \frac{\log x}{\log b}$

6. TRIGONOMETRIC VALUES

	0°	30°	45°	60°	90°
$\sin x$	0	$\frac{1}{2}$	$\frac{\sqrt{2}}{2}$	$\frac{\sqrt{3}}{2}$	1
$\cos x$	1	$\frac{\sqrt{3}}{2}$	$\frac{\sqrt{2}}{2}$	$\frac{1}{2}$	0
$\tan x$	0	$\frac{\sqrt{3}}{3}$	1	$\sqrt{3}$	undefined

7. TRIGONOMETRIC IDENTITIES

Pythagorean Identities	Reciprocal Identities	Quotient Identities
$\sin^2 x + \cos^2 x = 1$	$\csc x = \frac{1}{\sin x}$	$\tan x = \frac{\sin x}{\cos x}$
$\sin^2 x = 1 - \cos^2 x$	$\sec x = \frac{1}{\cos x}$	$\cot x = \frac{\cos x}{\sin x}$
$\cos^2 x = 1 - \sin^2 x$	$\cot x = \frac{1}{\tan x}$	
$1 + \cot^2 x = \csc^2 x$		
$\tan^2 x + 1 = \sec^2 x$		

8. TRANSFORMATIONS

Transformation ($a > 0$)	Description
$y = f(x + a)$ $y = f(x - a)$	Horizontal translation $(x + a)$ translates graph left $(x - a)$ translates graph right
$y = f(x) + a$ $y = f(x) - a$	Vertical translation $+a$ translates graph up $-a$ translates graph down
$y = f(-x)$	Reflection over the y-axis
$y = -f(x)$	Reflection over the x-axis
$y = af(x)$	Vertical dilation by a factor of a

9. COUNTING AND PROBABILITY

Quantity	Formula
Number of ordered arrangements of r items selected from n items	${}_nP_r = \frac{n!}{(n-r)!}$
Number of unordered groups of r members selected from n members	${}_nC_r = \frac{n!}{r!(n-r)!}$
Probability of r successes in n trials of a two-outcome experiment with probability, p, of success	${}_nC_r p^r (1-p)^{n-r}$

10. SERIES AND SEQUENCES

	Explicit	Recursive
Arithmetic	$a_n = a_1 + d(n - 1)$	Value for a_1 is specified; $a_n = a_{n-1} + d$ for $n > 1$
Geometric	$a_n = a_1 r^{n-1}$	Value for a_1 is specified; $a_n = a_{n-1} r$ for $n > 1$

Key Algebra 2/ Trigonometry Facts and Skills

1. EXPONENTS AND RADICALS

1.1 EXPONENTS

Exponents are used for repeated multiplication and roots. The base e (≈ 2.71828, an irrational number) is frequently used in exponential modeling.

Exponent Rules ($x > 0$):

- $x^0 = 1$
- $x^{-a} = \dfrac{1}{x^a}$ and $\dfrac{1}{x^{-a}} = x^a$
- $x^{\frac{a}{b}} = \sqrt[b]{x^a} = \left(\sqrt[b]{x}\right)^a$
- $x^a x^b = x^{a+b}$
- $\dfrac{x^a}{x^b} = x^{a-b} = \dfrac{1}{x^{b-a}}$
- $(x^a)^b = x^{ab}$

1.2 SIMPLIFYING RADICALS

Simplest form for a radical has no radicals in the denominator and no fractions or decimals in the radicand. To simplify a radical, reduce powers (prime factor trees can help) and rationalize denominators. To rationalize a monomial denominator, multiply the numberator and the denominator by the radical. To rationalize a binomial denominator, multiply by the conjugate. The conjugate of $\sqrt{a} + \sqrt{b}$ is $\sqrt{a} - \sqrt{b}$.

1.3 ARITHMETIC OPERATIONS WITH RADICALS

To add/subtract, simplify first and then add/subtract only like radicals. To multiply/divide, first multiply/divide outside and inside the radical sign separately and then simplify.

1.4 SOLVE RADICAL EQUATIONS

Isolate the radical on one side of the equation, raise each side of the equation to the reciprocal power to undo the radical, and solve for the variable. ALWAYS check the solutions in the original equation for extraneous roots.

Practice Exercises

1.1 The expression $(3c)^{-2}$, $c \neq 0$ is equivalent to

(1) $-6c^2$

(2) $\frac{1}{3c^2}$

(3) $\frac{1}{9c^2}$

(4) $\frac{3}{c^2}$

1.2 The expression $b^{-\frac{3}{2}}$, $b > 0$ is equivalent to

(1) $\frac{1}{\left(\sqrt[3]{b}\right)^2}$

(2) $\frac{1}{\left(\sqrt{b}\right)^3}$

(3) $-\left(\sqrt{b}\right)^3$

(4) $\left(\sqrt[3]{b}\right)^2$

1.3 Which expression represents the sum of $\frac{1}{\sqrt{3}} + \frac{1}{\sqrt{2}}$?

(1) $\frac{2\sqrt{3}+3\sqrt{2}}{6}$

(2) $\frac{2}{\sqrt{5}}$

(3) $\frac{\sqrt{3}+\sqrt{2}}{3}$

(4) $\frac{\sqrt{3}+\sqrt{2}}{2}$

1.4 When simplified, the expression $\left(\sqrt[3]{m^4}\right)\left(m^{-\frac{1}{2}}\right)$, $m > 0$ is equivalent to

(1) $\sqrt[3]{m^{-2}}$ (3) $\sqrt[5]{m^{-4}}$

(2) $\sqrt[4]{m^3}$ (4) $\sqrt[6]{m^5}$

1.5 Express the product of $3\sqrt{20}\left(2\sqrt{5}-7\right)$ in simplest radical form.

1.6 The expression $\dfrac{12}{3+\sqrt{3}}$ is equivalent to

(1) $12-\sqrt{3}$ (3) $4-2\sqrt{3}$

(2) $6-2\sqrt{3}$ (4) $2+\sqrt{3}$

1.7 The solution set of the equation $\sqrt{x+6}=x$ is

(1) {–2,3} (3) {3}
(2) {–2} (4) { }

1.8 Solve algebraically for x: $\sqrt{3x+1}+1=x$.

1.9 Express $\dfrac{7}{x-\sqrt{2}}$ as a radical in simplest form with a rational denominator.

1.10 The expression $\sqrt[4]{16a^6b^4}$ is equivalent to

(1) $2a^2b$ (3) $4a^2b$

(2) $2a^{\frac{3}{2}}b$ (4) $4a^{\frac{3}{2}}b$

Solutions

1.1 For a negative exponent, take the reciprocal of the base:

$$(3c)^{-2} = \left(\frac{1}{3c}\right)^2 = \frac{1}{(3c)^2} = \frac{1}{(3c)(3c)} = \frac{1}{9c^2}$$

CALC Check

```
Plot1 Plot2 Plot3
\Y1=(3X)^-2
\Y2=1/(9X²)
\Y3=
\Y4=
\Y5=
\Y6=
\Y7=
```

X	Y1	Y2
-3	.01235	.01235
-2	.02778	.02778
-1	.11111	.11111
0	ERROR	ERROR
1	.11111	.11111
2	.02778	.02778
3	.01235	.01235

X=-3

Make sure parentheses are around $9x^2$ in the calculator. Use the table to determine if the expressions are equivalent. ERROR means the expression is undefined at $x = 0$. Look at the original problem to see that $c \neq 0$.

The correct choice is (**3**).

1.2 For a negative exponent, take the reciprocal of the base. The fraction in the exponent means the expression will contain a radical. The numerator of the fraction will remain the exponent; the denominator will be the index of the radical.

$$b^{-\frac{3}{2}} = \left(\frac{1}{b}\right)^{\frac{3}{2}} = \frac{1}{b^{\frac{3}{2}}} = \frac{1}{\left(\sqrt{b}\right)^3}$$

CALC Check

```
Plot1 Plot2 Plot3
\Y1=X^(-3/2)
\Y2=1/(√(X))^3
\Y3=
\Y4=
\Y5=
\Y6=
\Y7=
```

X	Y1	Y2
-2	ERROR	ERROR
-1	ERROR	ERROR
0	ERROR	ERROR
1	1	1
2	.35355	.35355
3	.19245	.19245
4	.125	.125

X=4

The problem stated that $b > 0$. You can see why in the table.

The correct choice is (**2**).

1.3 Rationalize the monomial denominators by multiplying the numerator and denominator by the radical in the denominator.

$$\frac{1}{\sqrt{3}}+\frac{1}{\sqrt{2}}=\frac{1}{\sqrt{3}}\frac{\sqrt{3}}{\sqrt{3}}+\frac{1}{\sqrt{2}}\frac{\sqrt{2}}{\sqrt{2}}=\frac{\sqrt{3}}{3}+\frac{\sqrt{2}}{2}$$

Then rewrite each fraction with a common denominator and add the numerators.

$$\frac{\sqrt{3}}{3}+\frac{\sqrt{2}}{2}=\frac{\sqrt{3}}{3}\cdot\frac{2}{2}+\frac{\sqrt{2}}{2}\cdot\frac{3}{3}=\frac{2\sqrt{3}}{6}+\frac{3\sqrt{2}}{6}=\frac{2\sqrt{3}+3\sqrt{2}}{6}$$

CALC Check

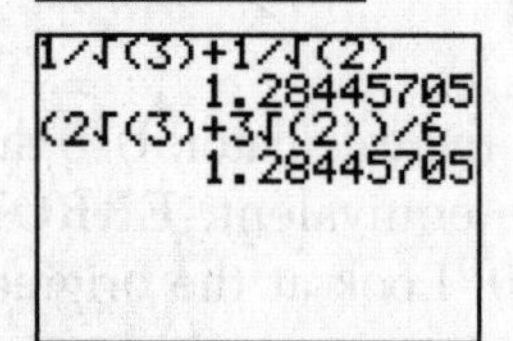

The correct choice is **(1)**.

1.4 Convert the radical to exponent form, $\sqrt[b]{x^a}=x^{\frac{a}{b}}$.

$$\left(\sqrt[3]{m^4}\right)\left(m^{-\frac{1}{2}}\right)=\left(m^{\frac{4}{3}}\right)\left(m^{-\frac{1}{2}}\right)$$

When multiplying powers, add exponents.

$$=m^{\left(\frac{4}{3}+-\frac{1}{2}\right)}=m^{\frac{5}{6}}$$

Convert back to radical form, $x^{\frac{a}{b}}=\sqrt[b]{x^a}$.

$$=\sqrt[6]{m^5}$$

Find the root symbol in the MATH menu.

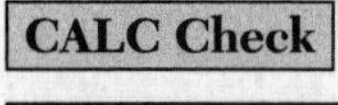


```
Plot1 Plot2 Plot3
\Y1■(³√(X^4))(X^
(-1/2)
\Y2■6*√(X^5)
\Y3=
\Y4=
\Y5=
\Y6=
```

X	Y1	Y2
-1	ERROR	ERROR
0	ERROR	0
1	1	1
2	1.7818	1.7818
3	2.498	2.498
4	3.1748	3.1748
5	3.8236	3.8236

X=5

The correct choice is **(4)**.

1.5 When multiplying radicals, first multiply coefficients and multiply the radicands. Then simplify.

$$3\sqrt{20}\left(2\sqrt{5}-7\right) = 6\sqrt{100} - 21\sqrt{20}$$

$$= 6(10) - 21\sqrt{4}\sqrt{5}$$

$$= 6(10) - 21(2)\sqrt{5}$$

$$= 60 - 42\sqrt{5}$$

Simplest radical form means the answer is not in decimal form. NEVER write a decimal answer if the problem specifies simplest radical form. Evaluating a radical on the calculator is best for multiple-choice questions.

CALC Check

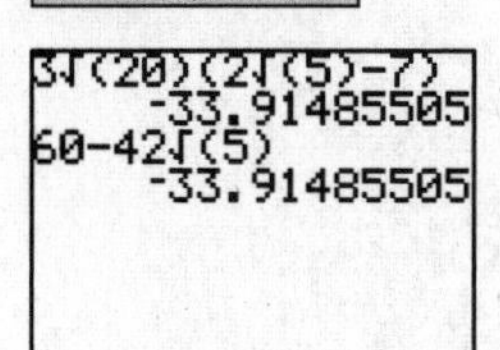

The correct answer is $\mathbf{60 - 42\sqrt{5}}$.

1.6 Rationalize the denominator by multiplying by the conjugate.

$$\frac{12}{3+\sqrt{3}} = \frac{12}{(3+\sqrt{3})} \cdot \frac{(3-\sqrt{3})}{(3-\sqrt{3})}$$

$$= \frac{12(3-\sqrt{3})}{9-3\sqrt{3}+3\sqrt{3}-3}$$

$$= \frac{12(3-\sqrt{3})}{6}$$

$$= 2(3-\sqrt{3}) = 6-2\sqrt{3}$$

CALC Check

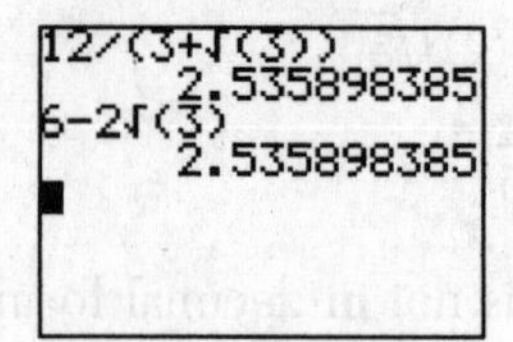

The correct choice is **(2)**.

1.7 <u>To solve a radical equation algebraically</u>:

Isolate the radical term (already done.)	$\sqrt{x+6} = x$
Square both sides.	$\left(\sqrt{x+6}\right)^2 = x^2$
Solve.	$x + 6 = x^2$
	$x^2 - x - 6 = 0$
	$(x + 2)(x - 3) = 0$
	$x = -2$ or $x = 3$

ALWAYS check answers in the *original* equation (see below for checking on the calculator).

Check $x = -2$: $\sqrt{(-2)+6} \stackrel{?}{=} -2$

$2 \stackrel{?}{=} -2$ No

Check $x = 3$: $\sqrt{3+6} \stackrel{?}{=} 3$

$3 = 3$ Yes

CALC Check

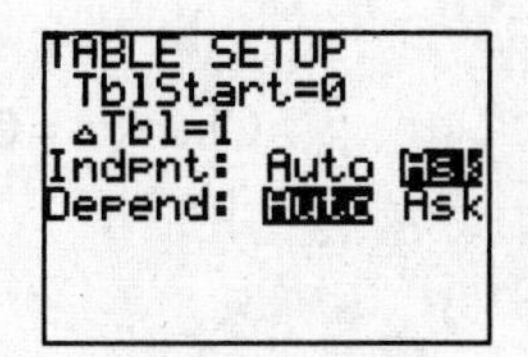

<u>To solve a radical equation graphically:</u>

Use your calculator to graph $y = \sqrt{x+6}$ and $y = x$ and find the point of intersection. Note that a graphical solution does not give extraneous roots.

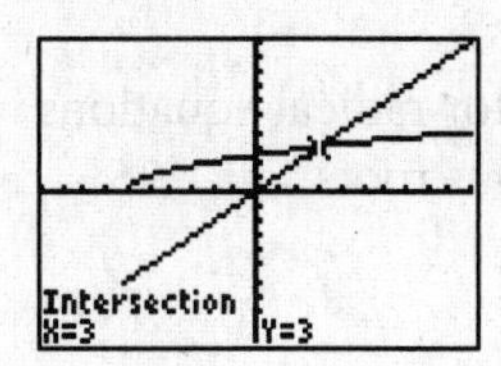

The correct choice is **(3)**.

1.8 Isolate the radical term. $\sqrt{3x+1}+1 = x$

$\sqrt{3x+1} = x-1$

Square both sides. $\left(\sqrt{3x+1}\right)^2 = (x-1)^2$

Solve.

$$\left(\sqrt{3x+1}\right)^2 = (x-1)(x-1)$$
$$3x + 1 = x^2 - x - x + 1$$
$$3x = x^2 - 2x$$
$$x^2 - 5x = 0$$
$$x(x - 5) = 0$$
$$x = 0 \text{ or } x = 5$$

Check answers in the *original* equation.

Check $x = 0$: $\sqrt{3(0)+1}+1 \stackrel{?}{=} 0$

$2 \stackrel{?}{=} 0$ No

Check $x = 5$: $\sqrt{3(5)+1}+1 \stackrel{?}{=} 5$

$5 = 5$ Yes

CALC Check

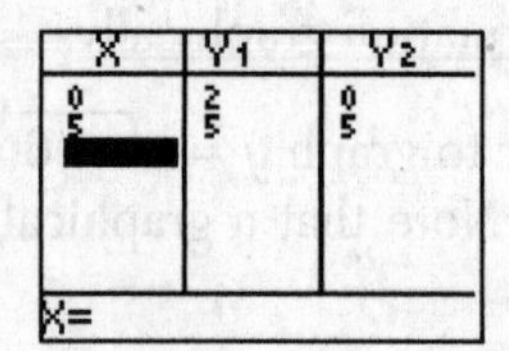

A calculator check is especially important for radical equations in case you have forgotten to check for extraneous roots.

The solution is $\boldsymbol{x = 5}$.

1.9 Rationalize the denominator by multiplying by the conjugate.

$$\frac{7}{x-\sqrt{2}} = \frac{7}{\left(x-\sqrt{2}\right)} \cdot \frac{\left(x+\sqrt{2}\right)}{\left(x+\sqrt{2}\right)} = \frac{7\left(x+\sqrt{2}\right)}{x^2-2} = \frac{7x+7\sqrt{2}}{x^2-2}$$

The rationalized form is $\boldsymbol{\frac{7x+7\sqrt{2}}{x^2-2}}$.

1.10 Convert the radical to exponent form, $\sqrt[m]{a^n} = a^{\frac{n}{m}}$.

$\sqrt[4]{16a^6b^4} = (16a^6b^4)^{\frac{1}{4}}$

Distribute the exponent. $= 16^{\frac{1}{4}}(a^6)^{\frac{1}{4}}(b^4)^{\frac{1}{4}}$

$16^{\frac{1}{4}} = 2$ and $(a^n)^m = a^{nm}$. $= 2a^{\frac{6}{4}}b^{\frac{4}{4}}$

Simplify the exponents. $= 2a^{\frac{3}{2}}b$

Find the root symbol in the MATH menu. Knowing $\sqrt[4]{16} = 2$ eliminates two of the choices.

CALC Check

```
4×√(16)
                2
```

The correct choice is **(2)**.

2. COMPLEX NUMBERS

2.1 COMPLEX NUMBERS

The imaginary unit is $i = \sqrt{-1}$ and has the property $i^2 = -1$. A complex number is $a + bi$, where a and b are real numbers. The conjugate of $a + bi$ is $a - bi$.

2.2 POWERS OF *i*

The first four powers of i are $i^0 = 1$, $i^1 = i$, $i^2 = -1$, and $i^3 = -i$. Successive powers just repeat the pattern: $1, i, -1, -i$. For any whole number power of i, you can find the result by raising i to the remainder obtained when the exponent is divided by 4.

2.3 OPERATIONS WITH COMPLEX NUMBERS

Radicals with a negative value in the radicand should be written in terms of i before performing any operations. To add/subtract complex numbers, add/subtract only like terms. To multiply, use the distributive property and simplify powers of i, such as $i^2 = -1$. Then combine like terms. To divide, multiply by the conjugate to rationalize the denominator, and then simplify.

Practice Exercises

2.1 The expression i^{25} is equivalent to
(1) 1 (3) i
(2) –1 (4) $-i$

2.2 What is the value of $i^{99} - i^3$?
(1) 1 (3) $-i$
(2) i^{96} (4) 0

2.3 What is the sum of $5 - 3i$ and the conjugate of $3 + 2i$?
(1) $2 + 5i$ (3) $8 + 5i$
(2) $2 - 5i$ (4) $8 - 5i$

2.4 What is the product of $\sqrt{-2}$ and $\sqrt{-18}$?
(1) 6 (3) $6i$
(2) –6 (4) $-6i$

2.5 The complex number $c + di$ is equal to $(3 + 2i)^2$. What is the value of c?
(1) 5 (3) 12
(2) 9 (4) 13

2.6 The expression $\dfrac{5-i}{7+3i}$ written in simplest $a+bi$ form is

(1) $\dfrac{32}{58}-\dfrac{22}{58}i$

(2) $\dfrac{16}{29}-\dfrac{11}{29}i$

(3) $\dfrac{38}{58}+\dfrac{8}{58}i$

(4) $\dfrac{19}{29}+\dfrac{4}{29}i$

2.7 The complex number –4 + 3*i* is shown in graphical form. Write and sketch its conjugate. What transformation relates the complex number and its conjugate?

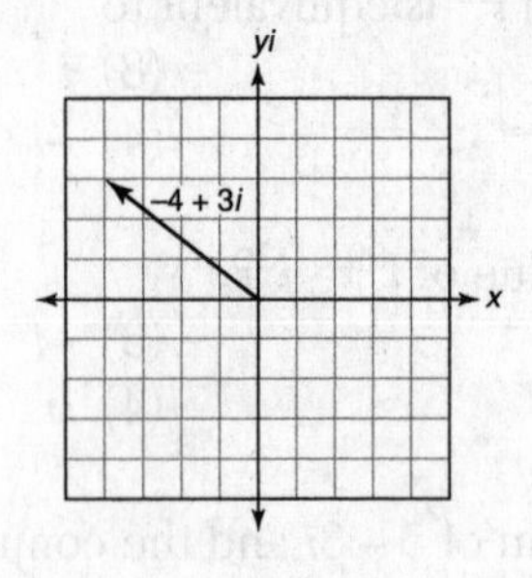

Solutions

2.1 The first four powers of i are $i^0 = 1$, $i^1 = i$, $i^2 = -1$, and $i^3 = -i$. Successive powers just repeat the pattern: 1, i, -1, $-i$. For any whole number power of i, you can find the result by raising i to the remainder obtained when the exponent is divided by 4.

$$\frac{25}{4} = 6 \text{ remainder } 1$$

$$i^{25} = i^1 = i$$

CALC Caution

```
NORMAL SCI ENG
FLOAT 0123456789
RADIAN DEGREE
FUNC PAR POL SEQ
CONNECTED DOT
SEQUENTIAL SIMUL
REAL a+bi re^θi
FULL HORIZ G-T
SET CLOCK 08/14/09 3:39PM
```

```
i^25
           -5E-13+i
i^26
          -1+4E-13i
```

The calculator has round-off errors on some calculations. You should know when to expect these and how to interpret the results. Round-off errors usually display in scientific notation. They often appear in arithmetic with complex numbers. On the calculator, $i^{25} = -5\text{E}{-13} + i$ where $-5\text{E}{-13}$ is calculator notation for -5×10^{-13}, which is very close to 0. This is calculator error, not calculator accuracy. The correct answer is $0 + 1i$ or just i. Similarly, on the calculator, $i^{26} = -1 + 4\text{E}{-13}i$. Again, interpret $4\text{E}{-13}$ as 0 to get the correct value of $-1 + 0i$ or -1.

The correct choice is **(3)**.

2.2

$$\frac{99}{4} = 24 \text{ remainder } 3$$
$$i^{99} = i^3 = -i$$
$$i^3 = -i$$
$$i^{99} - i^3 = -i - (-i) = 0$$

CALC Check

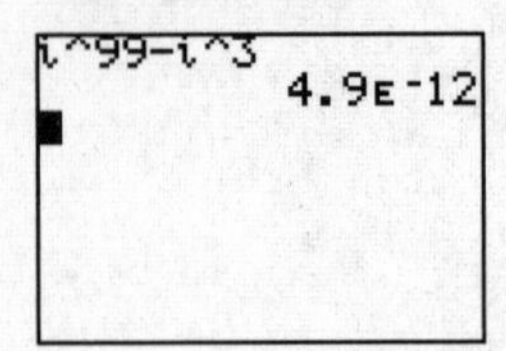

The calculator gives 4.9E–12. Allow for calculator round-off error. The answer is essentially 0.

The correct choice is **(4)**.

2.3 The conjugate of $a + bi$ is $a - bi$, so the conjugate of $3 + 2i$ is $3 - 2i$.

$$(5 - 3i) + (3 - 2i) = 8 - 5i$$

CALC Check

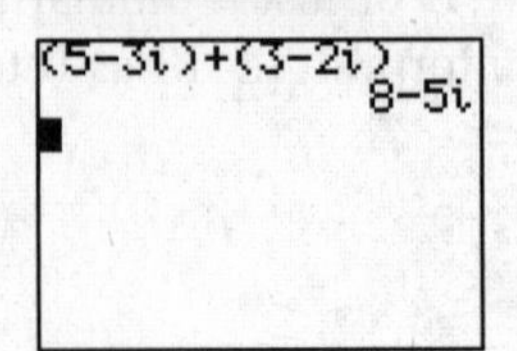

The correct choice is **(4)**.

2.4 Write radicals of negative numbers in terms of i before doing any operations.

$\sqrt{-2} = i\sqrt{2}$

$\sqrt{-18} = i\sqrt{18}$

$i^2 = -1$. Multiply the radicands and simplify.

$(i\sqrt{2})(i\sqrt{18})$

$i^2\sqrt{36}$

-6

CALC Caution

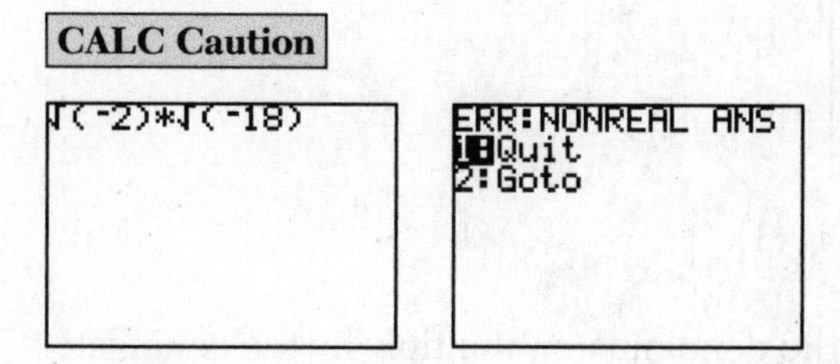

You must be in $a + bi$ mode to do operations with imaginary numbers.

CALC Check

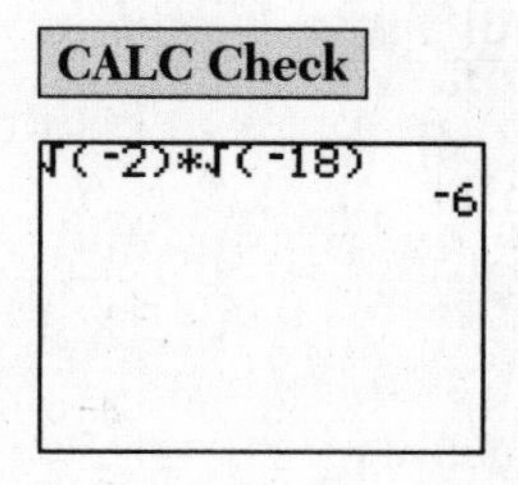

The correct choice is **(2)**.

2.5
$$
\begin{aligned}
(3+2i)^2 &= (3+2i)(3+2i) \\
&= 9 + 6i + 6i + 4i^2 \\
&= 9 + 12i + 4(-1) \\
&= 5 + 12i \\
&= c + di \\
c &= 5
\end{aligned}
$$

CALC Check

```
(3+2i)²
                5+12i
```

The correct choice is **(1)**.

2.6 To rationalize the denominator, multiply by the conjugate.

$$
\begin{aligned}
\frac{5-i}{7+3i} &= \frac{(5-i)}{(7+3i)} \cdot \frac{(7-3i)}{(7-3i)} \\
&= \frac{35-15i-7i+3i^2}{49-9i^2} \\
&= \frac{35-22i+(-3)}{49-(-9)} \\
&= \frac{32-22i}{58} \\
&= \frac{32}{58} - \frac{22}{58}i \\
&= \frac{16}{29} - \frac{11}{29}i
\end{aligned}
$$

CALC Check

```
(5-i)/(7+3i)
.5517241379-.37…
Ans▸Frac
   16/29-11/29i
```

On the calculator 11/29i means $\dfrac{11}{29}i$, not $\dfrac{11}{29i}$ (they are not the same).

The correct choice is **(2)**.

2.7 The conjugate of $-4 + 3i$ is $-4 - 3i$. It is depicted graphically below.

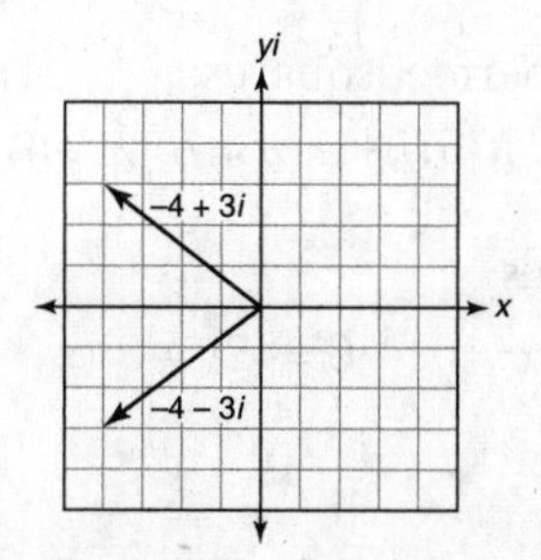

The graph of a complex number is reflected over the x-axis to form the graph of the conjugate.

3. POLYNOMIALS

3.1 OPERATIONS WITH POLYNOMIALS

To add/subtract polynomials, add/subtract only like terms. To multiply, use the distributive property. Remember to distribute negative terms completely. To divide, factor the numerator and denominator. Then reduce to lowest terms.

3.2 FACTORING POLYNOMIALS

- Greatest Common Factor:

$$3x^3 + 18x^2 - 9x = 3x(x^2 + 6x - 3)$$

- Factor by Grouping:

$$2x^3 - x^2 + 8x - 4 = x^2(2x - 1) + 4(2x - 1) = (2x - 1)(x^2 + 4)$$

- Difference of two perfect squares:

$$a^2 - b^2 = (a + b)(a - b)$$

- Quadratic trinomials:

$$2x^2 + x - 6 = (2x - 3)(x + 2)$$

CALC Tip

```
Plot1 Plot2 Plot3
\Y1=2X²+X-6
\Y2=
\Y3=
\Y4=
\Y5=
\Y6=
\Y7=
```

X	Y1
-4	22
-3	9
-2	0
-1	-5
0	-6
1	-3
2	4

X= -4

If you are having difficulty factoring a quadratic trinomial, try finding a root in the table on your calculator. In this example, $x = -2$ is a root, so $(x + 2)$ is a factor of $2x^2 + x - 6$. Note how the sign in the factor is the opposite of the sign of the root. Knowing one factor will help you find the other factor.

If you cannot find a root in the table, use CALC (press 2nd TRACE) and **2:zero**. Then return to the home screen (2nd MODE) and select MATH **1:frac** to see the root in fraction form. If the root is $x = 3/2$, then the factor is $(2x - 3)$. Note the placement the numerator and denominator and the opposite sign.

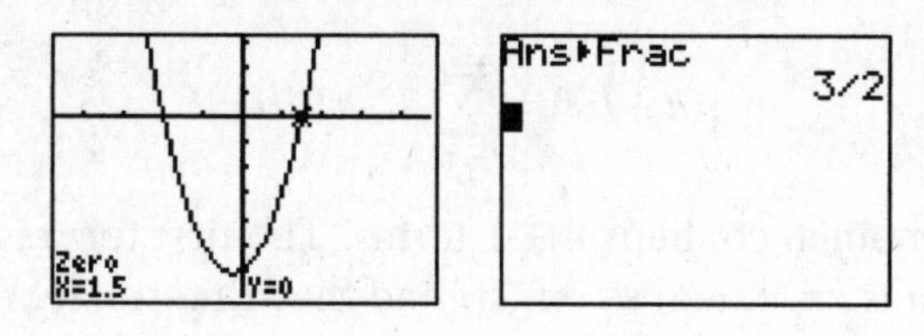

If the root found with CALC **2:zero** is irrational, the trinomial does not factor.

- Factor completely: Use any combination of factoring techniques, factoring out the common factor first, until all factors are prime.

$$2x^3 + 10x^2 + 12x = 2x(x^2 + 5x + 6) = 2x(x + 3)(x + 2)$$

CALC Check Caution

A table can check if you have factored correctly. However, the table cannot tell if you have factored completely. The table below shows that Y2 and Y1 both have the same values, but Y2 is not completely factored. Y3 also has the same values and is factored completely.

Plot1 Plot2 Plot3
\Y1■2X^3+10X²+12X
\Y2■2X(X²+5X+6)
\Y3■2X(X+3)(X+2)
\Y4=
\Y5=

X	Y1	Y2
-4	-16	-16
-3	0	0
-2	0	0
-1	-4	-4
0	0	0
1	24	24
2	80	80

X= -4

X	Y1	Y3
-4	-16	-16
-3	0	0
-2	0	0
-1	-4	-4
0	0	0
1	24	24
2	80	80

X= -4

3.3 BINOMIAL EXPANSION

Use these binomial theorem formulas to find powers of binomials or a specific term in the expanded power. They are given on the reference sheet.

$$(a+b)^n = {}_nC_0a^nb^0 + {}_nC_1a^{n-1}b^1 + {}_nC_2a^{n-2}b^2 + \cdots + {}_nC_na^0b^n$$

$$(a+b)^n = \sum_{r=0}^{n} {}_nC_ra^{n-r}b^r$$

The polynomial contains $n + 1$ terms. The first term is $r = 0$, the second term is $r = 1$, and so on. To find the kth term, let $r = k - 1$.

Practice Exercises

3.1 Subtract $\frac{3}{2}x^2 - \frac{1}{3}x + 3$ from $\frac{1}{2}x^2 + \frac{5}{3}x - 2$

(1) $-x^2 + 2x - 5$

(2) $-x^2 + \frac{4}{3}x - 5$

(3) $-x^2 + \frac{4}{3}x + 1$

(4) $x^2 - 2x + 5$

3.2 The expression $\left(\frac{1}{3}x - \frac{3}{2}\right)^2$ is equivalent to

(1) $\frac{1}{9}x^2 - \frac{9}{4}$

(2) $\frac{1}{9}x^2 + \frac{9}{4}$

(3) $\frac{1}{9}x^2 - x + \frac{9}{4}$

(4) $\frac{1}{9}x^2 - 3x + \frac{9}{4}$

3.3 When $\frac{2}{3}x^2 + \frac{1}{2}x$ is divided by $\frac{1}{2}x$ the result is

(1) $\frac{1}{3}x$

(2) $\frac{4}{3}x$

(3) $\frac{1}{3}x + 1$

(4) $\frac{4}{3}x + 1$

3.4 Factor completely: $x^2y^2 - 25$

3.5 Factor completely: $6x^2 - x - 12$

3.6 Factor completely: $x^3y - 4xy^3$

3.7 Factor completely: $2x^2 + 12x - 54$

3.8 What is the middle term in the expansion of $(x + y)^4$?

(1) x^2y^2
(2) $2x^2y^2$
(3) $6x^2y^2$
(4) $4x^2y^2$

3.9 What is the fourth term in the expansion of $(4x - 3)^5$?

(1) $-1620x$
(2) $1620x$
(3) $-4320x^2$
(4) $4320x^2$

3.10 Write out the expansion of $(2x - y)^4$.

3.11 Factor completely: $3x^3 - x^2 - 12x + 4$

Solutions

3.1 The expression "subtract A from B" is written algebraically as $B - A$, not $A - B$. If you are subtracting polynomials, use parentheses. Remember to change subtraction to addition of the opposite and distribute the negative through all terms of the subtracted polynomial.

Finally, combine like terms by adding coefficients.

$$\left(\frac{1}{2}x^2+\frac{5}{3}x-2\right)-\left(\frac{3}{2}x^2-\frac{1}{3}x+3\right)$$

$$=\left(\frac{1}{2}x^2+\frac{5}{3}x-2\right)+{}^{-}\left(\frac{3}{2}x^2-\frac{1}{3}x+3\right)$$

$$=\left(\frac{1}{2}x^2+\frac{5}{3}x-2\right)+\left(-\frac{3}{2}x^2+\frac{1}{3}x-3\right)$$

$$=\left(\frac{1}{2}-\frac{3}{2}\right)x^2+\left(\frac{5}{3}+\frac{1}{3}\right)x+(-2-3)$$

$$=-x^2+2x-5$$

The correct choice is **(1)**.

3.2 Write as a product of binomials and distribute.

$$\left(\frac{1}{3}x-\frac{3}{2}\right)^2=\left(\frac{1}{3}x-\frac{3}{2}\right)\left(\frac{1}{3}x-\frac{3}{2}\right)=\frac{1}{9}x^2-\frac{1}{2}x-\frac{1}{2}x+\frac{9}{4}=\frac{1}{9}x^2-x+\frac{9}{4}$$

The correct choice is **(3)**.

3.3 When dividing a polynomial by a monomial, split the problem into separate fractions with a common denominator.

$$\frac{\frac{2}{3}x^2+\frac{1}{2}x}{\frac{1}{2}x}=\frac{\frac{2}{3}x^2}{\frac{1}{2}x}+\frac{\frac{1}{2}x}{\frac{1}{2}x}=\left(\frac{\frac{2}{3}}{\frac{1}{2}}\right)x+1=\frac{4}{3}x+1$$

The correct choice is **(4)**.

3.4 This is a difference of two perfect squares, $a^2 - b^2 = (a+b)(a-b)$.

$$x^2y^2 - 25 = (xy+5)(xy-5)$$

3.5 $6x^2 - x - 12 = (3x+4)(2x-3)$

You should check factored trinomials by multiplying them.

$$(3x+4)(2x-3) = 6x^2 - 9x + 8x - 12 = 6x^2 - x - 12$$

Pay careful attention to the middle term, $-9x + 8x = -x$.

CALC Tip

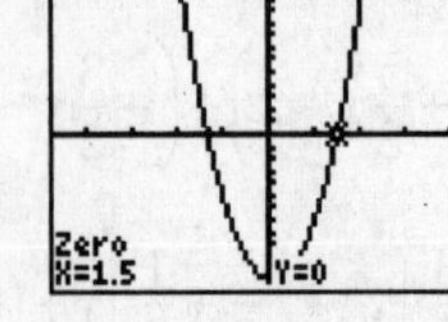

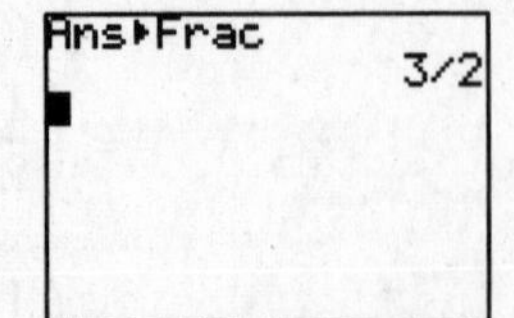

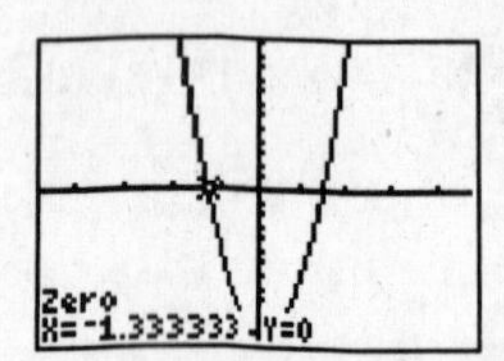

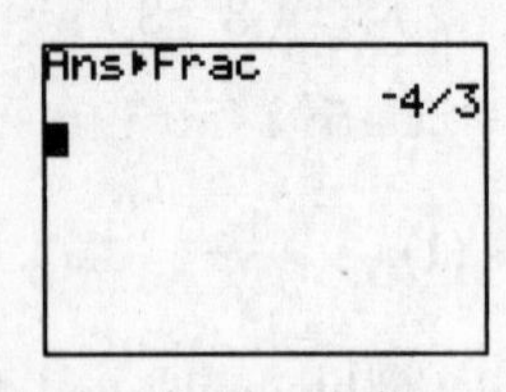

Since $x = 3/2$, $(2x - 3)$ is a factor. Since $x = -4/3$, $(3x + 4)$ is a factor.

3.6 Begin by factoring out the greatest common factor, and then factor the remaining polynomial. In this problem, the remaining polynomial is a difference of perfect squares.

$$\begin{aligned} x^3y - 4xy^3 &= xy(x^2 - 4y^2) \\ &= xy(x+2y)(x-2y) \end{aligned}$$

3.7 Begin by factoring out the greatest common factor, and then factor the remaining polynomial.

$$2x^2 + 12x - 54 = 2(x^2 + 6x - 27)$$
$$= 2(x + 9)(x - 3)$$

3.8 On the reference sheet, find the formula for expanding the power of a binomial: $(a+b)^n = \sum_{r=0}^{n} {}_nC_r a^{n-r} b^r$

Note two things:

(1) The summation starts at $r = 0$. This means the expansion of $(a + b)^n$ contains $n + 1$ terms. Since a fourth power will have five terms, the middle term will be the third term.

(2) Since the first term has $r = 0$ (not 1), the second term will be $r = 1$ and the third term will be $r = 2$.

In the given problem, $n = 4$, $r = 2$, $a = x$, and $b = y$. The third term of the expansion is

$${}_4C_2 x^{4-2} y^2 = 6x^2y^2$$

The correct choice is **(3)**.

3.9 Find the necessary formula on the reference sheet:

$$(a+b)^n = \sum_{r=0}^{n} {}_nC_r a^{n-r} b^r$$

In this problem, $n = 5$, $a = 4x$, and $b = -3$. The fourth term of the expansion will have $r = 4 - 1 = 3$. The fourth term is:

$${}_5C_3(4x)^{5-3}(-3)^3 = 10(4x)^2(-27) = -270(16x^2) = -4320x^2$$

The correct choice is **(3)**.

3.10 Find the necessary formula on the reference sheet:

$$(a+b)^n = \sum_{r=0}^{n} {}_nC_r a^{n-r} b^r$$

In this problem, $n = 4$, $a = 2x$, and $b = -y$:

$$\begin{aligned}(2x-y)^4 &= {}_4C_0(2x)^{4-0}(-y)^0 + {}_4C_1(2x)^{4-1}(-y)^1 + {}_4C_2(2x)^{4-2}(-y)^2 \\ &\quad + {}_4C_3(2x)^{4-3}(-y)^3 + {}_4C_4(2x)^{4-4}(-y)^4 \\ &= (1)(2x)^4(1) + (4)(2x)^3(-y) + (6)(2x)^2(y^2) + (4)(2x)^1(-y^3) \\ &\quad + (1)(1)(y^4) \\ &= (1)(16x^4) + (-4y)(8x^3) + (6y^2)(4x^2) + (-4y^3)(2x) + (1)(y^4)\end{aligned}$$

$$(2x-y)^4 = 16x^4 - 32x^3y + 24x^2y^2 - 8xy^3 + y^4$$

3.11 Begin with factor by grouping, then factor the difference of perfect squares. Be careful when factoring −4 from the second group to have the correct signs.

$$\begin{aligned}3x^3 - x^2 - 12x + 4 &= x^2(3x-1) - 4(3x-1) \\ &= (3x-1)(x^2-4) \\ &= (3x-1)(x-2)(x+2)\end{aligned}$$

4. QUADRATIC AND HIGHER-ORDER POLYNOMIALS

4.1 QUADRATIC FORMULA

If $ax^2 + bx + c = 0$, with $a \neq 0$, then $x = \dfrac{-b \pm \sqrt{b^2 - 4ac}}{2a}$.

Tip: If you have trouble remembering the quadratic formula accurately, learn to sing it! Look up "quadratic formula song" on the Internet. There are many different versions. Pick one and sing along. This works only if you sing out loud.

The discriminant is the expression under the radical sign: $b^2 - 4ac$. It can be used to determine the type of roots for the equation.

Discriminant $b^2 - 4ac$	positive, a perfect square	positive, not a perfect square	0	negative
Roots	rational and unequal	irrational and unequal	rational and equal	complex conjugates
Representative Graph ($a > 0$)	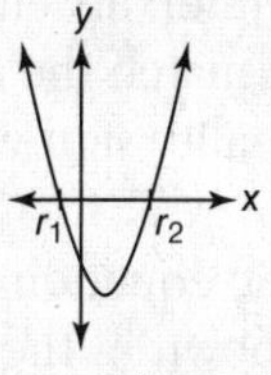 r_1, r_2, both rational	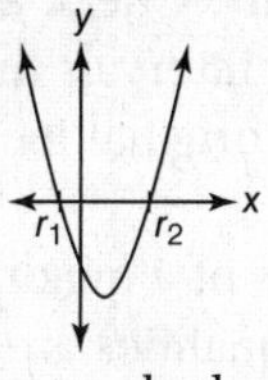 r_1, r_2, both irrational	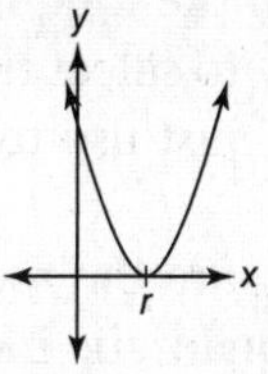	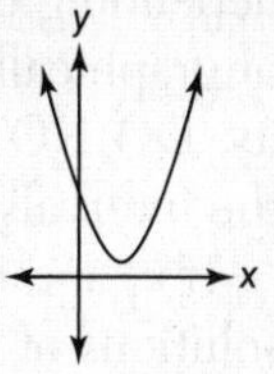

If r_1 and r_2 are the roots of the quadratic equation, then the sum of the roots is $r_1 + r_2 = -\dfrac{b}{a}$ and the product of the roots is $r_1 \cdot r_2 = \dfrac{c}{a}$.

4.2 COMPLETING THE SQUARE

The graph of $y = ax^2 + bx + c$ is a parabola with the equation written in standard form. The vertex form, $y = a(x - h)^2 + k$ of the parabola can be obtained by completing the square.

$$y = ax^2 + bx + c$$

$$y = a\left(x^2 + \frac{b}{a}x\right) + c$$

$$y = a\left(x + \frac{b}{2a}\right)^2 - a\left(\frac{b}{2a}\right)^2 + c$$

so that $h = -\dfrac{b}{2a}$ and $k = c - \dfrac{b^2}{4a}$.

4.3 SOLVING QUADRATIC EQUATIONS AND INEQUALITIES

To solve quadratic equations, set the equation equal to zero. Either use the quadratic formula or factor. Then set each factor equal to zero and solve (Zero Product Property.)

To solve quadratic inequalities, solve for the roots of the equation. Graph the roots on a number line with open or closed circles, depending on the inequality. Check all intervals either numerically or graphically to select the intervals that match the original inequality. DO NOT just use the original inequality sign without checking the intervals.

If r_1 and r_2 are the roots of a quadratic equation with $r_1 < r_2$, the solutions of quadratic inequalities are shown in the following table. For $ax^2 + bx + c \geq 0$ and $ax^2 + bx + c \leq 0$, the intervals are the same as shown in the table except they will be closed at the roots (solid circles).

	$ax^2 + bx + c > 0$	$ax^2 + bx + c < 0$
$a > 0$ 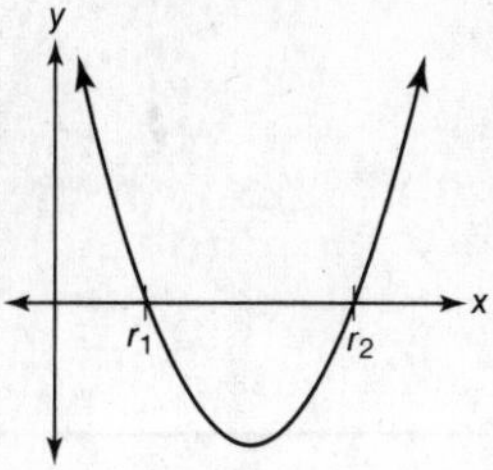	$x < r_1$ or $x > r_2$ 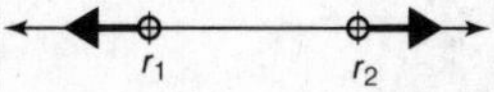	$r_1 < x < r_2$ 
$a < 0$ 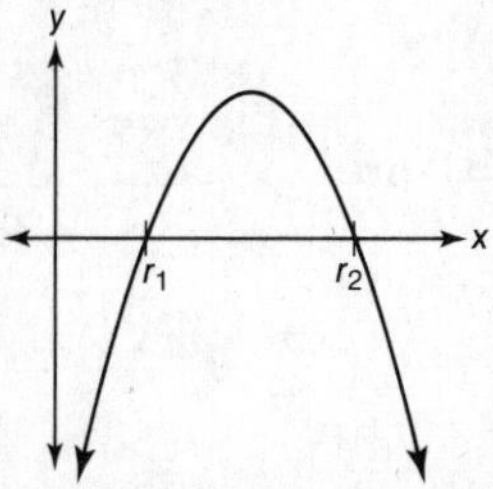	$r_1 < x < r_2$	$x < r_1$ or $x > r_2$ 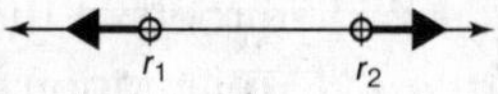

4.4 HIGHER-DEGREE POLYNOMIALS

If the polynomial factors, set each factor equal to zero and solve. Use the quadratic formula for trinomial factors. On a graph, the x-intercepts, if any, are the real roots of polynomials. These can be estimated from where the graph crosses the x-axis. A root is the x-value only, not the coordinates of the intercept.

Practice Exercises

4.1 Express, in simplest $a + bi$ form, the roots of the equation $x^2 + 5 = 4x$.

4.2 The roots of the equation $5x^2 - 2x + 1 = 0$ are
(1) real, rational, and unequal
(2) real, rational, and equal
(3) real, irrational, and unequal
(4) imaginary

4.3 If the equation $x^2 - kx - 36 = 0$ has $x = 12$ as one root, what is the value of k?
(1) 9 (3) 3
(2) –9 (4) –3

4.4 Which quadratic equation has the roots $3 + i$ and $3 - i$?
(1) $x^2 + 6x - 10 = 0$
(2) $x^2 + 6x + 8 = 0$
(3) $x^2 - 6x + 10 = 0$
(4) $x^2 - 6x - 8 = 0$

4.5 Ralph plans to solve the equation $x^2 - 10x = 36$ by completing the square. What number should he add to both sides of the equation?

4.6 If the equation of the parabola $y = 2x^2 - 12x + 5$ is written in the form $y = 2(x - h)^2 + k$, what are the values of h and k?

4.7 What is the solution set of the inequality $x^2 + 4x - 5 < 0$?

(1) $\{x|x < -1 \text{ or } x > 5\}$

(2) $\{x|x < -5 \text{ or } x > 1\}$

(3) $\{x|-1 < x < 5\}$

(4) $\{x|-5 < x < 1\}$

4.8 Sketch the graph of $y > x^2 - 3x - 4$.

4.9 Solve the following system of equations algebraically:

$$x^2 + y^2 = 25$$
$$2y - x = 5$$

4.10 Solve the following system of equations algebraically:

$$\frac{x+y}{x+3} = x - 2$$
$$y = x + 6$$

4.11 Solve for x: $(x^2 + 3x - 5)(2x + 7) = 0$. Express irrational roots in radical form.

4.12 Solve for x: $x^4 = 10x^2 + 24$. Express roots in simplest radical form or simplest $a + bi$ form as appropriate.

4.13 The graphs of $f(x) = -\frac{1}{3}x^3 + x^2 + 4x - 4$ and $y = 2$ are shown below. Use the graphs to estimate the solutions to the equation $-\frac{1}{3}x^3 + x^2 + 4x - 4 = 2$ to the nearest tenth.

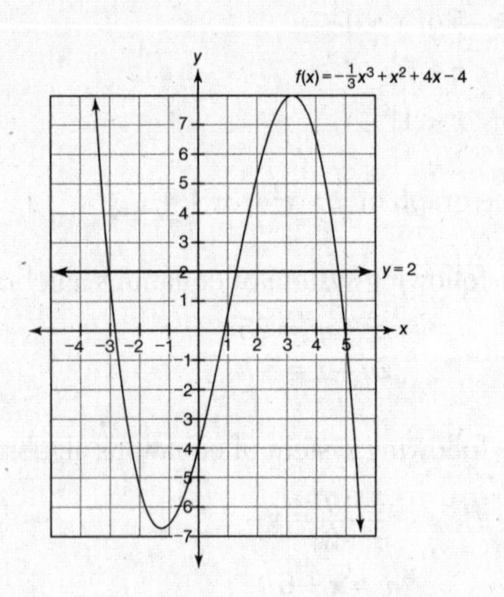

4.14 What is the solution set for the equation $(x + 3)^2(1 - x)(x^2 - 4) = 0$?

(1) $\{-3, 1, 2\}$
(2) $\{1, 2, 3\}$
(3) $\{-2, 1, 2, 3\}$
(4) $\{-3, -2, 1, 2\}$

Solutions

4.1 Rewrite the equation in standard form and use the quadratic formula. (Do not try to factor; the problem says the answers are imaginary.)

$$x^2 + 5 = 4x$$
$$x^2 - 4x + 5 = 0 \quad a = 1, b = -4, c = 5$$

$$x = \frac{-b \pm \sqrt{b^2 - 4ac}}{2a}$$

$$= \frac{-(-4) \pm \sqrt{(-4)^2 - 4(1)(5)}}{2(1)}$$

$$= \frac{4 \pm \sqrt{-4}}{2}$$

$$= \frac{4}{2} \pm \frac{2i}{2}$$

$$= 2 \pm i$$

Be very careful of signs, especially when using the calculator. Failure to use parentheses when squaring a negative number will result in a wrong answer.

CALC Caution

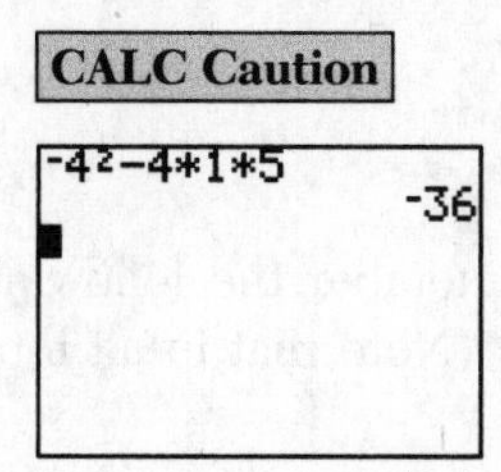

The solutions are $\mathbf{2 \pm i}$.

4.2 You could just solve the equation with the quadratic formula. Since the question asks only what kind of roots the equation has, it is quicker to evaluate the discriminant, $b^2 - 4ac$. In the problem, $a = 5$, $b = -2$, and $c = 1$. So $b^2 - 4ac = (-2)2 - 4(5)(1) = -16$. Since the negative discriminant is under the radical in the quadratic formula, the roots are imaginary.

The correct choice is **(4)**.

4.3 If $x = 12$ is a root, then the equation must check when 12 is substituted for x.

$$\begin{aligned} 12^2 - k(12) - 36 &= 0 \\ 144 - 12k - 36 &= 0 \\ 108 - 12k &= 0 \\ 108 &= 12k \\ k &= 9 \end{aligned}$$

You can also use sum and product of the roots formulas to solve. However, evaluating the root is easier. In case you did it with the sum and product of the roots formulas, here is that method. In this equation, $a = 1$, $b = -k$, and $c = -36$. The product of the roots is $r_1 \cdot r_2 = \frac{c}{a}$ or $12r_2 = -36$, so $r_2 = -3$. The sum of the roots is $r_1 + r_2 = -\frac{b}{a}$ or $12 + (-3) = k$, so $k = 9$.

The correct choice is **(1)**.

4.4 One way to do this problem is to remember the following facts about the roots of quadratic equations. (Note that in all four choices, $a = 1$.)

The sum of the roots is $-\frac{b}{a}$. $(3 + i) + (3 - i) = 6$, so $-\frac{b}{1} = 6$ and $b = -6$. This eliminates choices (1) and (2).

The product of the roots is $\frac{c}{a}$. $(3+i)(3-i)=9-i^2=9-(-1)=10$,

so $\frac{c}{1}=10$ and $c=10$. This is choice (3).

The correct choice is **(3)**.

4.5 To complete the square for a binomial of the form x^2+bx, you must add $\left(\frac{b}{2}\right)^2$.

In $x^2-10x=36$, $b=-10$. If Ralph added $\left(\frac{-10}{2}\right)^2=(-5)^2=25$ to both sides of his equation, he would have $x^2-10x+25=36+25$. The left side of this equation is now equivalent to $(x-5)^2$.

The correct answer is **25**.

4.6 This can be done by completing the square. First, factor 2 out of $2x^2-12x$.

$$\begin{aligned} y &= 2x^2-12x+5=2(x^2-6x)+5 \\ &= 2(\underbrace{x^2-6x+9}-9)+5 \\ &= 2((x-3)^2-9)+5 \\ &= 2(x-3)^2-18+5 \\ &= 2(x-3)^2-13 \end{aligned}$$

Therefore, $h=3$ and $k=-13$. Be careful with the sign of h. In the formula $y=2(x-h)^2+k$, the h is subtracted.

If you recognize $y=2(x-h)^2+k$ as the vertex form of the equation of a parabola where (h,k) are the coordinates of the vertex, you can find h and k by finding the vertex of the given parabola using the formula $x = -\frac{b}{2a}$, by graphing the parabola, or by making a table of values.

X	Y_1	
0	5	
1	-5	
2	-11	
3	-13	
4	-11	
5	-5	
6	5	

X=0

The vertex is at (3, –13). Those are the values of h and k. Be sure to show work on your paper to support calculator based solutions.

The correct answer is: $h = 3$ and $k = -13$

4.7 <u>To solve a quadratic inequality algebraically</u>:

Find the roots for equality.

$$x^2 + 4x - 5 = 0$$
$$(x + 5)(x - 1) = 0$$
$$x = -5 \text{ or } x = 1$$

Graph the roots on a number line using open circles because of the strict inequality.

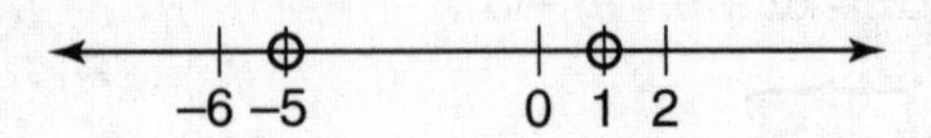

Check a value from each interval on the number line in the original inequality and shade intervals that check.

Check $x = -6$: $(-6)^2 + 4(-6) - 5 \overset{?}{<} 0$

$7 < 0$ No

Check $x = 0$: $(0)^2 + 4(0) - 5 \overset{?}{<} 0$

$-5 < 0$ Yes

Check $x = 2$: $(2)^2 + 4(2) - 5 \overset{?}{<} 0$

$7 < 0$ No

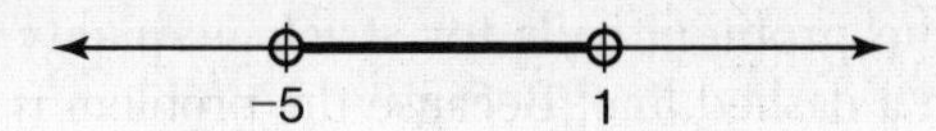

Read the solution off the number line:

$$-5 < x < 1$$

<u>To solve a quadratic inequality graphically:</u>

Graph $y = x^2 + 4x - 5$ on the calculator.

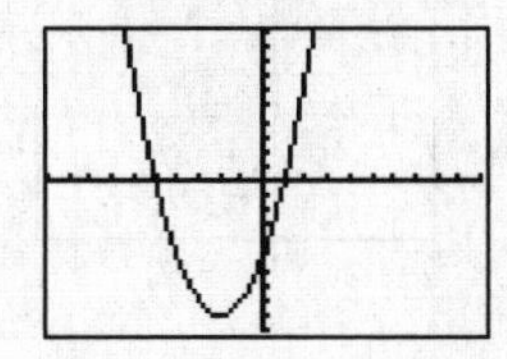

Since the problem states $x^2 + 4x - 5 < 0$, you want the x-values where the parabola is below the x-axis. These are $-5 < x < 1$.

The correct choice is **(4)**.

4.8 Make a table of values with your calculator.

X	Y1	
-2	6	
-1	0	
0	-4	
1	-6	
2	-6	
3	-4	
4	0	

X= -2

Find and label the vertex (1.5, 6.25).

Because the problem calls for strict inequality, >, graph the parabola with a dashed line. Because the problem is $y >$, shade *up* from the graph of the parabola.

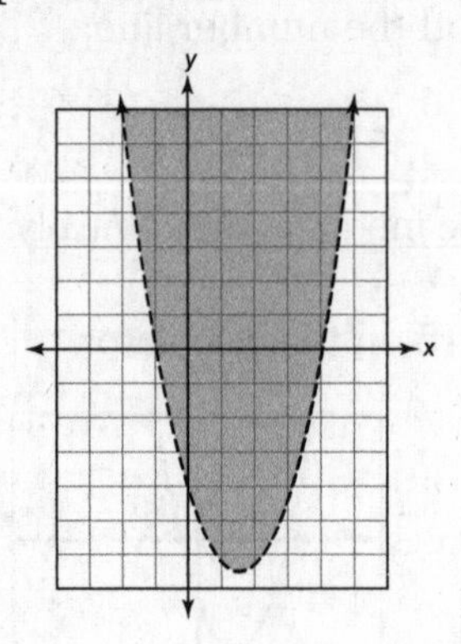

4.9 To solve a quadratic-linear system of equations algebraically, use substitution. Solve the linear equation for either variable, and then substitute that result into the quadratic equation.

Solve the linear equation for x.	$2y - x = 5 \quad \rightarrow \quad x = 2y - 5$
Substitute into the quadratic equation.	$(2y - 5)^2 + y^2 = 25$
	$(2y - 5)(2y - 5) + y^2 = 25$
Solve for y.	$4y^2 - 10y - 10y + 25 + y^2 = 25$
	$5y^2 - 20y = 0$
	$5y(y - 4) = 0$
	$5y = 0 \quad \text{or} \quad y - 4 = 0$
	$y = 0 \quad \text{or} \quad y = 4$
Find the x-value for each y-value.	$x = 2(0) - 5 \text{ or } x = 2(4) - 5$
	$x = -5 \quad \text{or} \quad x = 3$
Write the solutions as ordered pairs.	$(-5, 0)$ or $(3, 4)$

Unlike rational equations and absolute value equations, quadratic-linear systems will not lead to extraneous roots. Still, it is a good idea to check your answers in the original equations.

The solutions are **(–5, 0)** and **(3, 4)**.

4.10 To solve a nonlinear system of equations, use substitution. Since you know $y = x + 6$, substitute $x + 6$ in place of y in the first equation.

$$\frac{x+y}{x+3} = x-2 \quad \rightarrow \quad \frac{x+x+6}{x+3} = x-2 \quad \rightarrow \quad \frac{2x+6}{x+3} = x-2$$

Simplify the rational expression.

$$\frac{2\cancel{(x+3)}}{\cancel{x+3}} = x-2$$

$$2 = x-2$$

$$x = 4$$

$$y = 4+6 = 10$$

Check the solution in both original equations.

$$\frac{4+10}{4+3} \stackrel{?}{=} 4-2 \quad \text{and} \quad 10 \stackrel{?}{=} 4+6$$

$$2 = 2 \text{ Yes} \quad \text{and} \quad 10 = 10 \text{ Yes}$$

If you do not simplify the rational expression before solving, you will get a more difficult quadratic equation leading to two solutions, (4, 10) and (–3, 3). The second ordered pair does not check in the original equations.

The solution is **(4, 10)**.

4.11 This equation is already set equal to zero and factored. Apply the Zero Product Property:

$$(x^2 + 3x - 5)(2x + 7) = 0$$

$$x^2 + 3x - 5 = 0 \quad \text{or} \quad 2x + 7 = 0$$

$$x = \frac{-3 \pm \sqrt{3^2 - 4(1)(-5)}}{2(1)} \quad \text{or} \quad 2x = -7$$

$$x = \frac{-3 \pm \sqrt{29}}{2} \quad \text{or} \quad x = -\frac{7}{2}$$

The solution set is $\left\{\frac{-3 \pm \sqrt{29}}{2}, -\frac{7}{2}\right\}$.

4.12 If you let $x^2 = A$, then $x^4 = A^2$ and the equation can be written in quadratic form.

$$x^4 = 10x^2 + 24 \quad \rightarrow \quad A^2 = 10A + 24$$

Rewrite the equation in standard form and factor.

$$A^2 - 10A - 24 = 0$$

$$(A - 12)(A + 2) = 0$$

$$A = 12 \text{ or } A = -2$$

Since $A = x^2$, $x^2 = 12$ or $x^2 = -2$.

$$x = \pm\sqrt{12} \quad \text{or} \quad x = \pm\sqrt{-2}$$

$$x = \pm 2\sqrt{3} \quad \text{or} \quad x = \pm i\sqrt{2}$$

The solutions are $\{\pm 2\sqrt{3}, \pm i\sqrt{2}\}$.

4.13 Solving $-\frac{1}{3}x^3 + x^2 + 4x - 4 = 2$ is the same as finding where the function $f(x) = -\frac{1}{3}x^3 + x^2 + 4x - 4$ has a y-value of 2. This is where the graph of f intersects the horizontal line $y = 2$. This occurs at $x = -3$, $x \approx 1.3$, and $x \approx 4.7$. Since you are estimating a value from a graph, credit will be given for values that are within 0.1. For $x \approx 1.3$, you could receive credit for $x \approx 1.2$, 1.3, or 1.4. For $x \approx 4.7$, $x \approx 4.6$, 4.7, or 4.8 would all be considered correct.

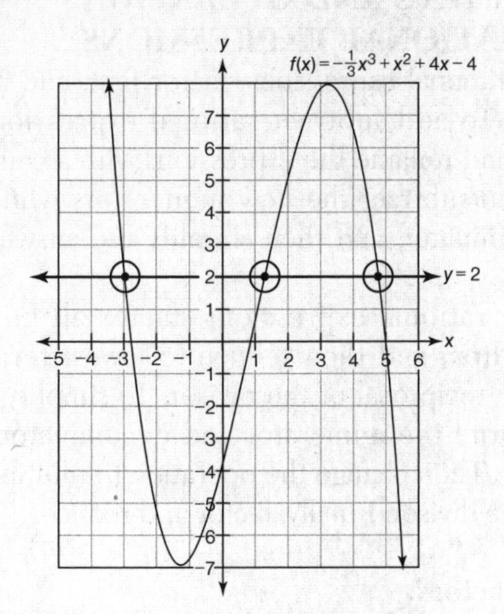

The solutions are $x = -3$, $x \approx 1.3$, and $x \approx 4.7$.

4.14 A polynomial partially factored is given; finish factoring and apply the Zero Product Property:

$(x + 3)^2(1 - x)(x^2 - 4) = 0$

$(x + 3)(x + 3)(1 - x)(x + 2)(x - 2) = 0$

$x + 3 = 0$ or $1 - x = 0$ or $x + 2 = 0$ or $x - 2 = 0$

$x = -3$ or $x = 1$ or $x = -2$ or $x = 2$. This is choice (4).

In the table, the roots are the x-values where the y-value is 0.

CALC Check

```
Plot1 Plot2 Plot3
\Y1=(X+3)²(1-X)(
X²-4)
\Y2=
\Y3=
\Y4=
\Y5=
\Y6=
```

X	Y1
-3	0
-2	0
1	0
2	0
3	-360

X=

The correct choice is **(4)**.

5. RATIONALS, ABSOLUTE VALUE, AND VARIATION

5.1 SIMPLIFYING AND OPERATIONS WITH RATIONAL EXPRESSIONS

To simplify rational expressions, factor first, and then reduce to lowest terms. To add/subtract rational expressions, factor the denominators and rename the expressions with a common denominator. Then add/subtract the new numerators while keeping the common denominator, and then simplify the answer by factoring and reducing.

To multiply rational expressions, factor all numerators and denominators first, and then reduce to lowest terms. To divide, multiply by the reciprocal of the divisor. To simplify complex fractions, first rewrite the numerator and denominator separately as single fractions. Then change the operation to multiplication by the reciprocal of the divisor. Finally, factor and reduce.

Opposite Factors:

$$\frac{a-b}{b-a}=-1$$

5.2 SOLVING RATIONAL EQUATIONS AND INEQUALITIES

To solve rational equations, rewrite the entire equation with a common denominator. Then equate the numerators and solve. This is equivalent to multiplying by the common denominator and then reducing. ALWAYS check solutions in the original equation for extraneous roots.

To solve rational inequalities, find the roots for the equation. Graph the roots of the equation and the roots of the denominators on a number line. Roots for the denominators will always be open circles. Roots for the numerator will be either open or closed circles, depending on the original inequality sign. Check all intervals either numerically or graphically to select the intervals that satisfy the original inequality. For nonlinear inequalities, do NOT just use the original inequality sign without checking the intervals.

5.3 SOLVING ABSOLUTE VALUE EQUATIONS AND INEQUALITIES

To solve absolute value equations, solve for two different cases. First assume that everything in the absolute value brackets is already positive. Just remove the brackets and solve. Next assume that everything inside the absolute value brackets is negative. Multiply everything inside the brackets by –1, remove the brackets, and solve. ALWAYS check the solutions in the original equation for extraneous roots.

To solve absolute value inequalities, solve for the roots to the equation. Graph the roots of the equation on a number line as open or closed circles, depending on the original inequality. Check all intervals either numerically or graphically to select the intervals that satisfy the original inequality. For nonlinear inequalities, do NOT just use the original inequality sign without checking the intervals.

5.4 DIRECT AND INVERSE VARIATION

Direct variation (proportion): If y varies directly with x, then $\frac{y_1}{x_1} = \frac{y_2}{x_2}$. This can also be written with a constant of variation $y = kx$ for some constant k. In a direct variation, x and y change by the same factor. For example, if x is doubled, y is doubled. The graph of a direct variation is a line through the origin.

Inverse variation: If y varies inversely with x, then $x_1y_1 = x_2y_2$. This can also be written with a constant of variation $y = \frac{k}{x}$ for some constant k. In an inverse variation, x and y change by reciprocal factors. For example, if x is doubled, y is halved. The graph of an inverse variation is a hyperbola.

Practice Exercises

5.1 What is the sum of $\frac{3}{x-3}$ and $\frac{x}{3-x}$?

(1) 1   (3) $\frac{x+3}{x-3}$

(2) –1   (4) 0

5.2 Express in simplest form: $\frac{12}{x^2-4} \cdot \frac{2-x}{3}$

5.3 If $f(x) = \frac{3x^2-27}{18x+30}$ and $g(x) = \frac{x^2-7x+12}{3x^2-7x-20}$, find $f(x) \div g(x)$ for all values of x for which the expression is defined. Express your answer in simplest form.

5.4 In simplest form, $\frac{\frac{1}{x^2}-\frac{1}{y^2}}{\frac{1}{y}+\frac{1}{x}}$, $x \neq 0, y \neq 0$, is equal to

(1) $\frac{x-y}{xy}$ (3) $x-y$

(2) $\frac{y-x}{xy}$ (4) $y-x$

5.5 Express $\dfrac{\frac{1}{x}-\frac{1}{x+1}}{\frac{2}{x-1}-\frac{1}{x}}$ as a fraction in simplest form.

5.6 What is the solution set of the equation $\dfrac{x+8}{x^2-4}-\dfrac{5}{x^2-2x}=\dfrac{2}{x}$?

(1) {1}
(2) {2}
(3) { }
(4) {1,2}

5.7 Solve for x and express your answer in simplest radical form:

$\dfrac{4}{x}-\dfrac{3}{x+1}=7.$

5.8 What is the solution set of the inequality $\dfrac{3x-2}{x+1}\geq 2$?

(1) $(-\infty, 4]$
(2) $[4, \infty)$
(3) $(-\infty, -1)\cup[4, \infty]$
(4) $(-1, 4]$

5.9 What is the solution set of the equation $|x^2-2x|=3x-6$?

(1) $\{2, \pm 3\}$
(2) $\{2\}$
(3) $\{\pm 3\}$
(4) $\{2, 3\}$

5.10 What is the solution set of the inequality $|2x-1|<9$?

(1) $\{x|-4<x<5\}$
(2) $\{x|x<-4 \text{ or } x>5\}$
(3) $\{x|x<5\}$
(4) $\{x|x<-4\}$

5.11 If a rectangular beam is supported at both ends and loaded in the middle, the distance the middle of the beam is deflected is directly proportional to the cube of the length of the beam. If a certain load produces a deflection of 18 millimeters at the center of a beam 3 meters long, what deflection, rounded to a *tenth of a millimeter,* would the same load produce if the beam were 5 meters long?

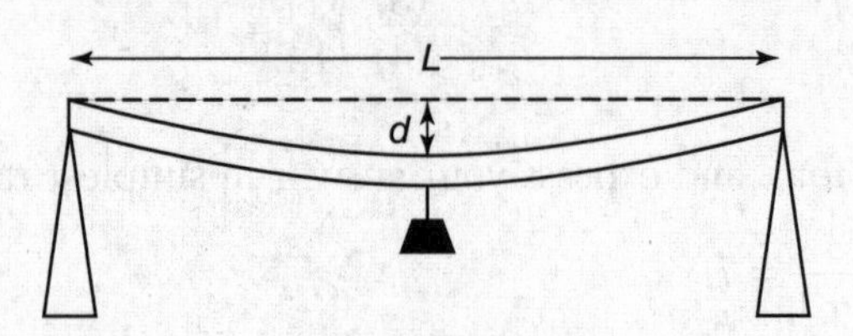

5.12 The speed of a truck varies inversely with the time to reach its destination. If the truck takes 3 hours to reach its destination while traveling at a constant speed of 50 miles per hour, how long will it take to reach the same location while traveling at a constant speed of 60 miles per hour?

(1) $2\frac{1}{3}$ hours

(2) 2 hours

(3) $2\frac{1}{2}$ hours

(4) $2\frac{2}{3}$ hours

Solutions

5.1 The easiest way to do this problem is to note that the denominators $x - 3$ and $3 - x$ are opposites.

$$\frac{3}{x-3}+\frac{x}{3-x}=\frac{3}{x-3}+\frac{x}{(3-x)}\cdot\left(\frac{-1}{-1}\right)=\frac{3}{x-3}+\frac{-x}{x-3}=\frac{3-x}{x-3}=-1$$

CALC Check

```
Plot1 Plot2 Plot3
\Y1■3/(X-3)+X/(3
-X)■
\Y2=
\Y3=
\Y4=
\Y5=
\Y6=
```

X	Y1
-3	-1
-2	-1
-1	-1
0	-1
1	-1
2	-1
3	ERROR

X=3

The correct choice is **(2)**.

5.2 To multiply fractions, factor all polynomial terms first and then simplify. Note that $\frac{2-x}{x-2}$ simplifies to -1, not 1.

$$\frac{12}{x^2-4}\cdot\frac{2-x}{3}=\frac{\overset{4}{\cancel{12}}}{\cancel{(x-2)}(x+2)}\cdot\frac{\overset{(-1)}{\cancel{2-x}}}{\cancel{3}}=\frac{-4}{x+2}$$

The correct answer is $-\frac{4}{x+2}$.

5.3 To divide fractions, multiply by the reciprocal of the divisor, factor, and reduce.

$$f(x)\div g(x)=\frac{3x^2-27}{18x+30}\div\frac{x^2-7x+12}{3x^2-7x-20}=\frac{3x^2-27}{18x+30}\cdot\frac{3x^2-7x-20}{x^2-7x+12}$$

$$=\frac{\cancel{3}\cancel{(x-3)}(x+3)}{2\,\cancel{6}\cancel{(3x+5)}}\cdot\frac{\cancel{(3x+5)}\cancel{(x-4)}}{\cancel{(x-3)}\cancel{(x-4)}}$$

$$=\frac{x+3}{2}$$

Note that for problems like this, you must work carefully and neatly. Do not try to do too many steps at once.

For values of x for which it is defined, $f(x) \div g(x) = \frac{x+3}{2}$.

5.4 The quickest way to solve this problem is to multiply both the numerator and denominator of the fraction by the least common denominator of all the individual fractions. The separate denominators are x, y, x^2, and y^2 so the LCD is x^2y^2.

$$\frac{\frac{1}{x^2}-\frac{1}{y^2}}{\frac{1}{y}+\frac{1}{x}} = \frac{\left(\frac{1}{x^2}-\frac{1}{y^2}\right)(x^2y^2)}{\left(\frac{1}{y}+\frac{1}{x}\right)(x^2y^2)}$$

$$= \frac{y^2-x^2}{x^2y+xy^2}$$

Now factor and reduce.

Note that $\frac{y+x}{x+y}$ simplifies to 1, not –1.

$$= \frac{(y-x)\cancel{(y+x)}}{xy\cancel{(x+y)}} = \frac{y-x}{xy}$$

The correct choice is **(2)**.

5.5 Because of all the different denominators involved, this problem is not most easily done by multiplying by the least common denominator. Instead, perform the necessary operations in the numerator and denominator to make each a single fraction.

$$\frac{\frac{1}{x}-\frac{1}{x+1}}{\frac{2}{x-1}-\frac{1}{x}}=\frac{\left(\frac{1}{x}\right)\left(\frac{x+1}{x+1}\right)-\left(\frac{x}{x}\right)\left(\frac{1}{x+1}\right)}{\left(\frac{x}{x}\right)\left(\frac{2}{x-1}\right)-\left(\frac{1}{x}\right)\left(\frac{x-1}{x-1}\right)}$$

$$=\frac{\frac{1}{x(x+1)}}{\frac{x+1}{x(x-1)}}$$

Then multiply the numerator by the reciprocal of the denominator and reduce.

$$=\frac{1}{\cancel{x}(x+1)}\cdot\frac{\cancel{x}(x-1)}{(x+1)}$$

$$=\frac{(x-1)}{(x+1)^2}$$

5.6 To solve a rational equation, first factor the denominators.

$$\frac{x+8}{x^2-4}-\frac{5}{x^2-2x}=\frac{2}{x}$$

$$\frac{x+8}{(x-2)(x+2)}-\frac{5}{x(x-2)}=\frac{2}{x}$$

Next get common denominators for all the terms in the equation.

$$\frac{x}{x}\frac{(x+8)}{(x-2)(x+2)}-\frac{5}{x(x-2)}\frac{(x+2)}{(x+2)}=\frac{2}{x}\frac{(x+2)(x-2)}{(x+2)(x-2)}$$

Equate the numerators.

$$x(x+8) - 5(x+2) = 2(x+2)(x-2)$$

OR

Multiply the original equation by the least common denominator of all the terms, in this case $x(x-2)(x+2)$. Careful simplificaiton clears all the fractions.

$$x(x-2)(x+2)\left(\frac{x+8}{(x-2)(x+2)} - \frac{5}{x(x-2)} = \frac{2}{x}\right)$$

$$\cancel{x}\cancel{(x-2)(x+2)}\frac{x+8}{\cancel{(x-2)(x+2)}} - \cancel{x(x-2)}(x+2)\frac{5}{\cancel{x(x-2)}}$$

$$= \cancel{x}(x-2)(x+2)\left(\frac{2}{\cancel{x}}\right)$$

Both methods yield the same equation to solve.

$$x(x+8) - 5(x+2) = 2(x-2)(x+2)$$

Put the equation in standard form.

$$x^2 + 8x - 5x - 10 = 2(x^2 - 4)$$
$$x^2 + 3x - 10 = 2x^2 - 8$$
$$x^2 - 3x + 2 = 0$$

Factor and solve.

$$(x-1)(x-2) = 0$$
$$x = 1 \quad \text{or} \quad x = 2$$

Finally, check the candidate solutions in the original equation. Use your calculator for the first check.

Check $x = 1$: $\dfrac{1+8}{1^2-4} - \dfrac{5}{1^2 - 2(1)} \stackrel{?}{=} \dfrac{2}{1}$

$2 = 2$ Yes

Check $x = 2$: $\dfrac{2+8}{2^2-4} - \dfrac{5}{2^2 - 2(2)} \stackrel{?}{=} \dfrac{2}{2}$

undefined $= 1$ No

Only $x = 1$ checks. The solution set is $\{1\}$.

Since this is a multiple-choice question, the easiest way to do it is to check each of the values from the choices, $x = 1$ and $x = 2$, in the equation. Only $x = 1$ checks.

The correct choice is **(1)**.

5.7 Because the answer must be in simplest radical form, you cannot do a graphical solution on your calculator. You must solve algebraically. First get common denominators for all terms in the equation.

$$\frac{4}{x} - \frac{3}{x+1} = 7$$

$$\frac{4}{x}\frac{(x+1)}{(x+1)} - \frac{3}{(x+1)}\frac{(x)}{(x)} = \frac{7}{1} \cdot \frac{x(x+1)}{x(x+1)}$$

Equate the numerators.

$$4x + 4 - 3x = 7x^2 + 7x$$

OR

Multiply through both sides of the original equation by the LCD of all the terms, in this case $x(x + 1)$.

$$x(x+1)\left(\frac{4}{x} - \frac{3}{x+1} = 7\right)$$

$$\cancel{x}(x+1)\left(\frac{4}{\cancel{x}}\right) - x\cancel{(x+1)}\left(\frac{3}{\cancel{x+1}}\right) = 7x(x+1)$$

$$4(x + 1) - 3x = 7x(x + 1)$$

Both methods yield the same equation to solve.

Get the equation in standard form.

$$4x + 4 - 3x = 7x^2 + 7x$$
$$x + 4 = 7x^2 + 7x$$
$$7x^2 + 6x - 4 = 0$$

Since the answers must be in radical form, you cannot factor. Use the quadratic formula.

$$x = \frac{-6 \pm \sqrt{(6)^2 - 4(7)(-4)}}{2(7)}$$

$$= \frac{-6 \pm \sqrt{148}}{14} = \frac{-6 + \sqrt{4}\sqrt{37}}{14}$$

$$= \frac{-6}{14} + \frac{2\sqrt{37}}{14} = -\frac{3}{7} \pm \frac{\sqrt{37}}{7}$$

Check for undefined roots. Checking these roots is complicated, so instead look at the original equation and identify which values of x make it undefined. Both $x = 0$ and $x = -1$ make the original equation undefined, and neither appeared as extraneous roots.

The solutions are $x = -\frac{3}{7} \pm \frac{\sqrt{37}}{7}$.

5.8 To solve an inequality with rational expressions algebraically: Find the roots by solving for the equality.

$$\frac{3x-2}{x+1} = 2$$
$$3x - 2 = 2(x + 1)$$
$$3x - 2 = 2x + 2$$
$$x = 4$$

Find x-values that make the problem undefined. This will occur when any denominator equals 0.

$$x + 1 = 0$$
$$x = -1$$

Graph roots and undefined values on a number line. Note that 4 is included (solid circle) because the original problem allows equality. However, –1 is not included (open circle) because –1 makes an expression in the problem undefined.

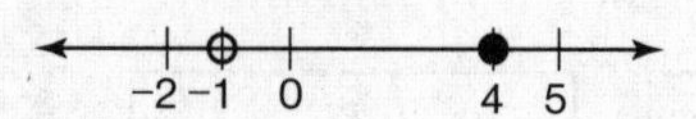

Check a value from each interval in the original inequality. Shade intervals that check on the number line.

Check $x = -2$: $\dfrac{3(-2)-2}{-2+1} \overset{?}{\geq} 2$

$8 \geq 2$ Yes

Check $x = 0$: $\dfrac{3(0)-2}{0+1} \overset{?}{\geq} 2$

$-2 \geq 2$ No

Check $x = 5$: $\dfrac{3(5)-2}{5+1} \overset{?}{\geq} 2$

$2.167 \geq 2$ Yes

The checking part of the solution can be made easier with the table feature on your calculator.

CALC Check

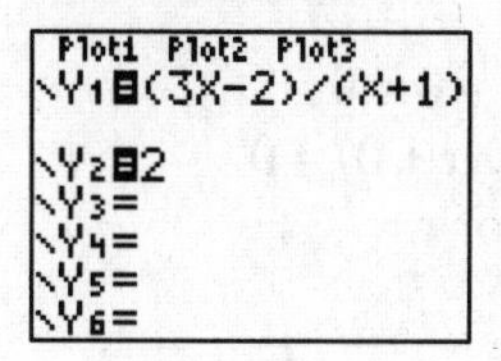

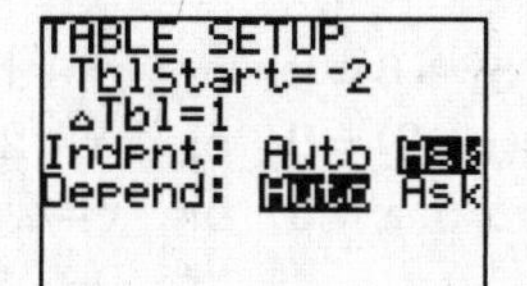

X	Y1	Y2
-2	8	2
0	-2	2
5	2.1667	2

X=

To solve an inequality with rational expressions graphically:

Use your calculator and graph $y = \dfrac{3x-2}{x+1}$ and $y = 2$ in separate equations.

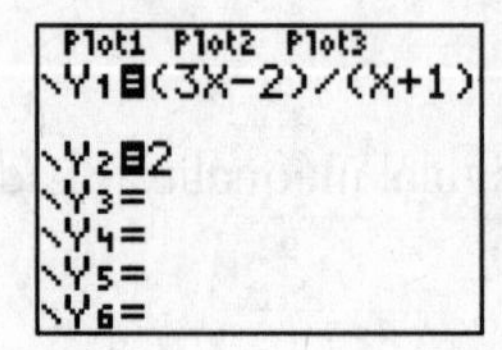

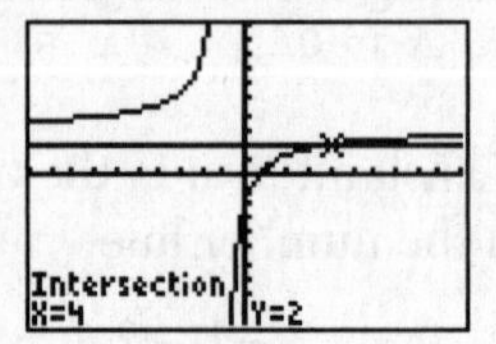

The only intersection is at $x = 4$. Check the original equation to see where the function is undefined; this happens at $x = -1$. This tells you the vertical asymptote in the graph occurs at $x = -1$. Finally, note that the curve is above (greater than) $y = 2$ on two intervals: $x < -1$ and $x \geq 4$. Note that -1 is not included because the equation is undefined there but 4 is included because the original problem allows equality.

The solution is $x < -1$ or $x \geq 4$. When written in interval notation, the solution is $(-\infty, -1) \cup [4, \infty]$.

The correct choice is **(3)**.

5.9 To solve an absolute value equation algebraically:

Write two separate equations to eliminate the absolute value.

$$x^2 - 2x = 3x - 6 \quad \text{or} \quad -(x^2 - 2x) = 3x - 6$$
$$-x^2 + 2x = 3x - 6$$

Solve both.

$$x^2 - 5x + 6 = 0 \quad \text{or} \quad x^2 + x - 6 = 0$$
$$(x-2)(x-3) = 0 \quad \text{or} \quad (x-2)(x+3) = 0$$
$$x = 2 \text{ or } x = 3 \quad \text{or} \quad x = 2 \text{ or } x = -3$$

Check all the candidate solutions, $x = -3$, $x = 2$, and $x = 3$, in the original problem.

Check $x = -3$: $|(-3)^2 - 2(-3)| \stackrel{?}{=} 3(-3) - 6$

$15 = -15$ No

Check $x = 2$: $|2^2 - 2(2)| \stackrel{?}{=} 3(2) - 6$

$0 = 0$ Yes

Check $x = 3$: $|3^2 - 2(3)| \stackrel{?}{=} 3(3) - 6$

$3 = 3$ Yes

Only $x = 2$ and $x = 3$ check.

To solve an absolute value equation graphically:

You can use a calculator. However, the separate equations method does not give an easy-to-read graph.

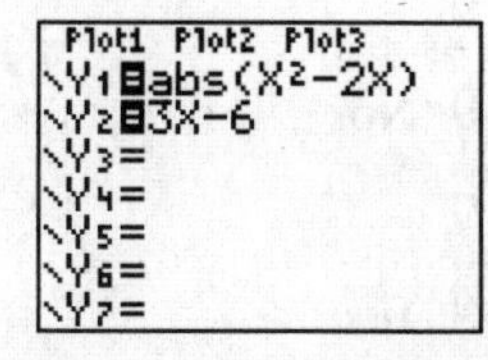

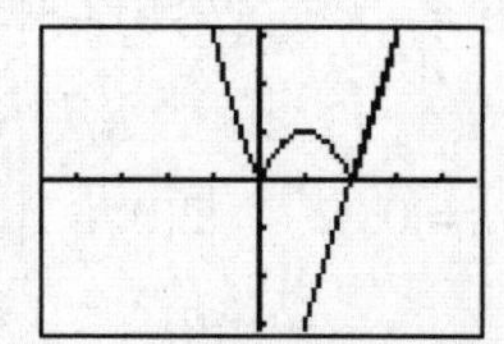

Instead, set the equation equal to 0, $|x^2 - 2x| - 3x + 6 = 0$

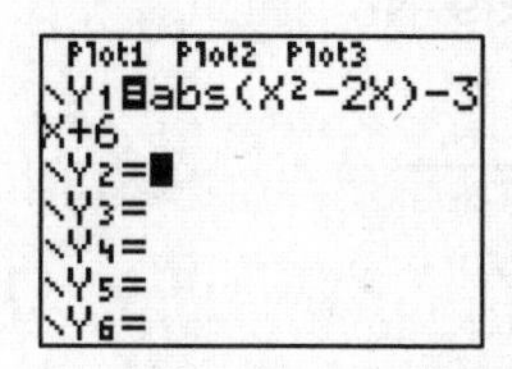

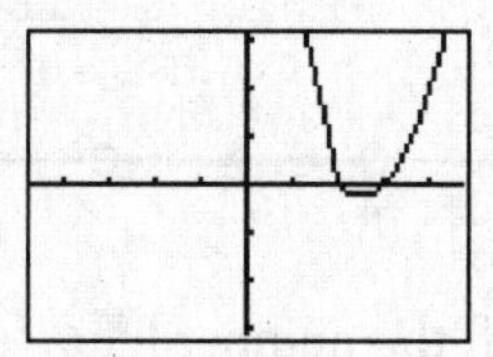

This graph shows solutions at $x = 2$ and $x = 3$.

The correct choice is **(4)**.

5.10 <u>To solve an absolute value inequality algebraically:</u>

Find the solutions for equality.

$$|2x - 1| = 9$$

$$2x - 1 = 9 \quad \text{or} \quad -(2x - 1) = 9$$
$$-2x + 1 = 9$$

$$x = 5 \quad \text{or} \quad x = -4$$

Both check.

Graph the roots on a number line with open circles because the original problem is a strict inequality. Check a value in each interval and shade the number line for intervals that check.

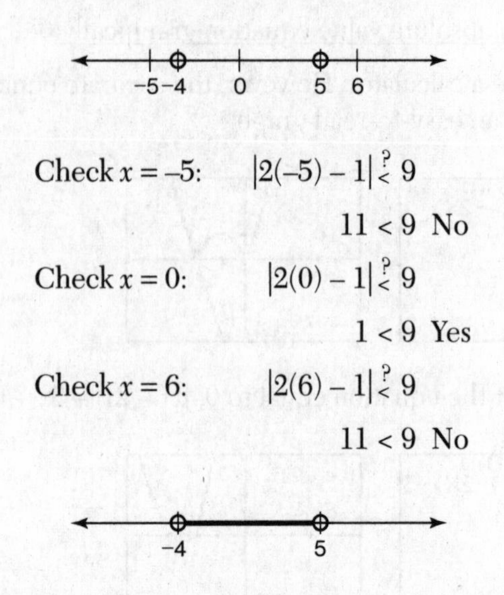

Read the solution off the number line.

$$-4 < x < 5$$

To solve an absolute value inequality graphically:

This problem can be solved on the calculator by graphing each side of the inequality separately.

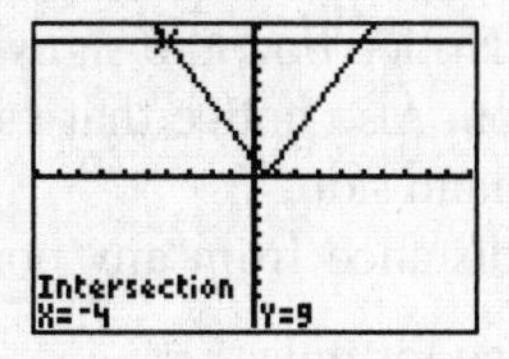

The roots can be found with CALC **5:intersect** (one is shown). The absolute value function is below (less than) $y = 9$ between the two roots: $-4 < x < 5$.

The correct choice is **(1)**.

5.11 Direct variation means d and l^3 always have the same ratio. Set up and solve a proportion.

$$\frac{\text{deflection}}{\text{length}^3} : \frac{18}{3^3} = \frac{d}{5^3}$$

$$d = \frac{\left(5^3\right)18}{3^3} = 83.333$$

The deflection of the beam when the length is 5 m is 83.3 mm.

5.12 Inverse variation means speed s and time t always have the same product: $s_1t_1 = s_2t_2$.

$$(50)(3) = 60t$$
$$150 = 60t$$
$$t = 2.5$$

The truck takes $2\frac{1}{2}$ hours while traveling at 60 miles per hour to reach the destination.

The correct choice is **(3)**.

6. CIRCLES

6.1 CENTER-RADIUS EQUATION OF A CIRCLE

$(x-h)^2+(y-k)^2=r^2$ is the equation of a circle with center at (h, k) and radius $= r$. Notice how the signs of h and k are opposite of those in the equation. Also notice that r is the square root of the constant on the right hand side.

The radius is the distance from any point on the circle to the center. Use the distance formula, $r=\sqrt{(x_P-h)^2+(y_P-k)^2}$, to find the radius if the center (h, k) and a point (x_P, y_P) on the circle are known.

6.2 COMPLETING THE SQUARE

The equation of a circle in general form $x^2+bx+y^2+cy+d=0$ can be transformed into center-radius form by completing the square twice.

$$x^2+bx+y^2+cy+d=0$$

$$(x^2+bx)+(y^2+cy)+d=0$$

$$\left(x+\frac{b}{2}\right)^2-\left(\frac{b}{2}\right)^2+\left(y+\frac{c}{2}\right)^2-\left(\frac{c}{2}\right)^2+d=0$$

$$\left(x+\frac{b}{2}\right)^2+\left(y+\frac{c}{2}\right)^2=\frac{b^2}{4}+\frac{c^2}{4}-d$$

Practice Exercises

6.1 Which is an equation for the circle whose graph is shown?

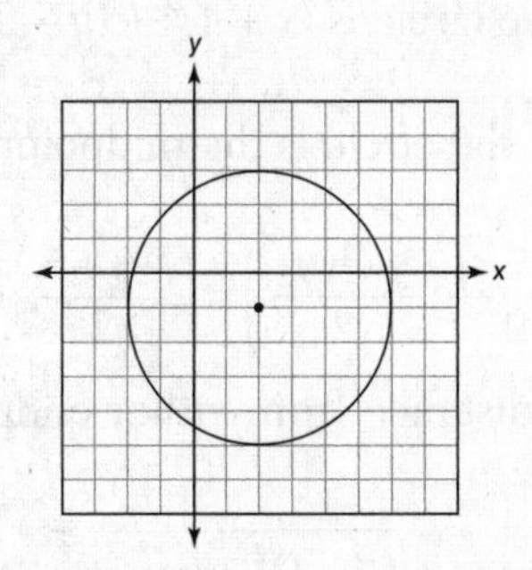

(1) $(x - 2)^2 + (y + 1)^2 = 16$ (3) $(x - 2)^2 + (y + 1)^2 = 8$
(2) $(x + 2)^2 + (y - 1)^2 = 16$ (4) $(x + 2)^2 + (y - 1)^2 = 8$

6.2 Write the equation of the circle with its center at (–4, 3) and that passes through point (2, 5).

6.3 Write an equation of a circle whose diameter has endpoints (–3, 3) and (5, 9).

6.4 Rewrite the equation of the circle $x^2 + y^2 + 6x - 12y + 20 = 0$ in center-radius form.

6.5 Find the coordinates of the center and the length of the radius of the circle that has the equation $x^2 + y^2 - 8x + 2y + 8 = 0$.

Solutions

6.1 If the center of a circle is (h, k) and its radius is r, then its equation is $(x - h)^2 + (y - k)^2 = r^2$. In the circle shown in the problem, the center is $(2, -1)$ and the radius is 4. The equation is $(x - 2)^2 + (y - (-1))^2 = 4^2$ or $(x - 2)^2 + (y + 1)^2 = 16$.

The correct choice is **(1)**.

6.2 The length of the radius is the distance from any point on the circle to the center. In this problem,

$$r = \sqrt{(x_P - h)^2 + (y_P - k)^2} = \sqrt{(2 - (-4))^2 + (5 - 3)^2} = \sqrt{40}.$$

The equation of the circle is $(x + 4)^2 + (y - 3)^2 = 40$.

6.3 The center of the circle is the midpoint of the diameter.

$$C(x, y) = \left(\frac{x_1 + x_2}{2}, \frac{y_1 + y_2}{2}\right) = \left(\frac{-3 + 5}{2}, \frac{3 + 9}{2}\right) = (1, 6)$$

The radius is the distance from either endpoint of the diameter to the center.

$$r = \sqrt{(5 - 1)^2 + (9 - 6)^2} = 5$$

The equation is $(x - 1)^2 + (y - 6)^2 = (5)^2$ or $(x - 1)^2 + (y - 6)^2 = 25$.

6.4 Rearrange the terms to get the x-terms together and the y-terms together. Then complete the square for each variable.

$$\begin{aligned}
&x^2 + 6x + y^2 - 12y + 20 = 0\\
&\underbrace{x^2 + 6x + 9}_{(x+3)^2} - 9 + \underbrace{y^2 - 12y + 36}_{(y-6)^2} - 36 + 20 = 0\\
&(x + 3)^2 - 9 + (y - 6)^2 - 36 + 20 = 0\\
&(x + 3)^2 + (y - 6)^2 = 9 + 36 - 20\\
&(x + 3)^2 + (y - 6)^2 = 25
\end{aligned}$$

The equation of the circle in center-radius form is

$$(x + 3)^2 + (y - 6)^2 = 25.$$

6.5 Rearrange the terms to get the x-terms together and the y-terms together. Then complete the square for each variable.

$$x^2 - 8x + \qquad y^2 + 2y \qquad + 8 = 0$$

$$\underbrace{x^2 - 8x + 16} - 16 + \underbrace{y^2 + 2y + 1} - 1 + 8 = 0$$

$$(x - 4)^2 \quad - 16 + \quad (y + 1)^2 \quad - 1 + 8 = 0$$

$$(x - 4)^2 + (y + 1)^2 = 9$$

Remember that the signs of the coordinates of the center are always the opposite of what appears in the equation.

The center of the circle is $(4, -1)$ and the radius is $\sqrt{9} = 3$.

7. FUNCTIONS

7.1 RELATIONS AND FUNCTIONS; DOMAIN AND RANGE

A relation is a set of ordered pairs (x, y). It can be represented by a table, a mapping diagram, a list of ordered pairs, an equation, a verbal description, or a graph.

A function is a relation where every x-value is associated with only one y-value. In other words, no two ordered pairs can have the same x-value but different y-values. In graph form, a function passes the vertical line test, where any vertical line intersects the graph at most once.

A one-to-one function is a function where every y-value is associated with only one x-value. In other words, no two ordered pairs can have the same y-value but different x-values. In graph form, a one-to-one function passes the horizontal line test, where any horizontal line intersects the graph at most once. The inverse of a one-to-one function is also a function. If a function is not one-to-one, its domain may be restricted to make it one-to-one so that its inverse is a function.

The domain is the set of all possible values for the independent variable, usually x. The domain can be determined from a graph by finding all the x-values where the graph exists. The following are three common rules for finding domain from an equation:

- Denominators cannot equal zero.
- Square root radicands must be greater than or equal to zero.
- Arguments of logs must be positive.

The range is the set of all possible values for the dependent variable, usually y. Range can be determined from a graph by finding all the y-values where the graph exists. Finding the range from the equation can be tricky; most students will get the best result by examining the graph.

A function from set A is onto set B if the range of the function is the same as set B.

7.2 COMPOSITIONS AND INVERSES

A composition of functions uses the output of one function as the input of another function. The two common ways of writing a composition are $f(g(x))$ and $(f \circ g)(x)$. Both mean evaluate g at x and then evaluate f at this result.

The inverse of a one-to-one function is found by switching the x and y in each ordered pair of the function and is written as $f^{-1}(x)$. Be careful! This is not a negative exponent and does not mean reciprocal. The inverse "undoes" the function so that $f^{-1}(f(x)) = f(f^{-1}(x)) = x$. The inverse is obtained algebraically by switching all x's and y's in the equation and then solving for y. Graphically, the inverse is found by reflecting the graph over the line $y = x$, which is equivalent to switching the x and y in the coordinates.

7.3 TRANSFORMATIONS OF FUNCTIONS

Transformation ($a > 0$)

$y = f(x + a)$
$y = f(x - a)$

Description:
Horizontal translation a units
$(x + a)$ translates graph left
$(x - a)$ translates graph right

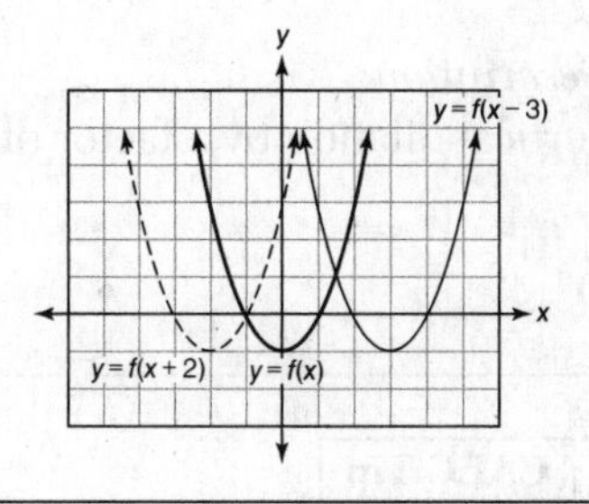

$y = f(x) + a$
$y = f(x) - a$

Description:
Vertical translation a units
$+a$ translates graph up
$-a$ translates graph down

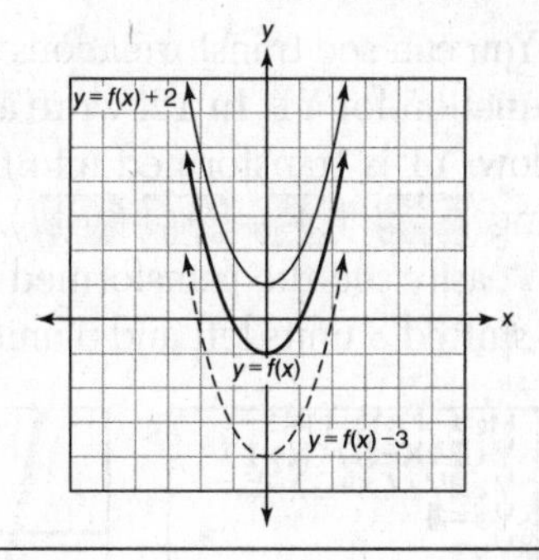

$y = f(-x)$

Description:
Reflection over the y-axis

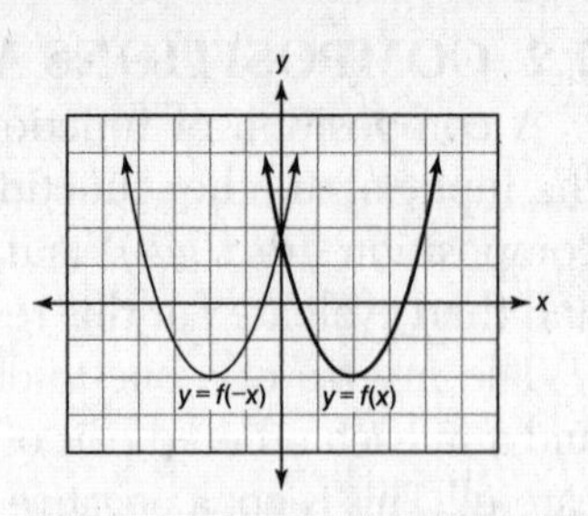

$y = f(x)$

Description:
Reflection over the x-axis

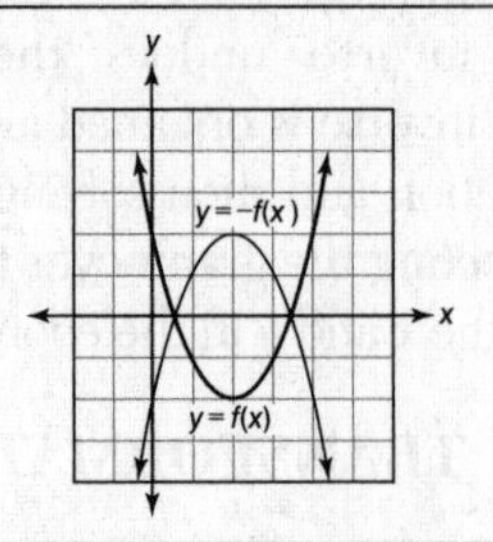

$y = af(x)$

Description:
Vertical dilation by a factor of a

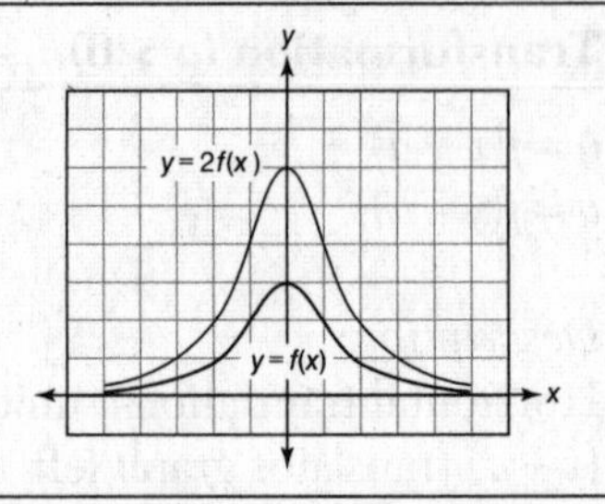

CALC Tip

You can see transformations of functions on your calculator. Enter a function for Y1. In Y2, write a transformation of Y1. In the example below, Y1 is transformed into $f(x+3) - 5$. (Y1 is found in Y-VARS by using VARS ❯ ENTER ENTER). Make Y2 have a thick curve so you can easily see the transformed graph. You can see that the graph of Y2 shifted 3 units left and 5 units down from the graph of Y1.

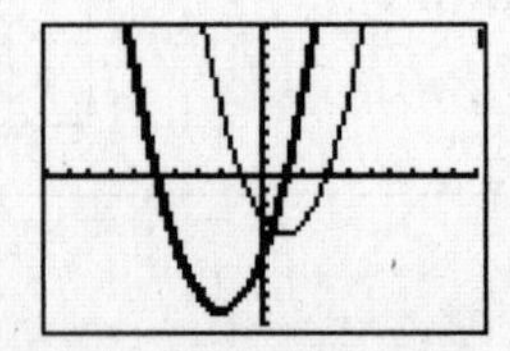

Practice Exercises

7.1 Which equation is *not* a function?

(1) $y = 3$
(2) $y = 3x^2 + 9$
(3) $2x + y = 9$
(4) $y^2 = 9 + x^2$

7.2 What is the domain of the function $f(x) = \dfrac{3x^2}{x^2 - 49}$?

(1) $\{x \mid x \in \text{real numbers}, x \neq 7\}$
(2) $\{x \mid x \in \text{real numbers}, x \neq \pm 7\}$
(3) $\{x \mid x \in \text{real numbers}\}$
(4) $\{x \mid x \in \text{real numbers}, x \neq 0\}$

7.3 What is the domain of $h(x) = \sqrt{x^2 - 4x - 5}$?

(1) $(-\infty, -5] \cup [1, \infty)$
(2) $(-\infty, -1] \cup [5, \infty)$
(3) $[-1, 5]$
(4) $[-5, 1]$

7.4 What is the range of the function $g(x) = 4\cos 2x$?

7.5 If $f(x) = x^2 + 1$, which expression represents $f(a + h)$?

(1) $a^2 + h^2 + 1$
(2) $a^2 + h^2 + 2$
(3) $a^2 + 1 + h$
(4) $a^2 + 2ah + h^2 + 1$

7.6 If $f(x) = x^2 - 3x$ and $g(x) = 2x - 4$, evaluate $(f \circ g)(5)$.

7.7 The graph below shows two functions, f and g. Evaluate $f(g(4))$.

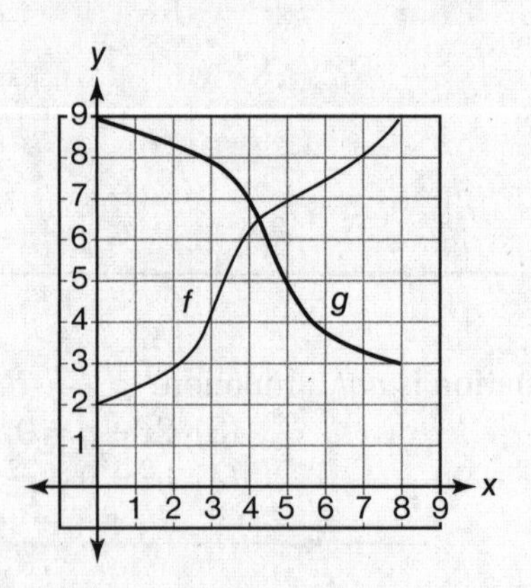

7.8 Which diagram represents a one-to-one function?

(1) 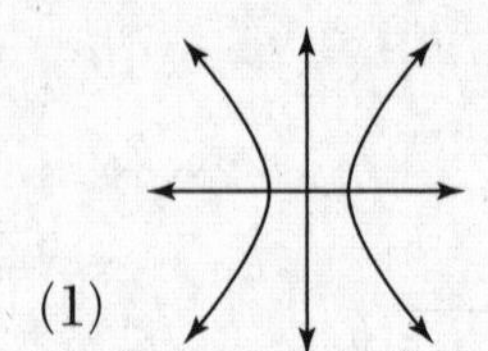

(3) 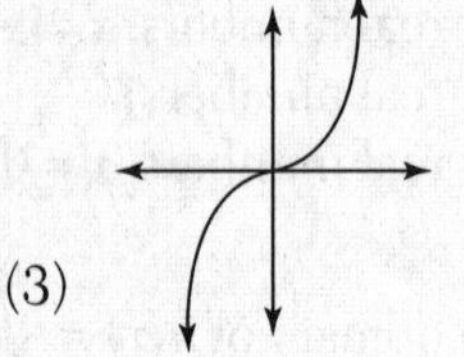

(2) 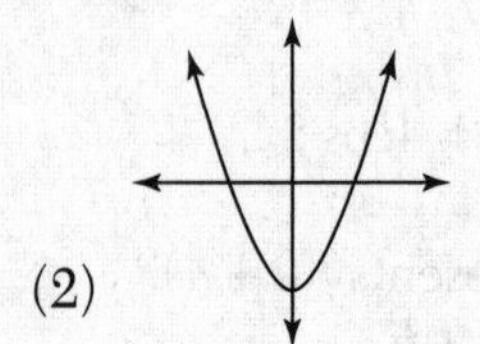

(4) 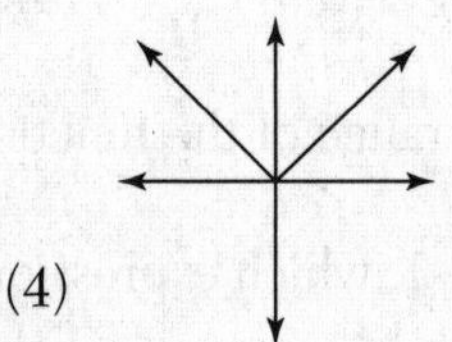

7.9 The accompanying graph shows the temperature in Albany over an 8-hour interval.

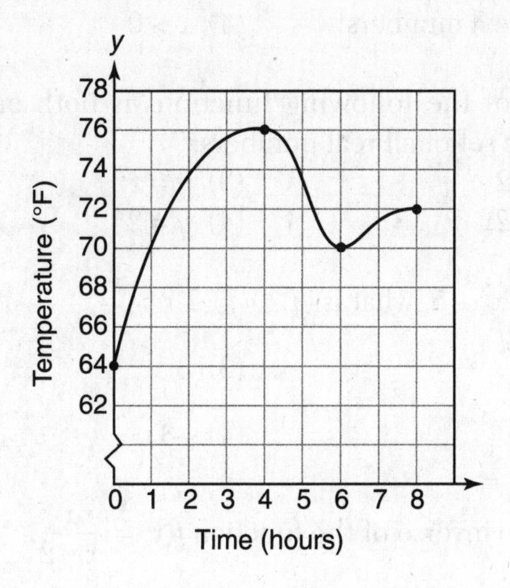

What is the range of the temperatures during this interval?

(1) $0 \leq y \leq 76$
(2) $62 \leq y \leq 78$
(3) $64 \leq y \leq 72$
(4) $64 \leq y \leq 76$

7.10 The effect of pH on the action of a certain enzyme is shown on the following graph.

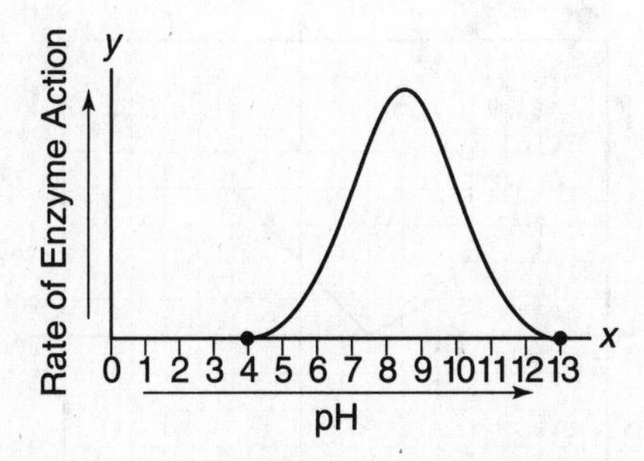

What is the domain of this function?

7.11 What is the domain of $f(x) = \ln x$?

(1) all integers
(2) all real numbers
(3) $x \geq 0$
(4) $x > 0$

7.12 Which of the following functions is both one-to-one and onto the set of all real numbers?

(1) $y = 2$
(2) $y = 2x$
(3) $y = x^2$
(4) $y = 2^x$

7.13 If $f(x) = x^3 - 5$, what is $(f^{-1} \circ f)(2)$?

(1) $\frac{1}{3}$
(2) 2
(3) 3
(4) –3

7.14 Find the inverse of the function $f(x) = \frac{x}{x-2}$.

7.15 The graph of $g(x)$ is obtained by translating the graph of $f(x) = x^3$ four units to the right and two units up. Find an equation for $g(x)$.

7.16 The graph of a function $f(x)$ is shown. On the same axes, draw and label the graph of $2f(-x)$.

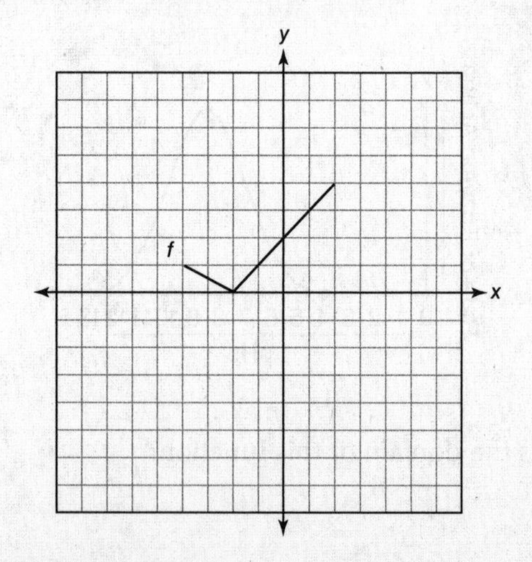

Solutions

7.1 The idea of a function is that for each input (x-value), there is only one output (y-value). Choice (4) has a y^2 term. For any x-value you substitute into the equation, you will get a quadratic equation for y which will have two solutions. Thus, each x-value has two y-values associated with it. So choice (4) is *not* a function.

Instead, you can eliminate choices to find the answer to this problem.

- Choice (1) is a horizontal line, which is a function.
- Choice (2) is a parabola, which is a function.
- Choice (3) is a slanted line, which is a function.

The correct choice is **(4)**.

7.2 If a domain is not given, it is assumed to be the largest set of real numbers for which the output of the function is defined and real. This means:

(A) Denominators must *not* equal 0 because division by 0 is undefined.
(B) Radicands (with an even index) must be greater than or equal to 0. Remember that even roots of negative numbers are imaginary.
(C) Arguments of logarithms must be greater than 0 because log (0) is undefined and the logarithm of a negative number is imaginary.

In this problem, only criterion (A) is relevant.

$$x^2 - 49 \neq 0$$
$$x^2 \neq 49$$
$$x \neq \pm 7$$

CALC Check

```
Plot1 Plot2 Plot3
\Y1=3X²/(X²-49)
\Y2=
\Y3=
\Y4=
\Y5=
\Y6=
\Y7=
```

X	Y1
-7	ERROR
0	0
7	ERROR

X=

The correct choice is **(2)**.

7.3 In this problem, only criterion (B) from solution 7.2 is relevant: $x^2 - 4x - 5 \geq 0$. To solve this algebraically, find the roots for equality, graph them on a number line, and then check each interval (see problem 5.7).

$$x^2 - 4x - 5 = 0$$
$$(x - 5)(x + 1) = 0$$
$$x = 5 \quad \text{or} \quad x = -1$$

−1 5

The domain is $x \leq -1$ or $x \geq 5$.
The function could also be graphed on a calculator.

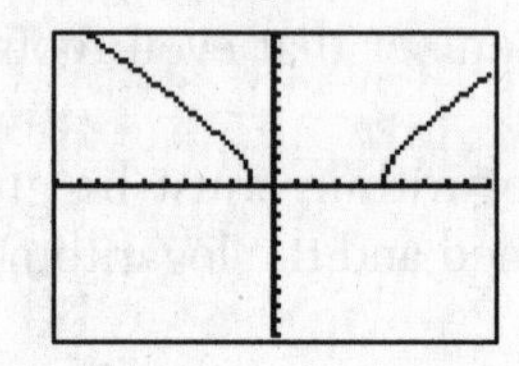

From the graph, the domain is $x \leq -1$ or $x \geq 5$. When written in interval notation, the domain is $(-\infty, -1] \cup [5, \infty)$.

CALC Check

```
Plot1 Plot2 Plot3
\Y1=√(X²-4X-5)
\Y2=
\Y3=
\Y4=
\Y5=
\Y6=
\Y7=
```

X	Y1
-1	0
0	ERROR
1	ERROR
2	ERROR
3	ERROR
4	ERROR
5	0

X= -1

The correct choice is **(2)**.

7.4 Range is the set of all possible outputs of a function. The cosine function $y = a \cos bx$ is centered on the x-axis. Its maximum and minimum values are determined by its amplitude, $|a|$. The given function has amplitude 4; it goes up and down 4 units from the x-axis. A graph of the function confirms this range.

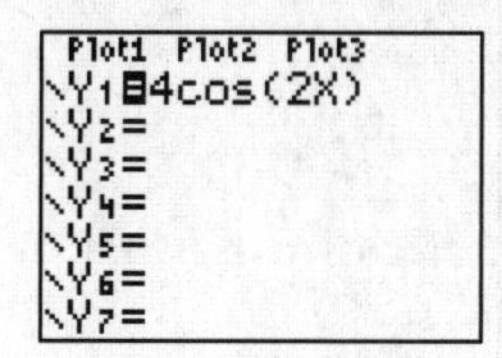

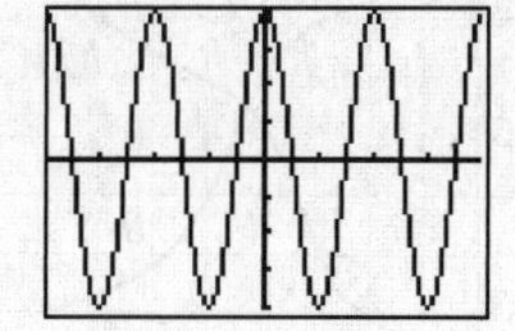

The range is $-4 \leq y \leq 4$.

7.5 To evaluate a function means to substitute a given value for x. In this problem, $f(a + h) = (a + h)^2 + 1$. Since that is not one of the choices, multiply the squared term.

$$f(a+h) = (a+h)^2 + 1 = (a+h)\,(a+h) + 1 = a^2 + ah + ah + h^2 + 1$$
$$= a^2 + 2ah + h^2 + 1$$

The correct choice is **(4)**.

7.6 The notation $(f \circ g)(5)$ means the same as $f(g(5))$. First evaluate $g(5)$. Then evaluate f at that answer.

$$g(5) = 2(5) - 4 = 6$$
$$f(6) = (6)^2 - 3(6) = 18$$

If you do the composition in the wrong order, you get $f(5) = 10$ and $g(10) = 16$, which are not correct.

CALC Check

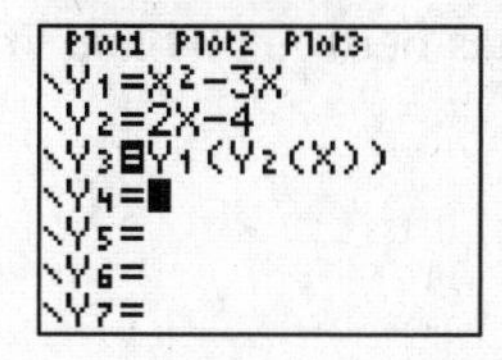

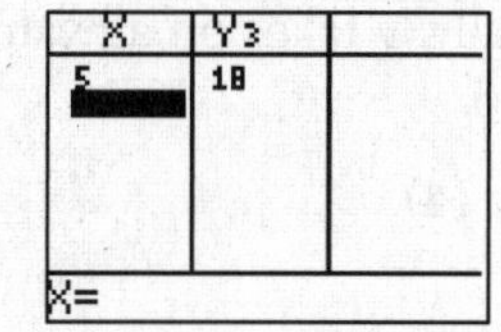

$(f \circ g)(5) = 18$

7.7 Evaluate $g(4)$ first. Then evaluate f at that answer. To evaluate $g(4)$ means to find the y-value of g when $x = 4$. From the graph, $g(4) = 7$. So $f(g(4)) = f(7) = 8$.

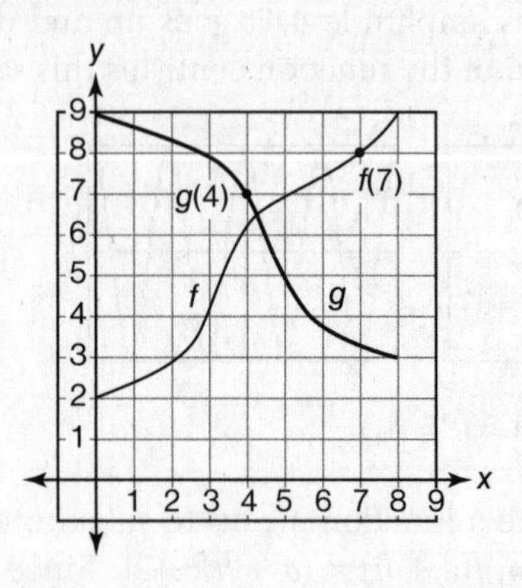

$f(g(4)) = 8$

7.8 In a function, each x-value is associated with (goes to) only one y-value. Graphically, that means any vertical line will intersect the graph of a function in at most one place. This eliminates choice (1); it is not a function.

In a one-to-one function, each y-value is associated with (comes from) only one x-value. Graphically, that means any horizontal line will intersect the graph of a one-to-one function in at most one place. This eliminates choices (2) and (4); they are functions but are not one-to-one.

The correct choice is **(3)**.

7.9 The range of a function is the set of all possible outputs, or y-values. The minimum y-value on the graph is 64. The maximum y-value is 76. Additionally, y takes on all values between 64 and 76. The range is $64 \leq y \leq 76$.

The correct choice is **(4)**.

7.10 The domain of a function is the set of all possible inputs, or x-values. The minimum x-value on the graph is 4. The maximum x-value is 13. Additionally, x takes on all values between 4 and 13.

The domain is $4 \le x \le 13$.

7.11 Arguments of logarithms, of any base, must be greater than 0 (see criterion (C) in solution 7.2). For $f(x) = \ln x$, the domain is $x > 0$.

The correct choice is **(4)**.

7.12 Graph all four functions.

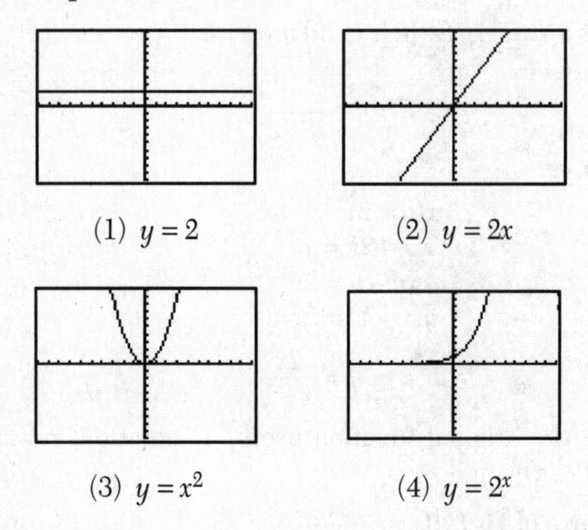

(1) $y = 2$ (2) $y = 2x$

(3) $y = x^2$ (4) $y = 2^x$

To be one-to-one, a function must pass the horizontal line test. Choices (1) and (3) are not one-to-one. To be onto the set of all real numbers, the outputs (y-values) of a function must take on every real number value. The y-values in choice (4) are all positive; this function will never give an output that is zero or negative. This function is *not* onto the set of real numbers. For choice (2), y can take on any real value, whether positive, zero, or negative. This function is both one-to-one and onto the real numbers.

The correct choice is **(2)**.

7.13 The concept of an inverse function is that it undoes whatever the original function does. Therefore, with appropriate domain restrictions when applicable, $(f^{-1} \circ f)(x) = x$. So $(f^{-1} \circ f)(2) = 2$. Be aware that $(f^{-1} \circ f)(x) = (f \circ f^{-1})(x) = x$.

Note that the problem could be done by first finding the formula for $f^{-1}(x)$ and then evaluating the composition $f^{-1}(f(2))$. This is extra work, which increases your chances of making a mistake.

The correct choice is **(2)**.

7.14 To find the inverse of a function defined by an equation $y = f(x)$, switch the x and the y and then solve for y. $f(x) = \dfrac{x}{x-2}$ is the same as $y = \dfrac{x}{x-2}$. Switch x and y.

$$x = \frac{y}{y-2}$$

Solve for y.

$$\begin{aligned} x(y-2) &= y \\ xy - 2x &= y \\ xy - y &= 2x \\ y(x-1) &= 2x \\ y &= \frac{2x}{x-1} \end{aligned}$$

Because the original function used $f(x)$ notation, rewrite the inverse as $f^{-1}(x)$.

The inverse of f is $f^{-1}(x) = \dfrac{2x}{x-1}$.

7.15 Any function $y = f(x)$ can be translated h units right and k units up by changing it to $y = f(x-h) + k$. Note that $(x-h)$ translates the function to the right. Many students mistakenly think $(x-h)$ translates to the left.

To translate $f(x) = x^3$ four units to the right and two units up, $h = 4$ and $k = 2$.

$$g(x) = f(x-4) + 2 = (x-4)^3 + 2$$

Note: do not multiply out $(x - 4)^3$. The directions do not ask for it to be written in standard form. Doing so increases your chances of making errors.

CALC Check

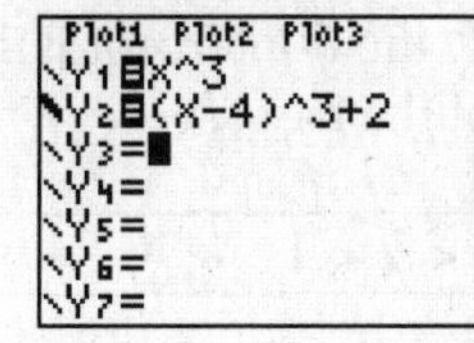

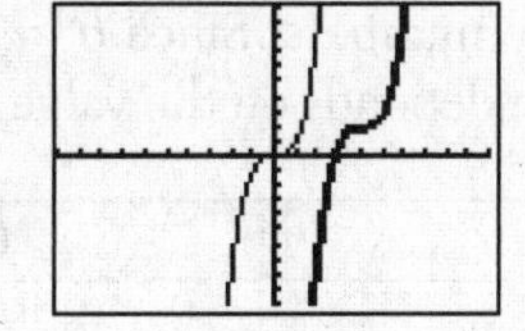

The correct answer is $g(x) = (x - 4)^3 + 2$.

7.16 The graph of $f(-x)$ is the reflection of the graph of f over the y-axis. The graph of $2f(x)$ is the vertical dilation of the graph of $f(x)$ by a factor of 2. The graph of $2f(-x)$ is both the reflection and vertical dilation.

You should do the transformations one at a time. To reflect over the y-axis, change each x-coordinate to its opposite while leaving the y-coordinates unchanged. The graph of the reflection is shown by dashed lines below. To dilate vertically by a factor of 2, multiply each y-coordinate by 2 while leaving the x-coordinates unchanged. The final answer is shown by the solid, heavy lines in the graph below. Remember to label your graphs.

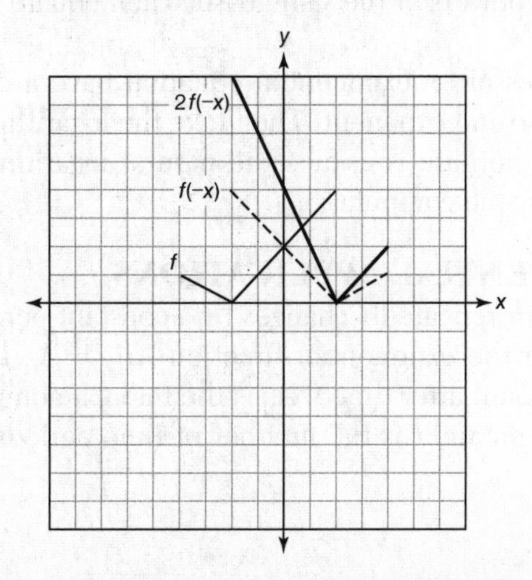

8. EXPONENTIAL AND LOGARITHMIC FUNCTIONS

8.1 EXPONENTIAL FUNCTIONS

The equation $y = b^x$, where b is positive and not equal to 1, is called an exponential function. The domain is all real numbers, and the range is all positive numbers. Since $b^0 = 1$, the y-intercept is 1. The shape of the graph depends on the value of the base, b.

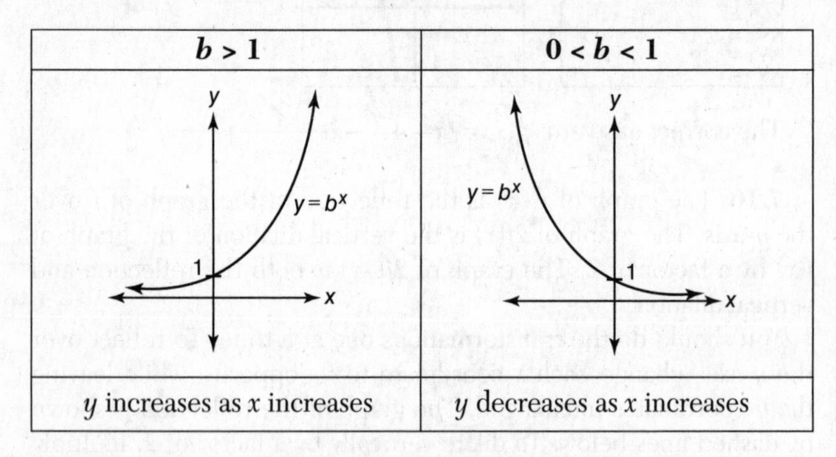

$b > 1$	$0 < b < 1$
y increases as x increases	y decreases as x increases

8.2 SOLVE EXPONENTIAL EQUATIONS

If both sides of an exponential equation have a common base, write them as powers of the same base. Then equate the exponents and solve.

If both sides of an exponential equation have a different base, isolate the base and exponent. Then take the logarithm of each side and solve. If the base is e, take the natural logarithm, ln, of each side to simplify the computations.

8.3 EXPONENTIAL APPLICATIONS

If an amount repeatedly changes by a constant percent, it can be modeled with the exponential function $A(t) = A_0(1 \pm r)^t$, where $A(t)$ is the amount after time t, A_0 is the initial amount, r is the rate in decimal form, and t is the number of time periods. Add r if the amount is increasing, such as for interest. Subtract r if the amount is decreasing, such as for depreciation.

8.4 LOGARITHMIC FUNCTIONS

A logarithm function is the inverse of the corresponding exponential function with the same base. This means that $y = \log_b x$ is equivalent to $x = b^y$. The graph of a logarithm function is the reflection of the graph of the corresponding exponential function over the line $y = x$. Two special bases have their own notation. The natural logarithm is $\log_e x = \ln x$. The common logarithm is $\log_{10} x = \log x$.

8.5 LOGARITHM FORMULAS

The following laws hold when x and y are both positive.

- $\log_b 1 = 0$
- $\log_b b = 1$
- $\log_b(xy) = \log_b x + \log_b y$
- $\log_b\left(\frac{x}{y}\right) = \log_b x - \log_b y$
- $\log_b x^c = c \log_b x$
- Change of base formula: $\log_b x = \frac{\log x}{\log b}$ or $\log_b x = \frac{\ln x}{\ln b}$

Practice Exercises

8.1 Growth of a certain strain of bacteria is modeled by the equation $G = Ae^{0.584t}$, where:

G = final number of bacteria
A = initial number of bacteria
t = time (in hours)

In how many hours will 4 bacteria first increase to approximately 2,500 bacteria? Round your answer to the *nearest hundredth of an hour.*

8.2 In the equation $y = 0.5(1.21)^x$, y represents the number of snowboarders in millions and x represents the number of years since 1998. Find the year in which the number of snowboarders will be 10 million for the first time. (Only an algebraic solution will be accepted.)

8.3 What is the value of b in the equation $4^{2b-3} = 8^{1-b}$?

(1) $-\frac{3}{7}$ (3) $\frac{9}{7}$

(2) $\frac{7}{9}$ (4) $\frac{10}{7}$

8.4 The amount A, in milligrams, of a 10-milligram dose of a drug remaining in the body after t hours is given by the formula $A = 10(0.8)^t$. Find, to the *nearest tenth of an hour*, how long it takes for half of the drug dose to be left in the body. What is the hourly percent decrease for the drug?

8.5 Solve for x: $\log_4(x_2 + 3x) - \log_4(x + 5) = 1$

8.6 If $\log_b x = y$ then x equals

(1) $y \cdot b$ (3) y^b

(2) $\dfrac{y}{b}$ (4) b^y

8.7 Sketch the graph of the functions $f(x) = 3^x$ and $g(x) = \log_3 x$. Considering the graphs, describe the relationship between $f(x)$ and $g(x)$. Specify the domain and the range of g.

8.8 The expression $\log_3(8 - x)$ is defined for all values of x such that

(1) $x > 8$ (3) $x < 8$

(2) $x \geq 8$ (4) $x \leq 8$

8.9 The expression $\frac{1}{2}\log m - 3 \log n$ is equivalent to

(1) $\log\sqrt{m} + \log n^3$ (3) $\log\dfrac{m^2}{\sqrt[3]{n}}$

(2) $\log\dfrac{1}{2}m - 3\log n$ (4) $\log\dfrac{\sqrt{m}}{n^3}$

8.10 Evaluate $\log_3 5$ to the *nearest thousandth*.

8.11 Solve for the value of x, rounded to the *nearest ten-thousandth*, that satisfies the equation $2 \ln (3x) = 5$.

Solutions

8.1 Substitute the given values into the equation. $2500 = 4e^{0.584t}$

Divide by 4 to isolate the exponential term. $625 = e^{0.584t}$

To get t down from the exponent, take the natural logarithm of both sides. $\ln(625) = \ln(e^{0.584t})$

Use a logarithm property to simplify. The logarithm of a power equals the exponent times the logarithm of the base. $\ln(625) = 0.584t \ln(e)$

Use $\ln(e) = 1$ to simplify. $\ln(625) = 0.584t$

Divide to solve for t. $t = \dfrac{\ln(625)}{0.584} = 11.0235$

Note that if you check the rounded answer, the check may not come out exact. Check before rounding for the most accurate check.

CALC Check

To the *nearest hundredth of an hour,* it will take 11.02 hours for the number of bacteria to increase to 2500.

8.2 Substitute $y = 10$ into the equation. $10 = 0.5(1.21)^x$

Divide by 0.5 to isolate the exponential term. $20 = (1.21)^x$

To get x down from the exponent, take the logarithm of both sides and simplify with a logarithm property.

$$\log(20) = \log(1.21)^x = x\log(1.21)$$

Divide to solve for x.

$$x = \frac{\log(20)}{\log(1.21)} = 15.715699$$

Add 15.7 to 1998 to find the year the number of snowboarders reaches 10 million. $1998 + 15.7 = 2013.7$

CALC Check

```
log(20)/log(1.21
)
         15.71569941
.5(1.21^Ans)
                  10
```

The number of snowboarders reaches 10 million sometime during the year 2013.

8.3 Since 4 and 8 are both powers of 2 ($4 = 2^2$ and $8 = 2^3$), rewrite the equation with common bases.

$$4^{2b-3} = 8^{1-b}$$

$$(2^2)^{2b-3} = (2^3)^{1-b}$$

When a power is raised to a power, exponents are multiplied. $2^{2(2b-3)} = 2^{3(1-b)}$

If two powers of the same base are equal, the exponents must be equal.

$$2(2b - 3) = 3(1 - b)$$
$$4b - 6 = 3 - 3b$$
$$7b = 9$$
$$b = \frac{9}{7}$$

CALC Check

Use TBLSET to set the table to Ask. Use TABLE to show the choices listed in order. Fractions were entered, but the calculator displays them in decimal form. Choice (3) has the same value for both expressions.

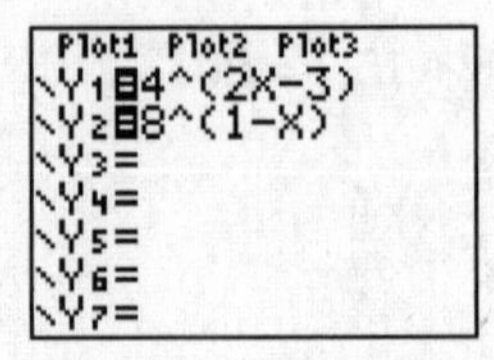

X	Y1	Y2
-.4286	.00476	19.504
.77778	.13501	1.5874
1.2857	.55204	.55204
1.4286	.82034	.41017

X=

The correct choice is **(3)**.

8.4 Half of 10 (the initial value) is 5, so substitute 5 for A.

$$5 = 10(0.8)^t$$

Divide both sides by 10 to isolate the exponential term.

$$0.5 = (0.8)^t$$

Take the logarithm of each side, and simplify with a logarithm property.

$$\log(0.5) = t \log(0.8)$$

Divide to solve for t.

$$t = \frac{\log(0.5)}{\log(0.8)} = 3.10628$$

CALC Check

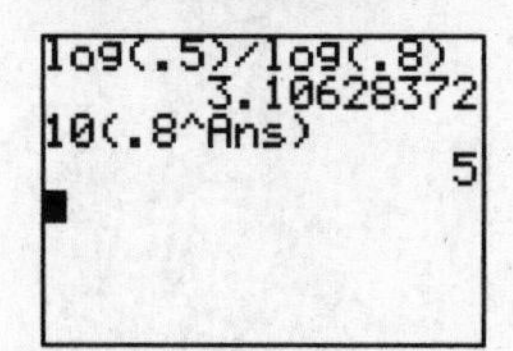

To the *nearest tenth of an hour,* half the drug will remain in the body after 3.1 hours.

To find the hourly percent decrease for the drug, look at the base of the exponential function, $b = 0.8$. If b were 1, there would be no change. So consider 1 to be the baseline of comparison. In the given equation, $b = 0.8$, which is 0.2 or 20% less than 1.

The hourly decrease is 20%.

8.5 To solve algebraically:

Use the properties of logarithms to combine the two logarithm terms into one. The difference of two logarithms is equal to the logarithm of the quotient.

$$\log_4(x^2 + 3x) - \log_4(x + 5) = 1$$

$$\log_4\left(\frac{x^2 + 3x}{x + 5}\right) = 1$$

Rewrite the equation in exponential form and solve.

$$\left(\frac{x^2 + 3x}{x + 5}\right) = 4^1$$

$$\left(\frac{x^2 + 3x}{x + 5}\right) = 4$$

$$x^2 + 3x = 4(x + 5)$$

$$x^2 + 3x = 4x + 20$$

$$x^2 - x - 20 = 0$$

$$(x - 5)(x + 4) = 0$$

$$x = 5 \quad \text{or} \quad x = -4$$

Always check the answers to logarithm equations in the *original equation*.

Check $x = 5$:

$$\log_4(5^2 + 3(5)) - \log_4(5 + 5) \stackrel{?}{=} 1$$
$$\log_4(40) - \log_4(10) \stackrel{?}{=} 1$$
$$\log_4\left(\frac{40}{10}\right) \stackrel{?}{=} 1$$
$$\log_4(4) = 1 \quad \text{Yes}$$

Check $x = -4$:

$$\log_4((-4)^2 + 3(-4)) - \log_4((-4) + 5) \stackrel{?}{=} 1$$
$$\log_4(4) - \log_4(1) \stackrel{?}{=} 1$$
$$1 - 0 = 1 \quad \text{Yes}$$

Notice that $x = -4$ is a solution. Do not reject negative roots without checking first.

<u>To solve graphically:</u>

This problem can be solved on the calculator if you remember the change of base formula: $\log_b x = \dfrac{\log x}{\log b}$. Graph $y = \log_4(x^2 + 3x) - \log_4(x + 5)$ and $y = 1$ together, and find the intersections. The graph, with $-8 \le x \le 8$ and $-4 \le y \le 4$, is shown along with one solution. The table shows a calculator check for both solutions.

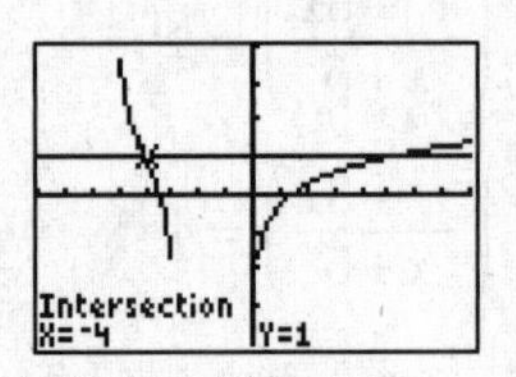

X	Y1	Y2
-4	1	1
5	1	1

X=

For full credit for a graphical solution, sketch the graphs and label the equations on your paper. Label the scale on both axes. Label the points of intersections with coordinates. The solutions to this equation are the x-values only. This particular graph is complicated and might be best used as a check for the algebraic solution.

The solution set is $\{-4, 5\}$.

8.6 Rewrite the equation in exponential form. Remember that a logarithm is an exponent; $y = \log_b x$ means "y is the exponent of base b that results in x." So $y = \log_b x$ is equivalent to $x = b^y$.

The correct choice is **(4)**.

8.7 Make a table of values for the function $f(x) = 3^x$ and plot the values. Sketch and label the graph. The function $g(x) = \log_3 x$ is the inverse of the function f. It can be graphed by switching the x- and y-coordinates in each ordered pair for f.

X	Y1	
-3	.03704	
-2	.11111	
-1	.33333	
0	1	
1	3	
2	9	
3	27	

X= -3

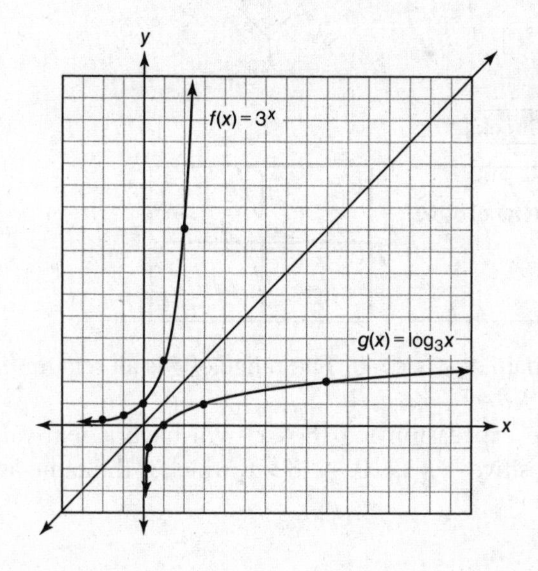

The graph of the function $g(x)$ is the reflection of the graph of the function $f(x)$ over the line $y = x$. If you rotate your graph 45°, it is easier to see the line of reflection. Try this whenever a reflection over $y = x$ may be required.

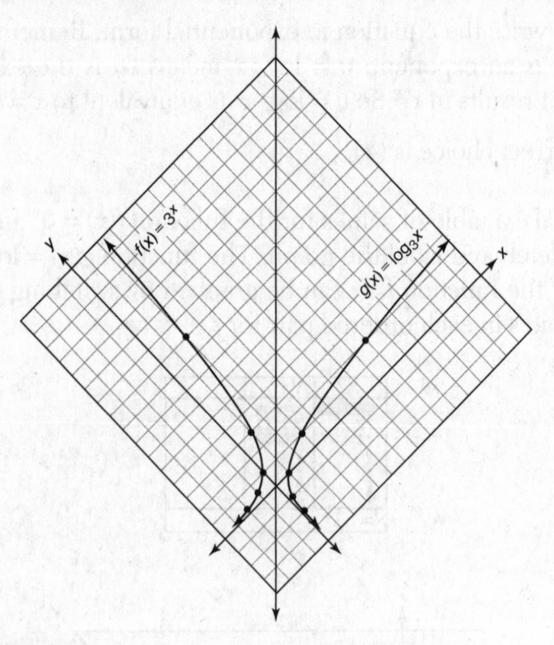

CALC Check

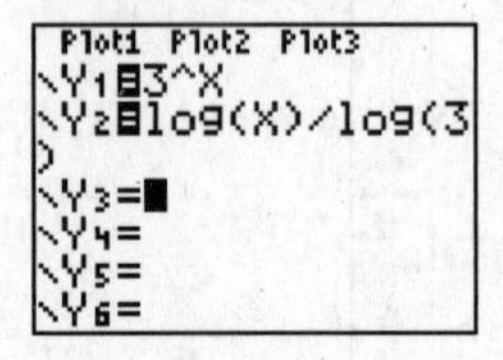

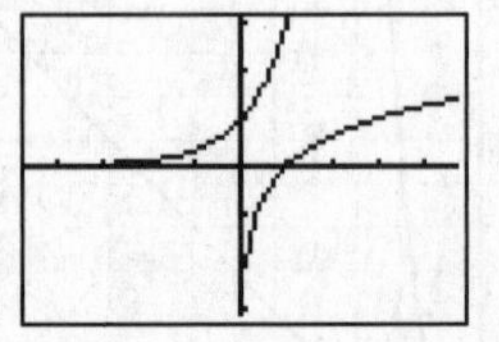

The domain of g is $x > 0$. The range of g is all real numbers.

8.8 The expression $\log_3(8 - x)$ will have a real value only if $(8 - x)$ is positive. $8 - x > 0 \rightarrow 8 > x$, which is the same as $x < 8$.

CALC Check

Finding a domain using a table or calculator graph can be tricky. This table suggests the domain may be $x \leq 7$. Look carefully at the graph to see that the function exists for $x < 8$.

Plot1 Plot2 Plot3
\Y1=log(8−X)/log(3)
\Y2=
\Y3=
\Y4=
\Y5=
\Y6=

X	Y1
4	1.2619
5	1
6	.63093
7	0
8	ERROR
9	ERROR
10	ERROR

X=10

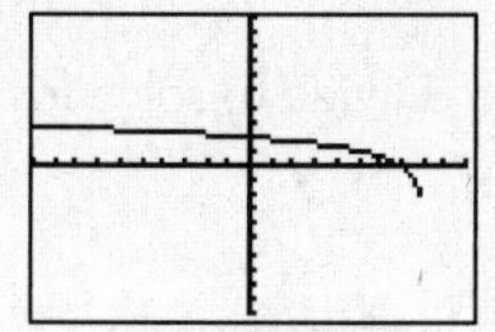

The correct choice is **(3)**.

8.9 Use the properties of logarithms and exponents.

$$\frac{1}{2}\log m - 3\log n$$

$a \log x = \log x^a$ $\qquad$ $= \log m^{\frac{1}{2}} - \log n^3$

$\log x - \log y = \log\left(\frac{x}{y}\right)$ $\qquad$ $= \log\left(\frac{m^{\frac{1}{2}}}{n^3}\right)$

$a^{\frac{1}{n}} = \sqrt[n]{a}$ $\qquad$ $= \log\left(\frac{\sqrt{m}}{n^3}\right)$

The correct choice is **(4)**.

8.10 If you remember the change of base formula, $\log_b x = \frac{\log x}{\log b}$, use it.

$$\log_3 5 = \frac{\log(5)}{\log(3)} = 1.46497$$

If you do not remember the formula, you can solve the equation $\log_3 5 = x$. First rewrite the equation in exponential form.

$$5 = 3^x$$

Take the logarithm of both sides.

$$\log(5) = \log(3^x)$$

Use the logarithm property $\log x^a = a \log x$

$$\log(5) = x \log(3)$$

Divide to solve for x.

$$x = \frac{\log(5)}{\log(3)} = 1.46497$$

CALC Check

To the *nearest thousandth*, $\log_3(5) = 1.465$

8.11 Divide both sides by 2 to isolate the natural logarithm term.

$$2\ln(3x) = 5$$

$$\ln(3x) = \frac{5}{2}$$

Rewrite in exponential form using base e.

$$3x = e^{\frac{5}{2}}$$

Divide by 3 and evaluate.

$$x = \frac{e^{\frac{5}{2}}}{3} = 4.06083$$

CALC Check

To the nearest *ten-thousandth*, the solution is 4.0608.

9. TRIGONOMETRIC FUNCTIONS

9.1 RADIAN MEASURE AND ARC LENGTH

To change from degree measure to radian measure, multiply the number of degrees by $\frac{\pi}{180°}$. To change from radian measure to degree measure, multiply the number of radians by $\frac{180°}{\pi}$.

Use $s = \theta r$ to find the length of an arc of a circle where s is the arc intercepted by a central angle, θ, measured in radians, and r is the radius of the circle. An angle that measures 1 radian intercepts an arc of length equal to the radius of the circle.

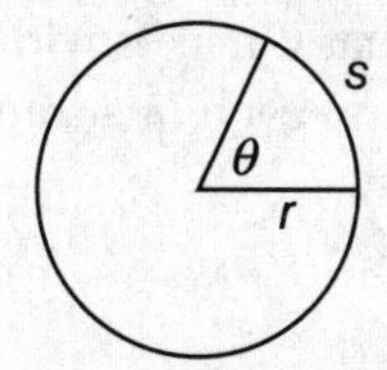

9.2 TRIGONOMETRIC FUNCTIONS OF ANY ANGLE

If angle θ is in standard position, $P(x, y)$ is a point on the terminal side of angle θ and $r = \sqrt{x^2 + y^2}$.

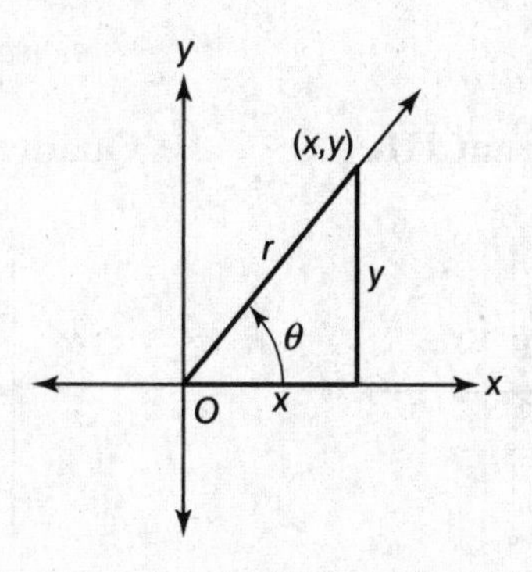

When angle θ is in standard position, the trigonometric functions of angle θ are defined:

$$\sin \theta = \frac{y}{r} \qquad \csc \theta = \frac{r}{y}$$

$$\cos \theta = \frac{x}{r} \qquad \sec \theta = \frac{r}{x}$$

$$\tan \theta = \frac{y}{x} \qquad \cot \theta = \frac{x}{y}$$

The signs of the trigonometric values depend on the signs of x and y in the quadrant in which the terminal side of angle θ is found. Note that r is always positive.

For any angle θ in standard position, the reference angle θ_{ref} is the positive acute angle that θ makes with the x-axis. The following table shows θ_{ref} for angles in each quadrant.

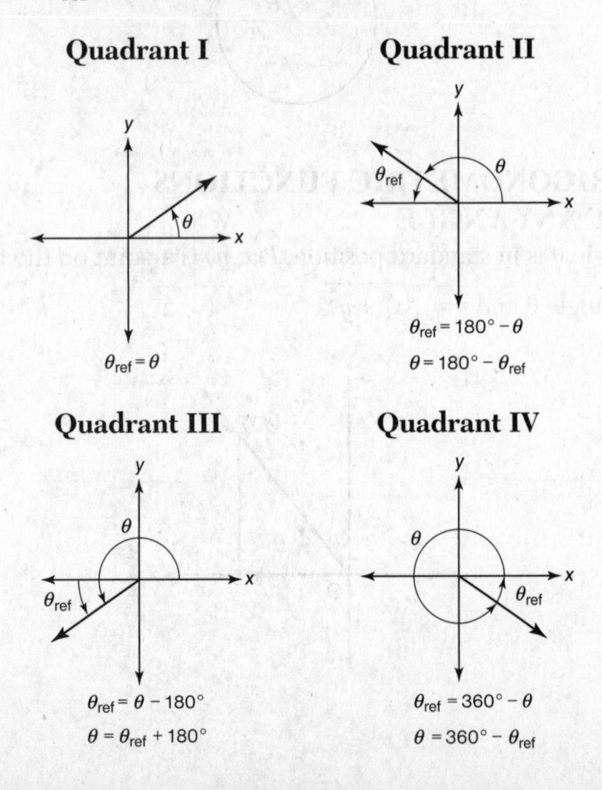

The next table shows how to express the trigonometric function of an angle, x, in any quadrant as a function of its reference angle, x_{ref}.

	Quadrant I: All functions are +	**Quadrant II: Only sin *x* is +**
$\sin x$ (and $\csc x$) $\cos x$ (and $\sec x$) $\tan x$ (and $\cot x$)	$\sin x = \sin x_{ref}$ $\cos x = \cos x_{ref}$ $\tan x = \tan x_{ref}$	$\sin x = \sin x_{ref}$ $\cos x = -\cos x_{ref}$ $\tan x = -\tan x_{ref}$
	Quadrant III: Only tan *x* is +	**Quadrant IV: Only cos *x* is +**
$\sin x$ (and $\csc x$) $\cos x$ (and $\sec x$) $\tan x$ (and $\cot x$)	$\sin x = -\sin x_{ref}$ $\cos x = -\cos x_{ref}$ $\tan x = \tan x_{ref}$	$\sin x = -\sin x_{ref}$ $\cos x = \cos x_{ref}$ $\tan x = -\tan x_{ref}$

9.3 TRIGONOMETRIC VALUES FOR 30º, 45º, AND 60º

Know the exact values of the trigonometric functions for these angles.

x	**30º**	**45º**	**60º**
$\sin x$	$\frac{1}{2}$	$\frac{\sqrt{2}}{2}$	$\frac{\sqrt{3}}{2}$
$\cos x$	$\frac{\sqrt{3}}{2}$	$\frac{\sqrt{2}}{2}$	$\frac{1}{2}$
$\tan x$	$\frac{\sqrt{3}}{3}$	1	$\sqrt{3}$

Notice the pattern in the numerators for $\sin x$ and $\cos x$. In Quadrant I, $\sin x$ increases and $\cos x$ decreases.

To find the other three trigonometric functions, use the reciprocal identities:

$$\csc x = \frac{1}{\sin x} \qquad \sec x = \frac{1}{\cos x} \qquad \cot x = \frac{1}{\tan x}$$

Know how to find a trigonometric value for any angle using your calculator. In most instances, use the exact value (not the decimal approximation) of the trigonometric values for 30°, 45°, and 60°.

9.4 COFUNCTIONS

The prefix "*co-*" on a trigonometric function indicates that two functions are cofunctions. For example, tangent and cotangent are cofunctions. Cofunctions of two angles have the same value if the angles are complementary (sum to 90°). In the previous table, sin 30° = cos 60° because sine and cosine are cofunctions and 30° + 60° = 90°. The third pair of cofunctions is secant and cosecant.

9.5 UNIT CIRCLE AND TRIGONOMETRIC VALUES FOR THE QUADRANTAL ANGLES: 0°, 90°, 180°, 270°

In the unit circle, radius = 1, so trigonometric ratios are simplified.

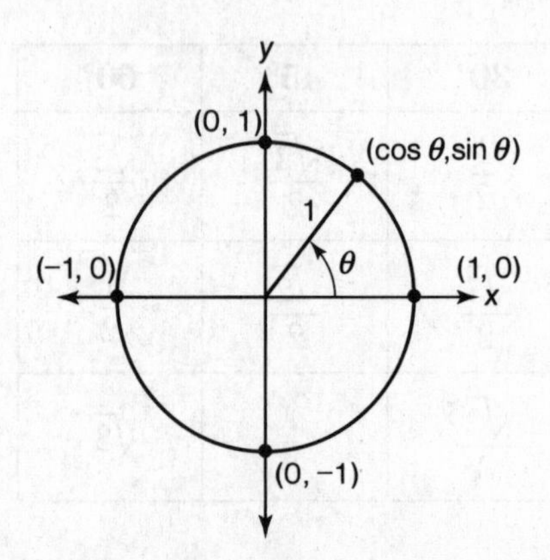

If $P(x, y)$ is the point at the intersection of the unit circle and the terminal side of angle θ, $\cos \theta = x$ and $\sin \theta = y$. The coordinates of the points where the unit circle intersects the x- and y-axes can be used to find the trigonometric values for angles with terminal sides on the axes (quadrantal angles). These are shown in the following table.

θ	0°	90°	180°	270°
$\sin \theta$	0	1	0	–1
$\cos \theta$	1	0	–1	0
$\tan \theta$	0	undefined	0	undefined

9.6 INVERSE TRIGONOMETRIC FUNCTIONS

To find an angle for a known trigonometric value, use the inverse trigonometric function.

One solution to $\sin \theta = k$ is $\sin^{-1} k = \theta$, sometimes written as $\arcsin k = \theta$. There may be a second solution in another quadrant.

Practice Exercises

9.1 What is 135° expressed in radian measure?

(1) 135π
(2) $\frac{\pi}{135}$
(3) $\frac{3\pi}{4}$
(4) $\frac{4\pi}{3}$

9.2 What is the number of degrees, rounded to the *nearest tenth*, of an angle whose radian measure is 2?

9.3 Tami is at point A on a circular track that has a radius of 150 feet, as shown in the accompanying diagram. She runs counterclockwise along the track from A to S, a distance of 247 feet. Find, to the *nearest degree*, the measure of the central angle that subtends minor arc AS.

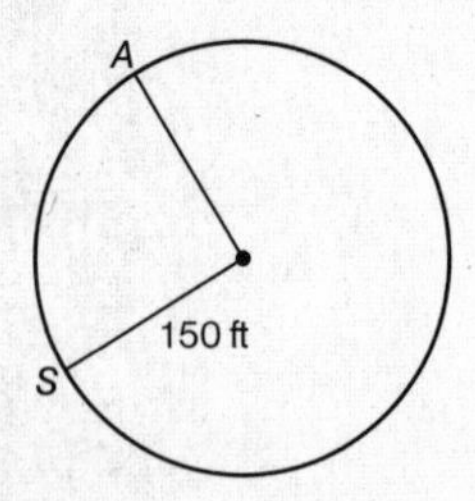

9.4 If csc $\theta = -2$, what is the value of sin θ?

(1) -2 (3) $-\frac{1}{2}$

(2) 2 (4) $\frac{1}{2}$

9.5 If $\tan (2x + 18)^\circ = \cot (4x - 12)^\circ$, then the value of x could be

(1) 14 (3) 29

(2) 15 (4) 46

9.6 If θ is an angle in standard position and its terminal side passes through the point $\left(\frac{1}{2}, \frac{\sqrt{3}}{2}\right)$ on a unit circle, a possible value of θ is

(1) 30° (3) 120°

(2) 60° (4) 150°

9.7 In the unit circle shown in the accompanying diagram, what are the coordinates of (x, y)?

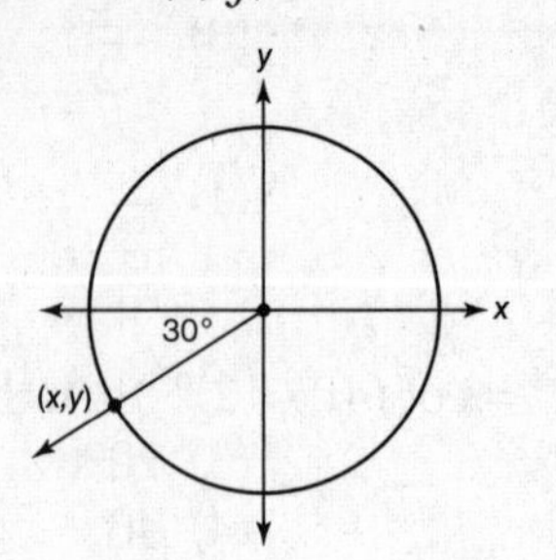

(1) $\left(-\frac{\sqrt{3}}{2}, -\frac{1}{2}\right)$ (3) $(-30, -210)$

(2) $\left(-\frac{1}{2}, -\frac{\sqrt{3}}{2}\right)$ (4) $\left(-\frac{\sqrt{2}}{2}, -\frac{\sqrt{2}}{2}\right)$

9.8 If $\sin\theta > 0$ and $\sec\theta < 0$, in which quadrant does the terminal side of angle θ lie?

(1) I (3) III
(2) II (4) IV

Solutions

9.1 To convert degrees to radians, multiply by $\frac{\pi}{180°}$:

$$135° \cdot \frac{\pi}{180°} = \frac{135\pi}{180} = \frac{3\pi}{4}$$

The correct choice is **(3)**.

9.2 To convert radians to degrees, multiply by $\frac{180°}{\pi}$:

$$2 \cdot \frac{180°}{\pi} = \frac{360°}{\pi} = 114.591559$$

To convert decimal degress to degrees and minutes, you can subtract the whole number of degrees and multiply the remaining decimal by 60.

```
2(180/π)
       114.591559
Ans-114
       .5915590262
Ans*60
      35.49354157
```

To the nearest minute, this gives 114°35′.

Alternatively, you can use the ▶DMS function in the ANGLE menu to convert to degrees-minutes-seconds.

```
2(180/π)
       114.591559
Ans▶DMS
   114°35'29.612"
```

To the nearest minute, 2 radians is 114°35′.

9.3 You must know the formula for arc length on a circle $s = \theta r$, where s is the arc length, r is the radius of the circle, and θ is the central angle *in radians*. In the problem, you know $r = 150$ feet and $s = 247$ feet.

$$247 = 150\theta$$
$$\theta = \frac{247}{150} = 1.646666667 \text{ radians}$$

To convert radians to degrees, multiply by $\frac{180°}{\pi}$.

$$1.646666667\left(\frac{180°}{\pi}\right) = 94.34705°$$

To the *nearest degree*, the angle measures 94°.

9.4 Sine and cosecant are reciprocals of each other.

$$\sin\theta = \frac{1}{\csc\theta} = \frac{1}{-2} = -\frac{1}{2}$$

The correct choice is **(3)**.

9.5 Cofunctions, such as tangent and cotangent, of two angles will be equal when the angles are complementary (sum to 90°).

$$\begin{aligned}(2x + 18) + (4x - 12) &= 90\\ 6x + 6 &= 90\\ 6x &= 84\\ x &= 14\end{aligned}$$

The correct choice is **(1)**.

9.6 If θ is an angle in standard position and its terminal side passes through the point (x, y) on the unit circle, then $\cos\theta = x$ and $\sin\theta = y$. In this problem, x and y are both positive, so the angle is in Quadrant I.

$$\cos\theta = \frac{1}{2} \quad \rightarrow \quad \theta = \cos^{-1}\left(\frac{1}{2}\right) = 60°$$

The correct choice is **(2)**.

9.7 If θ is an angle in standard position and its terminal side passes through the point (x, y) on the unit circle, then $\cos\theta = x$ and $\sin\theta = y$. Since 30° is the reference angle, not θ, you need to choose the correct signs for x and y. In Quadrant III, both sine and cosine are negative. The coordinates are given by

$$x = -\cos 30° = -\frac{\sqrt{3}}{2}$$

$$y = -\sin 30° = -\frac{1}{2}$$

The correct choice is **(1)**.

9.8 Sine is positive in Quadrants I and II. Secant is the reciprocal of cosine, and both are negative in Quadrants II and III. The only quadrant where sine is positive and secant is negative is Quadrant II.

The correct choice is **(2)**.

10. TRIGONOMETRIC GRAPHS

10.1 BASIC GRAPHS OF TRIGONOMETRIC FUNCTIONS

Make sure the calculator is in radian mode. ZoomTrig is a commonly used window on the graphing calculator for trigonometric functions.

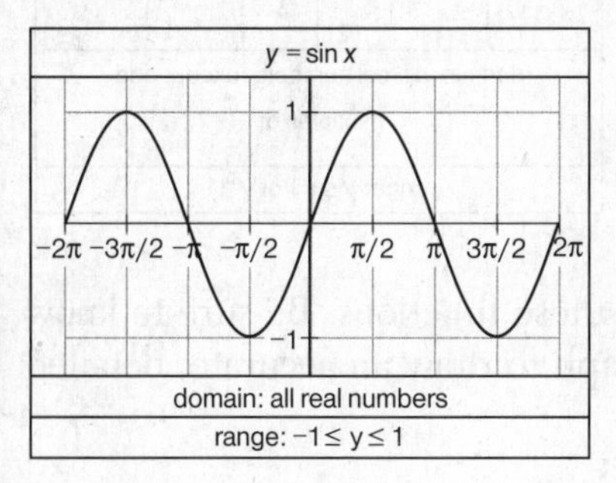

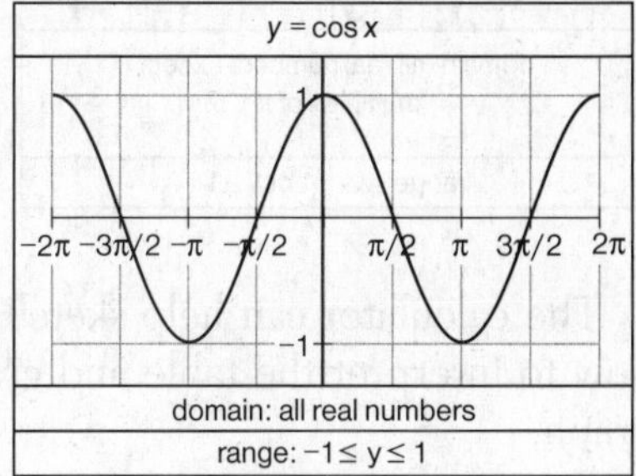

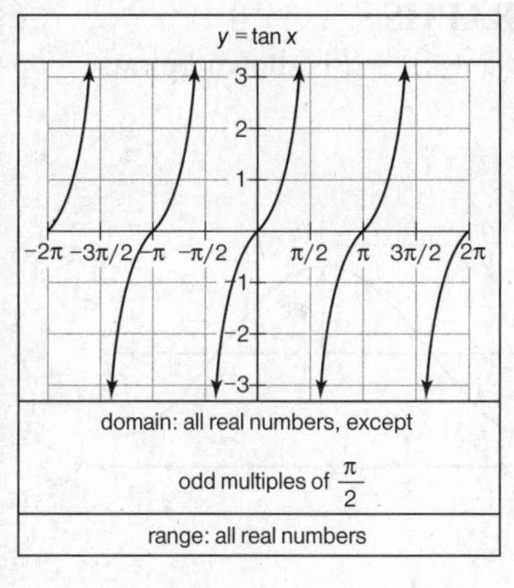

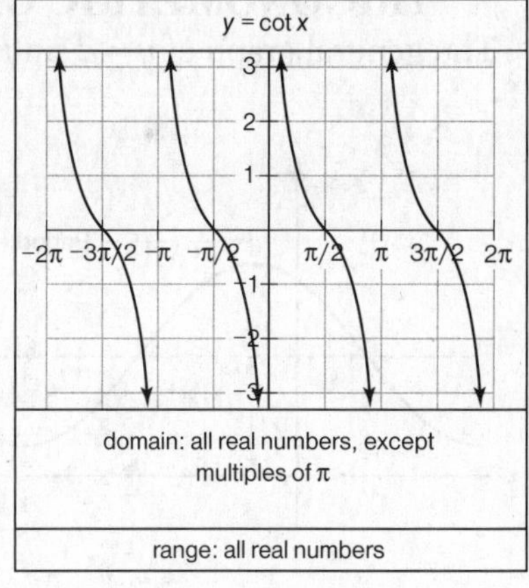

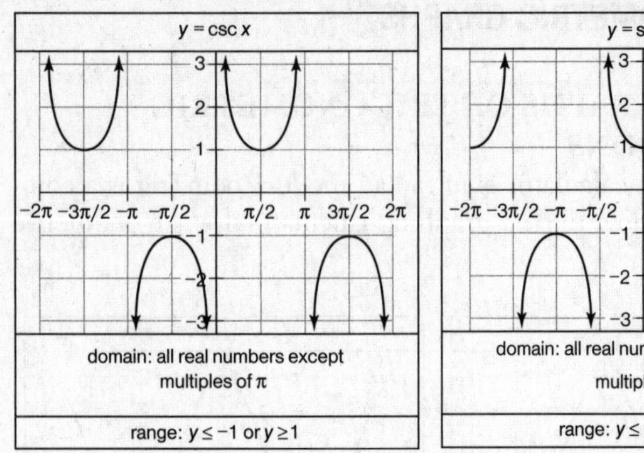

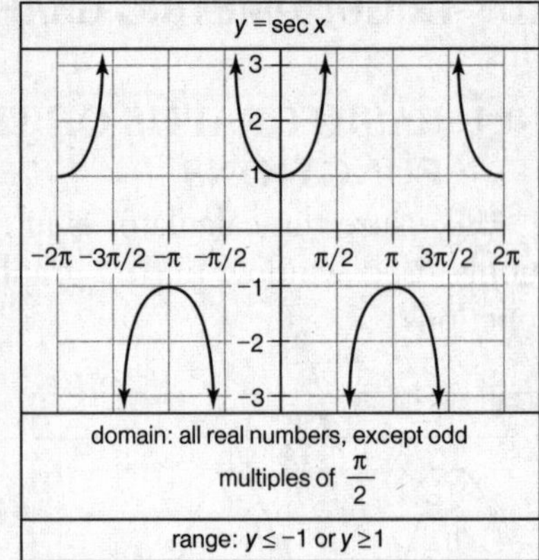

The calculator can help sketch these functions. Be sure to know how to interpret the table and graph to draw an accurate, detailed graph.

10.2 TRANSFORMATIONS OF TRIGONOMETRIC GRAPHS

The general graph of $y = a\sin(b(x + c)) + d$ is shown below.

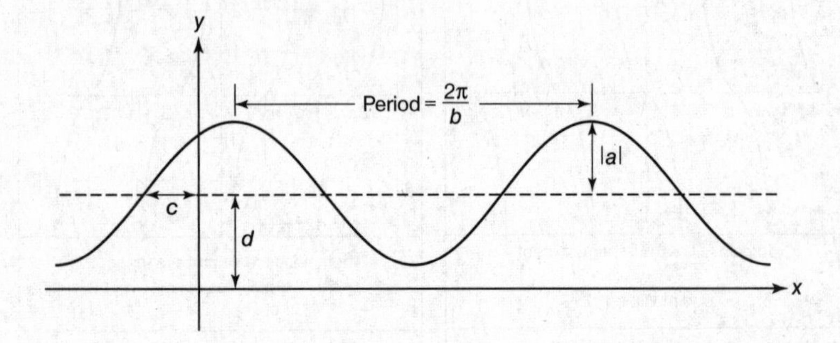

- Amplitude = $|a|$ = half the distance from the maximum value to the minimum value.
- Period = $\frac{2\pi}{|b|}$ = the length of one complete cycle of the graph.
- Frequency = $|b|$ = the number of complete cycles in the interval $0 \leq x \leq 2\pi$. From a graph, find the period first, and then use $\frac{2\pi}{\text{period}} = b$ to find the frequency. Another form of the formula is $b \cdot \text{period} = 2\pi$.
- Phase shift = c = how far the beginning of the cycle has been translated horizontally. When $c = 0$, a cycle begins at $x = 0$. The sign of c in the equation indicates the direction of the horizontal translation. A positive value of c represents a translation to the left, and a negative value of c represents a translation to the right.
- Vertical shift = d = how far the midline of the graph has been translated vertically. When $d = 0$, the midline for the graph is the x-axis. Positive values of d translate the function up; negative values of d translate it down. The midline is the average of the maximum and minimum values for the function.

These transformation concepts apply to both sine and cosine functions. When writing the equation from a graph, multiple responses may be correct.

10.3 GRAPHS OF INVERSE TRIGONOMETRIC FUNCTIONS

In order to define inverse functions for sine, cosine, and tangent, the domain of each must be restricted to create one-to-one functions.

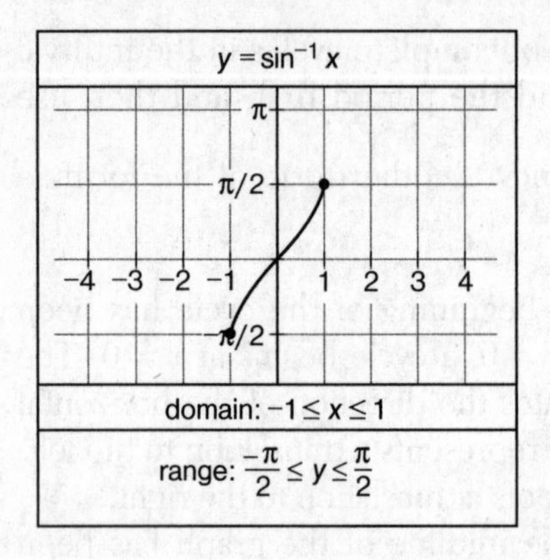

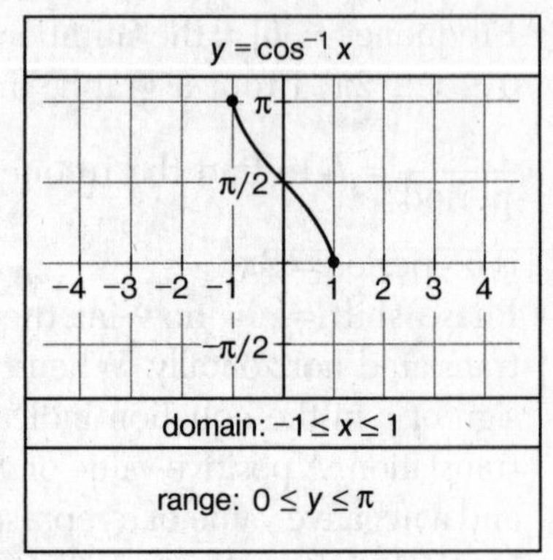

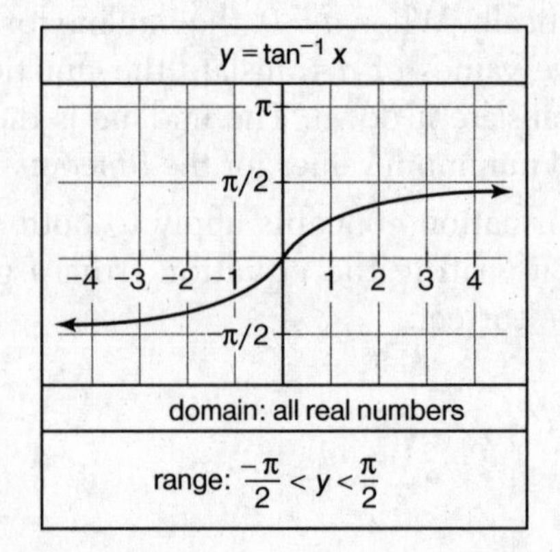

The calculator can help sketch these functions. Be sure to know how to interpret the table and graph to draw an accurate, detailed graph. ZoomDecimal is a better window than ZoomTrig for the inverse trigonometric functions.

Practice Exercises

10.1 A monitor displays the graph $y = -3 \sin 5x$. What is the amplitude?

(1) –5
(2) –3
(3) 3
(4) 5

10.2 A certain radio wave is represented by the equation $y = 5 \sin 2x$. What is the period of this wave?

(1) 5
(2) 2
(3) π
(4) 2π

10.3 What is the phase shift of the function $y = 5 \sin 3(x - 0.6)$?

(1) 5 up
(2) 5 down
(3) 0.6 to the right
(4) 0.6 to the left

10.4 The accompanying diagram shows a section of a sound wave as displayed on an oscilloscope

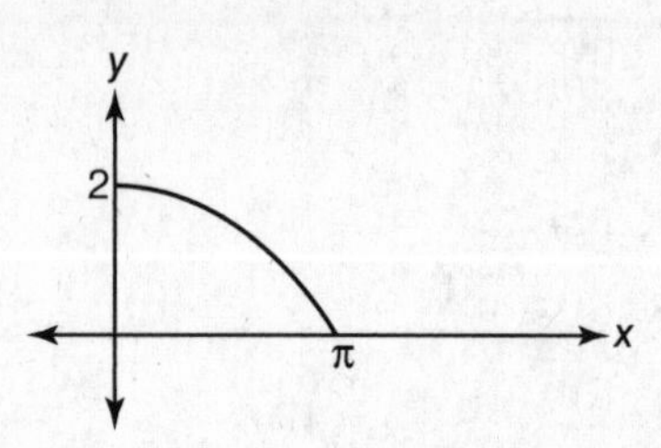

Which equation could represent this graph?

(1) $y = 2\cos\frac{x}{2}$ (3) $y = \frac{1}{2}\cos 2x$

(2) $y = 2\sin\frac{x}{2}$ (4) $y = \frac{1}{2}\sin 2x$

10.5 A student attaches one end of a rope to a wall at a fixed point 3 feet above the ground, as shown in the accompanying diagram. The student moves the other end of the rope up and down, producing a wave described by the equation $y = a \sin bx + c$. The range of the rope's height above the ground is between 1 foot and 5 feet. The period of the wave is 4π. Write an equation that represents this wave.

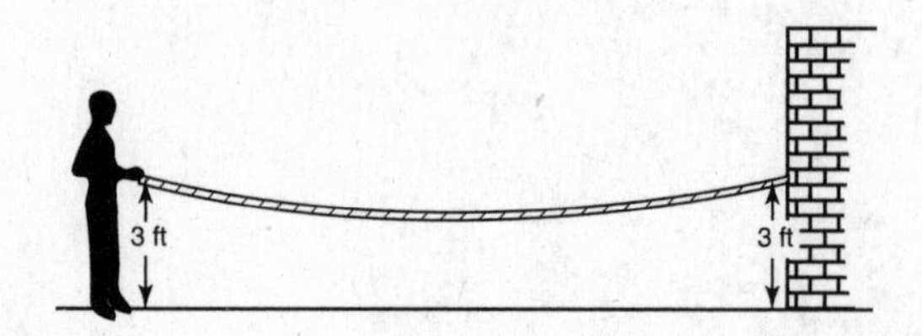

10.6 A radio wave has an amplitude of 3 and a wavelength (period) of π meters. Using the interval $0 \leq x \leq 2\pi$, draw a possible sine curve for this wave that passes through the origin. Write an equation for your curve.

10.7 Sketch the graph of $y = \cos^{-1} x$.

10.8 Sketch the graph of $y = \csc x$ over the interval $-2\pi \leq x \leq 2\pi$.

Solutions

10.1 In the equation $y = a \sin bx$, the amplitude is $|a|$. In this problem, the amplitude is $|-3| = 3$.

The correct choice is **(3)**.

10.2 In the equation $y = a \sin bx$, the frequency is b and the period is given by $\frac{2\pi}{b}$. In this problem, $b = 2$ and the period is $\frac{2\pi}{2} = \pi$.

The correct choice is **(3)**.

10.3 In the equation $y = a \sin b(x + c)$, the value of c gives the phase shift and the sign tells the direction. Positive means left; negative means right. In this problem, the phase shift is 0.6 units to the right.

The correct choice is **(3)**.

10.4 It is helpful to extend the graph to the right.

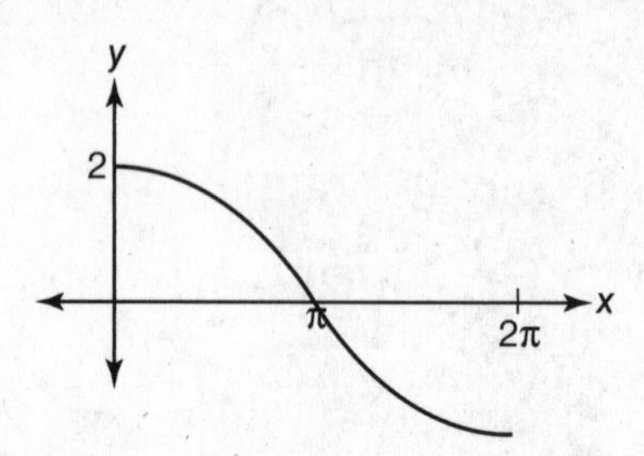

When $x = 0$, the graph is at its maximum. This is part of a cosine wave. The amplitude, the distance from the center line to the maximum value, is 2. Since exactly half a cycle fits in the interval $0 \le x \le 2\pi$, the frequency is $\frac{1}{2}$. The equation is $y = 2 \cos \frac{1}{2}x$, which can also be written as $y = 2 \cos \frac{x}{2}$.

The correct choice is **(1)**

10.5 Draw a diagram. Note that once the student begins moving the end of the rope up and down, it will no longer look like the original picture.

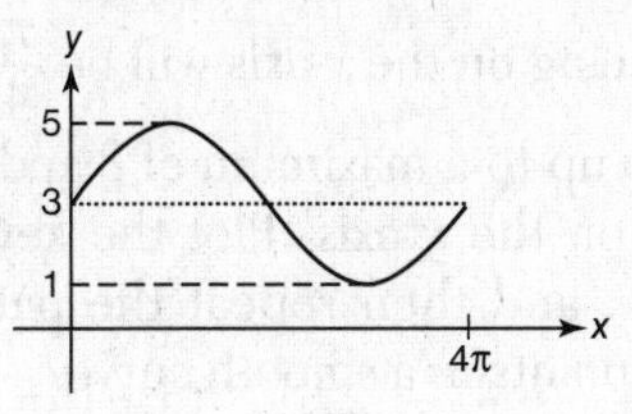

To write the equation in $y = a \sin bx + c$ form, find a, b and c.

Caution: Because c is a constant added to the function, it represents the value of the vertical shift, not the phase shift as is common usage. The letters used in each position are arbitrary. Other letters can be and are sometimes used instead.

$\|a\|$ = **Amplitude**	b = **Frequency**	c = **Vertical Shift**
Half the distance from the minimum value to the maximum value. The minimum is 1, and the maximum is 5.	$b = \frac{2\pi}{\text{period}}$ Period, the length of one complete cycle, is 4π.	The vertical distance between the x-axis and the centerline of the graph; positive if the centerline has been translated up from the x-axis. This sine graph has been translated up 3 feet.
$a = \frac{5-1}{2} = 2 \rightarrow$ $a = 2$	$b = \frac{2\pi}{4\pi} = \frac{1}{2} \rightarrow$ $b = \frac{1}{2}$	$c = 3$.

One possible equation for the wave is $y = 2 \sin\left(\frac{1}{2}x\right) + 3$.

10.6 Since the period is π and the curve is graphed on the interval 0 to 2π, you need two complete cycles of a sine wave. Label π and 2π on the x-axis. Then divide each of those intervals into four equal parts; the spacing on the x-axis will be $\frac{\pi}{4}$. The amplitude is 3, so the graph will go up to a maximum of 3 and down to a minimum of –3. Mark these on the y-axis. Plot the key points of one cycle shown in the table, and then repeat the pattern for the second cycle. Connect the points in a smooth curve.

x	y
0	0
$\frac{\pi}{4}$	3
$\frac{\pi}{2}$	0
$\frac{3\pi}{4}$	–3
π	0

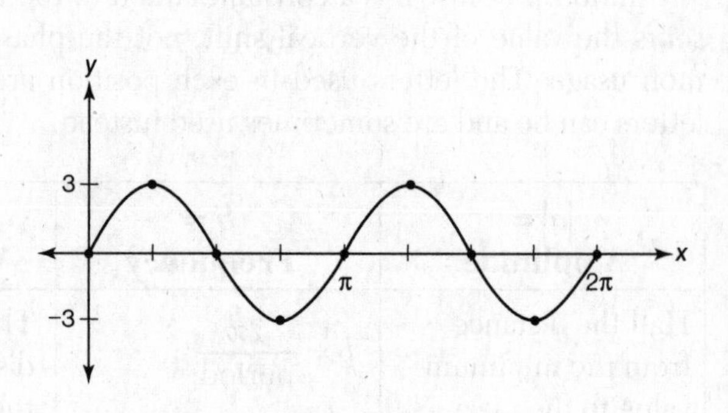

The equation will be in $y = a \sin bx$ form, where a is the amplitude and b is the frequency. The amplitude is 3, and the period is found by $b = \frac{2\pi}{\text{period}} = \frac{2\pi}{\pi} = 2$.

One equation of the graph is $y = 3\sin 2x$.

10.7 Remember that the range for $\cos^{-1}x$ is $0 \le x \le \pi$. Then make a table of values for $y = \cos x$ with that domain. It is convenient to rewrite all the x-values with the same denominator.

Since $\cos^{-1}x$ is the inverse of $\cos x$, to graph $\cos^{-1}x$, reverse the coordinates of all the points in the table for $\cos x$: (0, 1) in the table is graphed as (1, 0), etc.

x	$\cos x$
0	1
$\frac{\pi}{6} = \frac{2\pi}{12}$	$\frac{\sqrt{3}}{2} \approx 0.87$
$\frac{\pi}{4} = \frac{3\pi}{12}$	$\frac{\sqrt{2}}{2} \approx 0.71$
$\frac{\pi}{3} = \frac{4\pi}{12}$	$\frac{1}{2} = 0.5$
$\frac{\pi}{2} = \frac{6\pi}{12}$	0
$\frac{2\pi}{3} = \frac{8\pi}{12}$	$-\frac{1}{2} = -0.5$
$\frac{3\pi}{4} = \frac{9\pi}{12}$	$-\frac{\sqrt{2}}{2} \approx -0.71$
$\frac{5\pi}{6} = \frac{10\pi}{12}$	$-\frac{\sqrt{3}}{2} \approx -0.87$
$\pi = \frac{12\pi}{12}$	-1

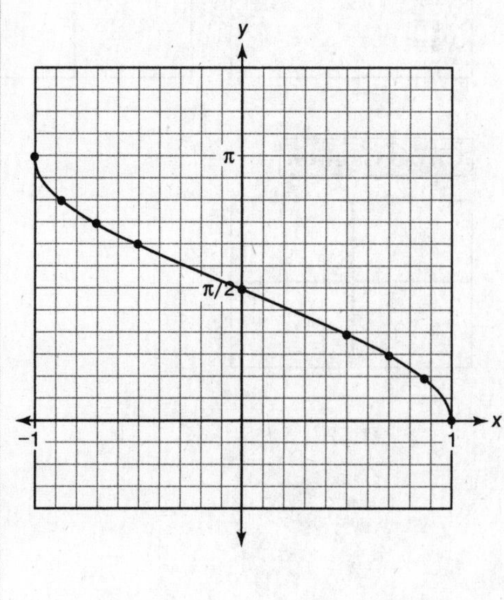

CALC Check

Be sure you are in Radian Mode. ZoomDecimal gives a better graph for inverse trig functions than ZoomTrig, shown far below. Make sure you know what the graphs of the basic functions look like so you can determine if your calculator is representing them well. Using the table can help if you are unsure.

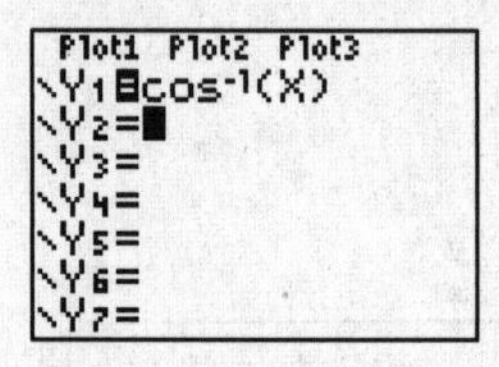

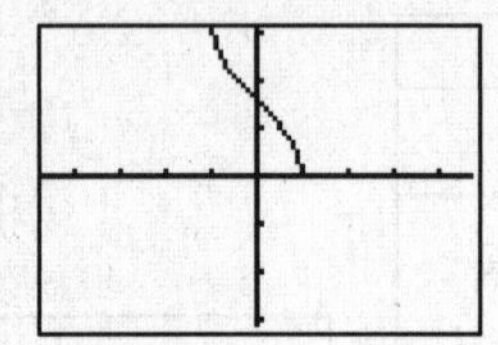

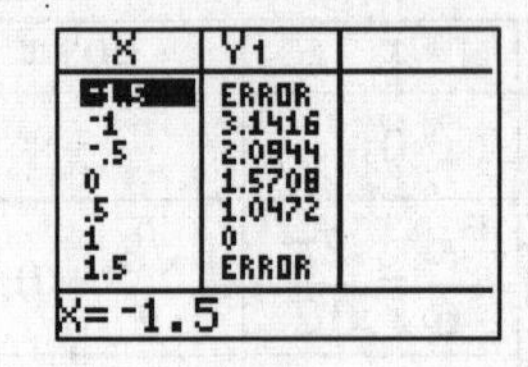

CALC Caution

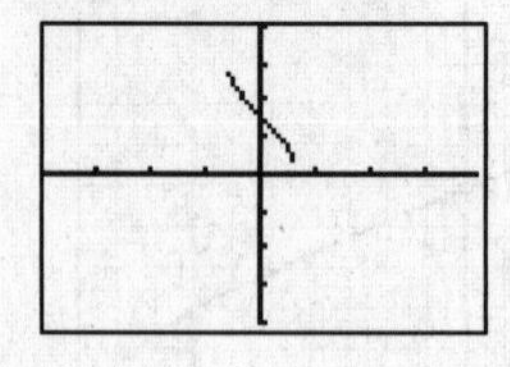

10.8 The easiest way to sketch $y = \csc x$ is first to sketch $y = \sin x$. Remember that cosecant is the reciprocal of sine. At every point where $\sin x = 0$ (including the y-axis), $\csc x$ is undefined and has a vertical asymptote, usually represented by a dashed line on a graph. At every point where $\sin x = 1$, $\csc x = 1$. The graphs of the two functions intersect at those points. Complete the graph of $y = \csc x$ by drawing an alternating series of ∪s and ∩s as shown in the graph below.

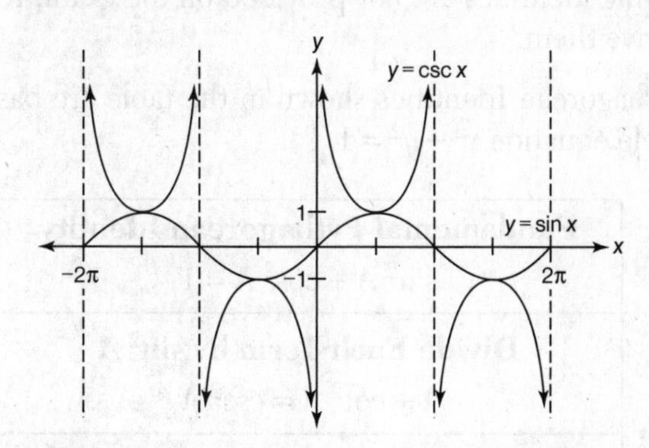

Radian mode and ZoomTrig are best for graphing reciprocal trigonometric functions.

CALC Check

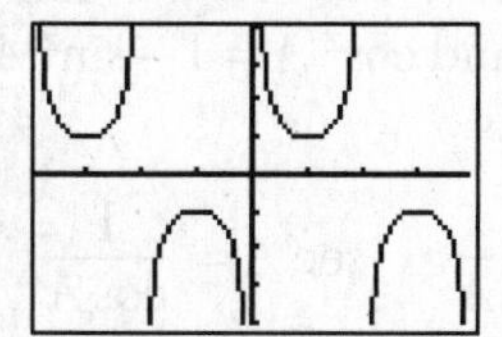

11. TRIGONOMETRIC IDENTITIES, FORMULAS, AND EQUATIONS

11.1 BASIC TRIGONOMETRIC IDENTITIES

The relationships among the trigonometric functions can be very helpful when evaluating trigonometric functions, simplifying trigonometric expressions, and solving trigonometric equations. The following identities are *not* provided on the exam. Know or be able to derive them.

- The Pythagorean Identities shown in the table are based on the unit circle equation $x^2 + y^2 = 1$.

Fundamental Pythagorean Identity $\sin^2 A + \cos^2 A = 1$
Divide Each Term by $\sin^2 A$ $1 + \cot^2 A = \csc^2 A$
Divide Each Term by $\cos^2 A$ $\tan^2 A + 1 = \sec^2 A$

- The identity $\sin^2 A + \cos^2 A = 1$ can be rearranged to give $\sin^2 A = 1 - \cos^2 A$ and $\cos^2 A = 1 - \sin^2 A$.
- Reciprocal identities:

$$\csc A = \frac{1}{\sin A} \qquad \sec A = \frac{1}{\cos A} \qquad \cot A = \frac{1}{\tan A}$$

- Quotient identities:

$$\tan A = \frac{\sin A}{\cos A} \qquad \cot A = \frac{\cos A}{\sin A}$$

11.2 TRIGONOMETRIC FORMULAS

Formulas for finding trigonometric values of the sum and difference of angles, half angle, and double angle are provided on the reference sheet during the exam. Know which formulas are provided and when and how to use them.

Functions of the Sum of Two Angles

$\sin(A+B) = \sin A \cos B + \cos A \sin B$

$\cos(A+B) = \cos A \cos B - \sin A \sin B$

$$\tan(A+B) = \frac{\tan A + \tan B}{1 - \tan A \tan B}$$

Functions of the Difference of Two Angles

$\sin(A-B) = \sin A \cos B - \cos A \sin B$

$\cos(A-B) = \cos A \cos B + \sin A \sin B$

$$\tan(A-B) = \frac{\tan A - \tan B}{1 + \tan A \tan B}$$

Functions of the Half Angle

$$\sin \frac{1}{2}A = \pm\sqrt{\frac{1-\cos A}{2}}$$

$$\cos \frac{1}{2}A = \pm\sqrt{\frac{1+\cos A}{2}}$$

$$\tan \frac{1}{2}A = \pm\sqrt{\frac{1-\cos A}{1+\cos A}}$$

Functions of the Double Angle

$\sin 2A = 2 \sin A \cos A$

$\cos 2A = \cos^2 A - \sin^2 A$

$\cos 2A = 2 \cos^2 A - 1$

$\cos 2A = 1 - 2 \sin^2 A$

$$\tan 2A = \frac{2 \tan A}{1 - \tan^2 A}$$

The double angle formulas are sometimes used to rewrite an equation in a single variable before solving. You have three choices for $\cos 2A$; choose one that that matches the other trigonometric functions in the equation.

11.3 TRIGONOMETRIC EQUATIONS

When solving trigonometric equations, use the following techniques. Use trigonometric identities to rewrite equations in terms of a single trigonometric function of a common variable. Factor completely or use the quadratic formula. Use inverse trigonometric functions to solve for the angle. Remember to consider all appropriate quadrants.

Practice Exercises

11.1 If $\sin A = \frac{4}{5}$, $\tan B = \frac{5}{12}$, and angles A and B are in Quadrant I, what is the value of $\sin(A + B)$?

(1) $\frac{63}{65}$

(2) $-\frac{63}{65}$

(3) $\frac{33}{65}$

(4) $-\frac{33}{65}$

11.2 The expression $\cos 40° \cos 10° + \sin 40° \sin 10°$ is equivalent to

(1) $\cos 30°$

(2) $\cos 50°$

(3) $\sin 30°$

(4) $\sin 50°$

11.3 If x is an acute angle and $\cos x = \frac{4}{5}$ then $\cos 2x$ is equal to

(1) $\frac{6}{25}$

(2) $-\frac{1}{25}$

(3) $\frac{2}{25}$

(4) $\frac{7}{25}$

11.4 If x is a positive acute angle and $\sin x = \frac{1}{2}$ what is $\sin 2x$?

(1) $-\frac{1}{2}$

(2) $\frac{1}{2}$

(3) $-\frac{\sqrt{3}}{2}$

(4) $\frac{\sqrt{3}}{2}$

11.5 If θ is a positive acute angle and $\sin 2\theta = \frac{\sqrt{3}}{2}$ then $(\cos\theta + \sin\theta)^2$ equals

(1) 1

(2) $1 + \frac{\sqrt{3}}{2}$

(3) 30°

(4) 60°

11.6 Solve algebraically for all values of θ in the interval $0° \leq \theta < 360°$ that satisfy the equation $\frac{\sin^2\theta}{1+\cos\theta} = 1$.

11.7 Find all values of x in the interval $0° \leq \theta < 360°$ that satisfy the equation $3\cos 2x = \cos x + 2$. Express your answers to the *nearest degree*. [The use of a grid is optional.]

11.8 If $(\sec x - 2)(2\sec x - 1) = 0$, then x terminates in

(1) Quadrant I, only
(2) Quadrants I and II, only
(3) Quadrants I and IV, only
(4) Quadrants I, II, III, and IV

11.9 If $\tan A = 8$ and $\tan B = \frac{1}{2}$, what is the value of $\tan(A + B)$?

(1) $\frac{4}{3}$

(2) $\frac{17}{10}$

(3) $-\frac{15}{6}$

(4) $-\frac{17}{6}$

11.10 If $\cos\theta = \frac{1}{8}$ and θ is an acute angle, then the value of $\sin\frac{\theta}{2}$ is

(1) $\frac{3}{2}$

(2) $\frac{\sqrt{7}}{4}$

(3) $\frac{9}{16}$

(4) $\frac{3}{4}$

11.11 If $\cos x = -\frac{4}{5}$ and x is in Quadrant III, find the value of $\tan\left(\frac{1}{2}x\right)$.

Solutions

11.1 $\sin(A + B)$ can be found using the angle sum formula shown on the reference sheet: $\sin(A + B) = \sin A \cos B + \cos A \sin B$. You need to know $\sin A$, $\sin B$, $\cos A$, and $\cos B$ to use this formula.

For angle A, you are given $\sin A = \frac{4}{5}$ and that A is in Quadrant I.

You need to determine $\cos A$. For angle B, you are given $\tan B = \frac{5}{12}$ and that B is in Quadrant I. You need to determine $\sin B$ and $\cos B$.

Draw separate triangles for angles A and B. Label the sides implied by the given trigonometric function values and use the Pythagorean Theorem to find the unknown side of each triangle. Use ratios of right triangles to determine the missing trigonometric values.

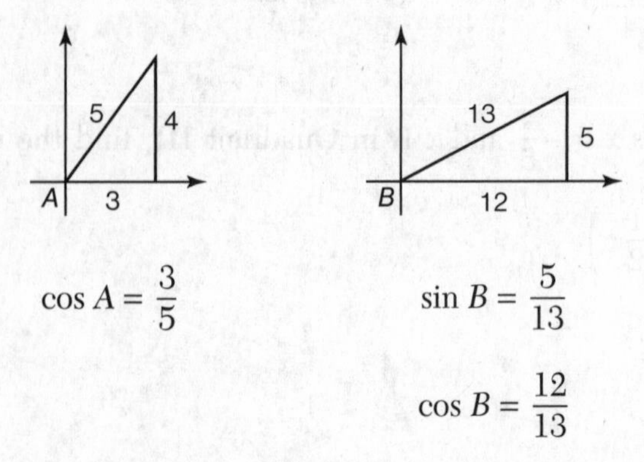

$$\cos A = \frac{3}{5} \qquad \sin B = \frac{5}{13}$$

$$\cos B = \frac{12}{13}$$

Substitute values into the formula $\sin(A + B) = \sin A \cos B + \cos A \sin B$.

$$= \left(\frac{4}{5}\right)\left(\frac{12}{13}\right) + \left(\frac{3}{5}\right)\left(\frac{5}{13}\right) = \frac{48}{65} + \frac{15}{65} = \frac{63}{65}$$

You can do this on the calculator accurately and easily because both angles are in Quadrant I. Using a calculator is trickier if the angles are not in Quadrant I.

CALC Check

```
sin(A+B)
        .9692307692
Ans▸Frac
              63/65
```

The correct choice is **(1)**.

11.2 The expression is of the form cos A cos B + sin A sin B, which is the formula for cos(A – B) shown on the reference sheet. With $A = 40°$ and $B = 10°$, $\cos(A - B) = \cos(40° - 10°) = \cos 30°$.

CALC Check

```
cos(40)cos(10)+s
in(40)sin(10)
        .8660254038
cos(30)
        .8660254038
```

The correct choice is **(1)**.

11.3 The reference sheet shows that $\cos 2A = 2\cos^2 A - 1$. So

$$\cos 2x = 2\left(\frac{4}{5}\right)^2 - 1 = 2\left(\frac{16}{25}\right) - 1 = \left(\frac{32}{25}\right) - \left(\frac{25}{25}\right) = \left(\frac{7}{25}\right)$$

CALC Check

```
cos⁻¹(4/5)
        36.86989765
cos(2Ans)
                .28
Ans▸Frac
               7/25
```

The correct choice is **(4)**.

11.4 The reference sheet shows that $\sin 2A = 2\sin A\cos A$. To find $\cos x$, label a diagram with the values from the sine ratio. Then use the Pythagorean Theorem to find the missing side.

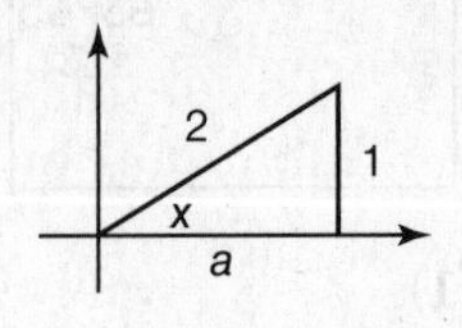

$$a^2 + 1^2 = 2^2$$
$$a^2 + 1 = 4$$
$$a^2 = 3$$
$$a = \sqrt{3}$$

From the diagram, $\cos x = \dfrac{\sqrt{3}}{2}$. Substitute values into the formula.

$$\sin 2x = 2\left(\frac{1}{2}\right)\left(\frac{\sqrt{3}}{2}\right) = \frac{\sqrt{3}}{2}$$

You can save time on this problem if you recognize the value for sin 30° and know the exact value of sin 60°.

Again, this problem can be easily worked out on the calculator because the angle is in Quadrant I.

CALC Check

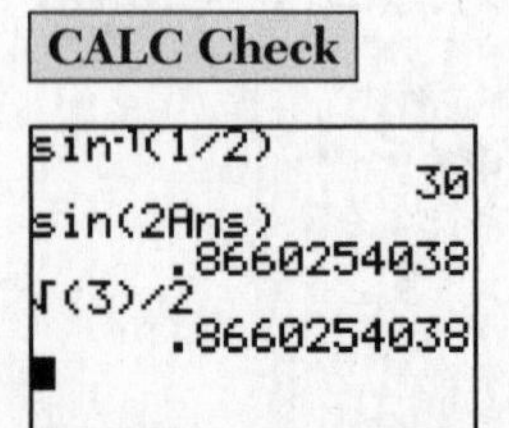

The correct choice is **(4)**.

11.5 $(\cos\theta + \sin\theta)^2 = (\cos\theta + \sin\theta)(\cos\theta + \sin\theta) = \cos^2\theta + 2\cos\theta\sin\theta + \sin^2\theta$. The reference sheet shows that $2\sin A\cos A = \sin 2A$. Also, you should recognize that $\cos^2\theta + \sin^2\theta = 1$.

$$(\cos\theta + \sin\theta)^2 = 1 + \sin 2\theta = 1 + \frac{\sqrt{3}}{2}$$

CALC Check

```
sin⁻¹(√(3)/2)
                60
Ans/2
                30
(cos(Ans)+sin(An
s))²
       1.866025404
```

The correct choice is **(2)**.

11.6 The problem involves two different trigonometric functions. From the Pythagorean Identity $\sin^2\theta + \cos^2\theta = 1$, you get $\sin^2\theta = 1 - \cos^2\theta$. This is an important variation used to change $\sin^2\theta$ into a function of $\cos\theta$. First, substitute.

$$\frac{\sin^2\theta}{1+\cos\theta} = 1 \rightarrow \frac{1-\cos^2\theta}{1+\cos\theta} = 1$$

Then cross-multiply and solve.

$$\begin{aligned} 1 - \cos^2\theta &= 1 + \cos\theta \\ \cos^2\theta + \cos\theta &= 0 \\ \cos\theta(\cos\theta + 1) &= 0 \\ \cos\theta &= 0 \quad &&\text{or } \cos\theta + 1 = 0 \\ \theta = 0° \text{ or } \theta &= 270° &&\text{or } \cos\theta = -1 \\ & &&\theta = 180° \end{aligned}$$

The answer $\theta = 180°$ does not check in the original equation because it makes the denominator equal to 0. There are two solutions: $\theta = 90°$ or $\theta = 270°$.

The extraneous root could have been avoided in this problem by simplifying the rational expression after substituting:

$$\frac{1-\cos^2\theta}{1+\cos\theta}=1 \rightarrow \frac{(1-\cos\theta)\cancel{(1+\cos\theta)}}{\cancel{(1+\cos\theta)}}=1$$

$$1 - \cos\theta = 1$$
$$\cos\theta = 0$$
$$\theta = 90° \text{ or } \theta = 270°$$

CALC Check

Use degree mode for graphical solutions to trigonometric equations in the interval $0° \leq x < 360°$. Set the x-values in WINDOW to $0° \leq x \leq 360°$, and then select ZoomFit for the best graph. It is easy to see on the graph that 90º and 270º are solutions, but you can use CALC (press 2nd TRACE) **5:intersect** to confirm these values.

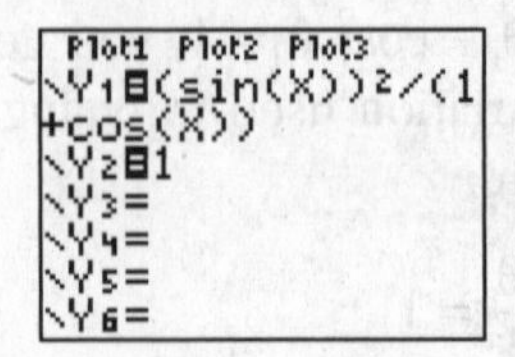

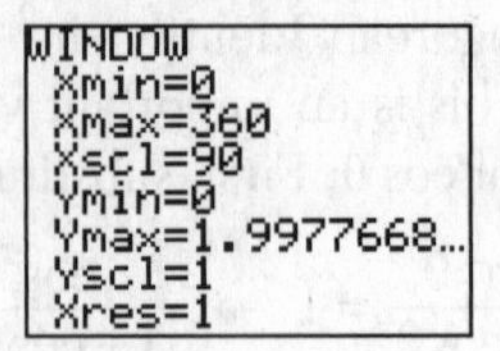

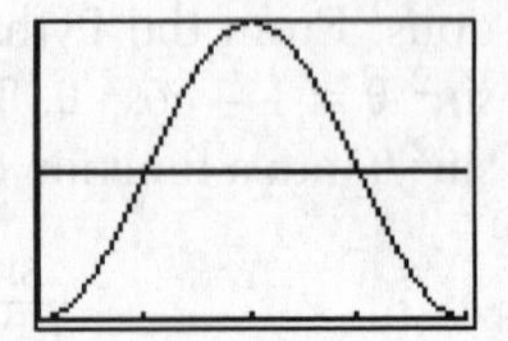

The solution set is {90°, 270°}.

11.7 The equation contains both x and $2x$. The only trigonometric function in the equation is cosine so use the formula $\cos 2A = 2\cos^2 A - 1$ from the reference sheet to rewrite $\cos 2x$ in terms of $\cos x$ only.

$$3\cos 2x = \cos x + 2$$
$$3(2\cos^2 x - 1) = \cos x + 2$$
$$6\cos^2 x - 3 = \cos x + 2$$

Rewrite in standard form.

$$6\cos^2 x - \cos x - 5 = 0$$

Factor and solve for x.

$(6 \cos x + 5)(\cos x - 1) = 0$

$$\cos x = -\frac{5}{6} \qquad \text{or } \cos x = 1$$

$$x_{\text{ref}} = \cos^{-1}\left(\frac{5}{6}\right) = 33.5573° \quad \text{or } x = \cos^{-1}(1) = 0°$$

In the interval $0° \leq x < 360°$, cosine is negative in Quadrants II and III. Since 33.5573° is in Quadrant I, it is not a solution. It is a reference angle. In Quadrant II, $x = 180° - 33.5573° = 146.4427°$. In Quadrant III, $x = 180° + 33.5573° = 213.5573°$.

The solution set is {0°, 146°, 214°}.

Since the use of a grid is optional, you can solve the equation graphically if you choose. You must sketch a detailed graph on scaled axes, labeled with equations and points of intersection. Only the x-values of these points are the solutions.

Graph $y = 3\cos 2x$ (use the left side of the original equation) and $y = \cos x + 2$ in degree mode. In [WINDOW], set Xmin = 0 and Xmax = 360. Select ZoomFit. Use CALC (press 2nd TRACE) **5:intersect** to find the three points of intersection. Note that the graph shows a fourth solution, $x = 360°$, which is not in the given interval.

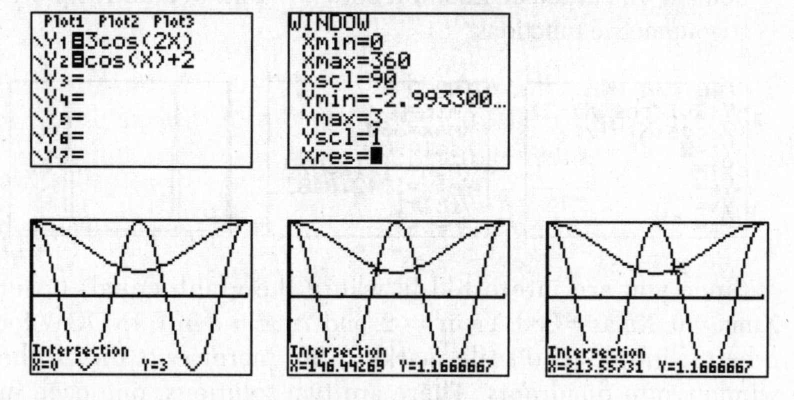

The solution set is {0°, 146°, 214°}.

11.8 Set each factor equal to 0 and solve.

$$(\sec x - 2)(2\sec x - 1) = 0$$

$$\sec x = 2 \quad \text{or} \quad \sec x = \frac{1}{2}$$

Remember that $\sec x = \dfrac{1}{\cos x}$.

$$\frac{1}{\cos x} = 2 \quad \text{or} \quad \frac{1}{\cos x} = \frac{1}{2}$$

Take the reciprocals.

$$\cos x = \frac{1}{2} \quad \text{or} \quad \cos x = 2$$

Cosine can be $\frac{1}{2}$ in Quadrants I or IV because cosine is positive in only those quadrants. Cosine can never equal 2. So there are two solutions, one each in Quadrants I and IV.

CALC Caution

This graph is the result of using ZoomFit. ZoomFit does not work well if there are undefined values in the WINDOW domain. In particular, ZoomFit does not work well for reciprocal trigonometric functions.

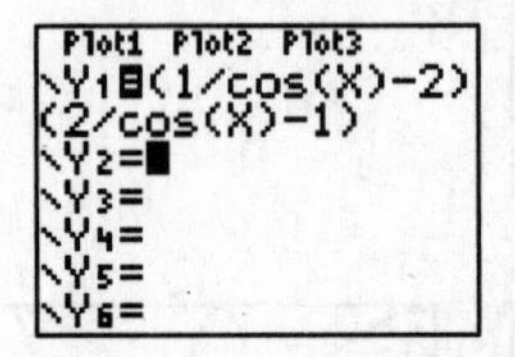

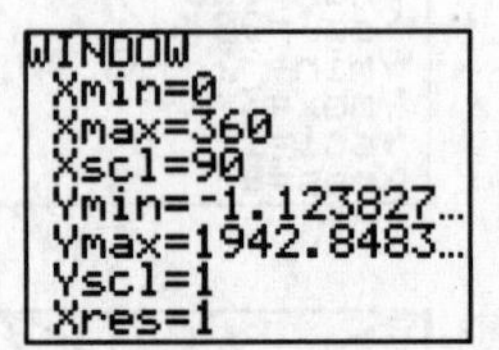

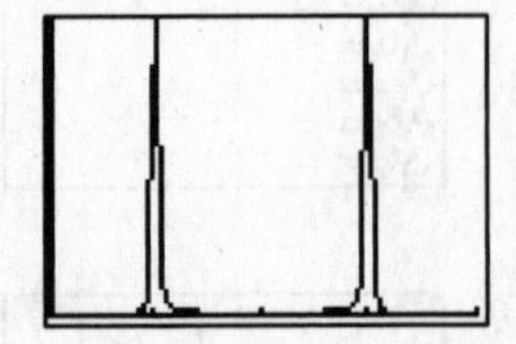

Since you are interested in where the graph equals 0, set Xmin = 0, Xmax = 360, Ymin = –2, and Ymax = 2 in WINDOW for a better graph. The scale marks at 90° increments divide the window into quadrants. There are two solutions, one each in Quadrants I and IV.

CALC Check

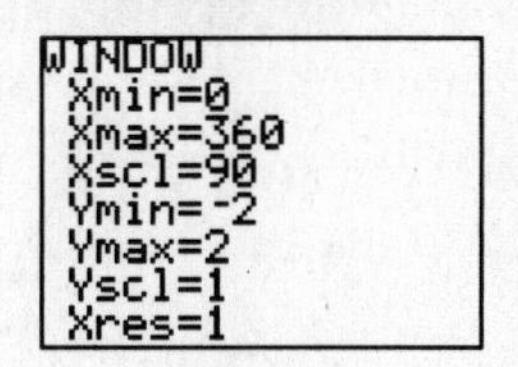

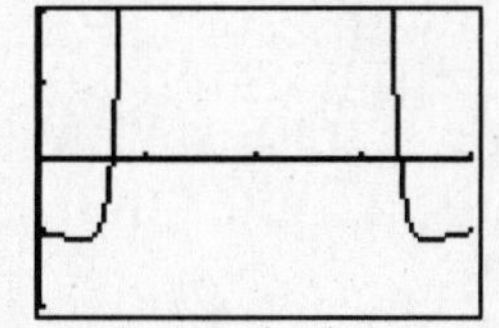

The correct choice is **(3)**.

11.9 From the reference sheet,

$$\tan(A+B)=\frac{\tan A+\tan B}{1-\tan A\tan B}=\frac{8+\frac{1}{2}}{1-(8)\left(\frac{1}{2}\right)}=\frac{\frac{17}{2}}{-3}=-\frac{17}{6}$$

CALC Check

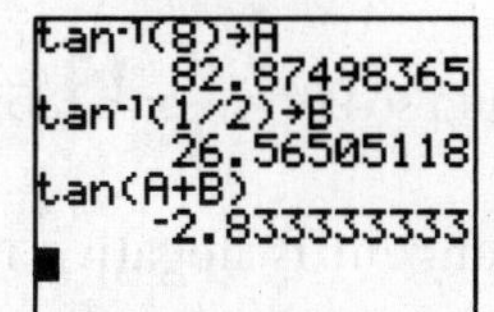

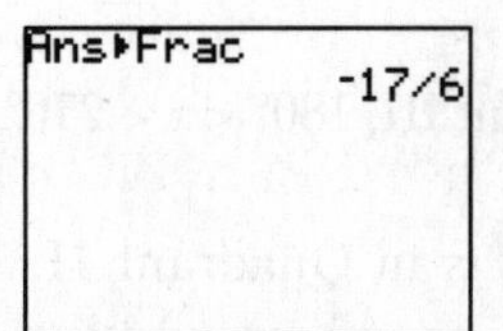

The correct choice is **(4)**.

11.10 From the reference sheet, $\sin\frac{1}{2}A=\pm\sqrt{\frac{1-\cos A}{2}}$.

$$\sin\frac{\theta}{2}=\pm\sqrt{\frac{1-\left(\frac{1}{8}\right)}{2}}=\pm\sqrt{\frac{\frac{7}{8}}{2}}=\pm\sqrt{\frac{7}{16}}=\pm\frac{\sqrt{7}}{4}$$

Since θ is acute, $\frac{1}{2}\theta$ is acute. The correct sign is positive.

CALC Check

```
cos⁻¹(1/8)
        82.81924422
sin(Ans/2)
        .6614378278
√(7)/4
        .6614378278
```

The correct choice is **(2)**.

11.11 From the reference sheet, $\tan\frac{1}{2}A = \pm\sqrt{\frac{1-\cos A}{1+\cos A}}$.

$$\tan\frac{1}{2}x = \pm\sqrt{\frac{1-\left(-\frac{4}{5}\right)}{1+\left(-\frac{4}{5}\right)}} = \pm\sqrt{\frac{\frac{9}{5}}{\frac{1}{5}}} = \pm\sqrt{9} = \pm 3$$

Since x is in Quadrant III, $180° < x < 270°$ and so $90° < \frac{1}{2}x < 135°$.

This means that $\frac{1}{2}x$ is in Quadrant II. Tangent is negative in Quadrant II, so the correct choice of sign is (–).

CALC Caution

```
cos⁻¹(-4/5)
        143.1301024
tan(.5Ans)
                  3
```

The calculator gives the principal value for inverse cosine. For a negative value, inverse cosine gives a Quadrant II angle. However, this problem specifies the angle is in Quadrant III. Redo the check by finding the reference angle first. Then add 180° to find the correct Quadrant III angle before proceeding.

CALC Check

```
cos⁻¹(4/5)
     36.86989765
Ans+180
     216.8698976
tan(.5Ans)
              -3
```

$$\tan\left(\frac{1}{2}x\right)=-3$$

12. TRIGONOMETRIC LAWS AND APPLICATIONS

The three formulas below are provided during the exam on the reference sheet.

12.1 AREA OF A TRIANGLE

The area, K, of any triangle can be found using any two sides of the triangle, a and b, and their included angle, C.

$$K = \frac{1}{2}ab \sin C$$

12.2 LAW OF SINES

In a triangle, the sines of the angles are in proportion to the sides directly opposite them. Use the Law of Sines to find a side if one side and two angles are known or to find an angle if two sides and one nonincluded angle are known (see "12.4 Ambiguous Case.") In any $\triangle ABC$,

$$\frac{a}{\sin A} = \frac{b}{\sin B} = \frac{c}{\sin C}$$

12.3 LAW OF COSINES

Use the Law of Cosines to find a side of a triangle if any two sides and the included angle are known or to find an angle if all three sides are known. In any $\triangle ABC$,

$$a^2 = b^2 + c^2 - 2bc \cos A$$

12.4 AMBIGUOUS CASE

If two sides and one nonincluded angle of a triangle are given, zero, one, or two triangles may be drawn to fit the given data. This problem is known as the ambiguous case. Use the Law of Sines to find the possible solutions for a second angle in the triangle, considering both the Quadrant I and Quadrant II (180° – original angle) solutions. Determine if it is possible to have a triangle with either or both of these two angle measures. Remember that the three angles in the triangle must sum to 180°.

Practice Exercises

12.1 The accompanying diagram shows the floor plan for a kitchen. The owners plan to carpet the entire kitchen except the work space, which is represented by scalene triangle ABC. Find the area of this work space to the *nearest tenth of a square foot.*

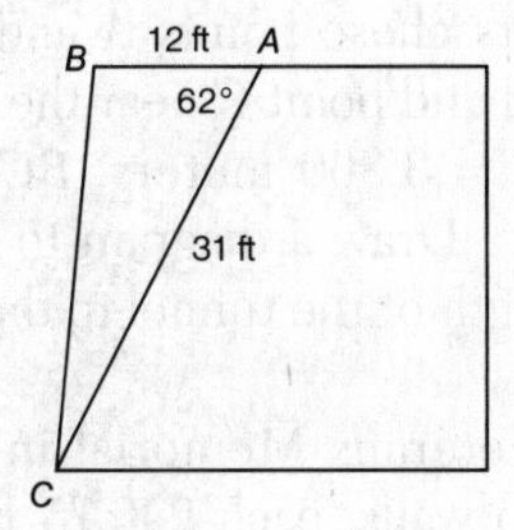

12.2 In the accompanying diagram of parallelogram $ABCD$, $m\angle A = 30°$, $AB = 10$, and $AD = 6$. What is the area of parallelogram $ABCD$?

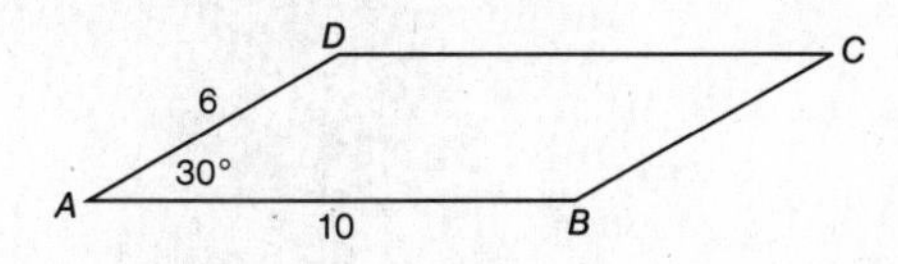

12.3 Kieran is traveling from city A to city B. As the accompanying map indicates, Kieran could drive directly from A to B along County Route 21 at an average speed of 55 miles per hour or could travel on the interstates, 45 miles along I-85 and 20 miles along I-64. The two interstates intersect at an angle of 150° at C and have a speed limit of 65 miles per hour. How much time, rounded to the *nearest minute*, will Kieran save by traveling along the interstates at an average speed of 65 miles per hour?

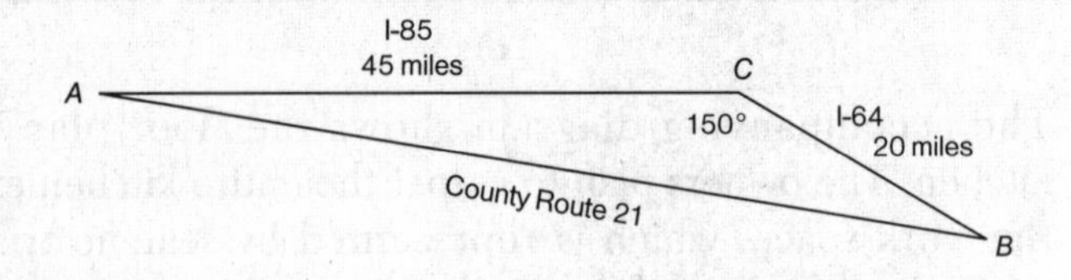

12.4 To measure the distance through a mountain for a proposed tunnel, surveyors chose points A and B at each end of the proposed tunnel and point C near the mountain. They determined that AC = 3,800 meters, BC = 2,900 meters, and $m\angle ACB = 110°$. Draw a diagram to illustrate this situation and find the length of the tunnel to the *nearest meter*.

12.5 The Vietnam Veterans Memorial in Washington, D.C., is made up of two walls, each 246.75 feet long, that meet to form an angle. The distance between the far ends of the walls is 438.14 feet. Find, to the *nearest tenth of a degree*, the angle between the walls where they meet.

12.6 In the accompanying diagram of $\triangle ABC$, $m\angle A = 30°$, $m\angle C = 50°$, and $AC = 13$.

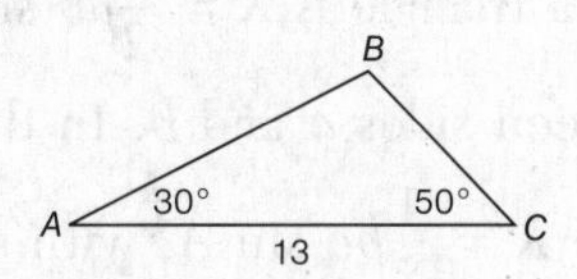

What is the length of side $\overline{AB}$ to the *nearest tenth*?

(1) 6.6 (3) 11.5
(2) 10.1 (4) 12.0

12.7 How many distinct triangles can be formed if $m\angle A = 30°$, side $b = 12$, and side $a = 8$?

(1) 1 (3) 3
(2) 2 (4) 0

12.8 Sam is designing a triangular piece for a metal sculpture. He tells Martha that two of the sides of the piece are 40 inches and 15 inches, and the angle opposite the 40-inch side measures 120°. Martha decides to sketch the piece that Sam described. How many different triangles can she sketch that match Sam's description?

(1) 1 (3) 3
(2) 2 (4) 0

12.9 How many distinct triangles can be formed if $m\angle A = 40°$, side $b = 15$, and side $a = 9$?

(1) 1 (3) 3
(2) 2 (4) 0

12.10 Two equal forces act on a body at an angle of 70°. If the resultant force is 100 newtons, find the value of one of the two equal forces to the *nearest hundredth of a newton*.

Solutions

12.1 The area of a triangle is $K = \frac{1}{2}ab \sin C$, where C is the included angle between sides a and b. In this triangle, A is the included angle. Use $K = \frac{1}{2}bc \sin A$, with $b = 31$, $c = 12$, and $m\angle A = 62°$.

$$\text{Area} = \frac{1}{2}(31)(12) \sin 62° = 164.2282$$

The area of the work space to the *nearest tenth of a foot* is 164.2 square feet.

12.2 The area of parallelogram $ABCD$ is twice the area of triangle ADB. The area of a triangle is $K = \frac{1}{2}ab \sin C$.

$$\text{Area}_{ABCD} = (2)\frac{1}{2}(6)(10)\sin 30° = 30$$

The area of parallelogram $ABCD$ is 30.

12.3 You need to know the distance from A to B along County Route 21. You know two sides and the included angle of a triangle and need the third side; use the Law of Cosines. To match the diagram as given, it is convenient both to switch a and c and to replace A with C in the formula given on the reference sheet: $c^2 = a^2 + b^2 - 2ab \cos C$. In the given triangle, $a = 20$, $b = 45$, and $m\angle C = 150°$.

$$c^2 = 20^2 + 45^2 - 2(20)(45)\cos 150° = 3983.845727$$

$$c = \sqrt{3983.845727} = 63.11771326 \text{ miles}$$

Use time $= \frac{\text{distance}}{\text{rate}}$ to find the amount of time each route takes. If traveling at 55 miles per hour, the trip along County Road 21 takes $\frac{63.11771326}{55} = 1.147594786$ hours. If traveling at 65 miles per hour, the trip on the interstates takes $\frac{45+20}{65} = 1$ hour. The difference in the times is $1.14759476 - 1 \approx 0.147594786$ hours. Convert hours to minutes.

$$0.147594786 \cancel{\text{hours}} \cdot \frac{60 \text{ minutes}}{1 \cancel{\text{hour}}} \approx 8.856 \text{ minutes}$$

To the *nearest minute*, Kieran will save 9 minutes by traveling along the interstates.

12.4 Draw a diagram.

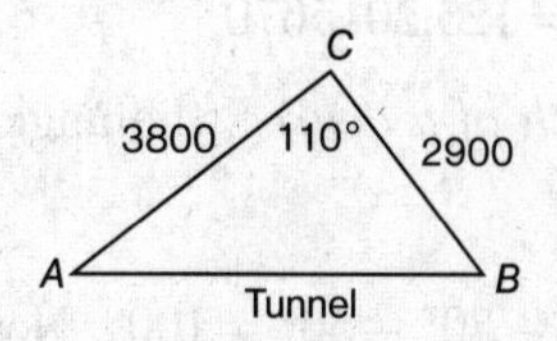

You know two sides and the included angle of a triangle and need the third side; use the Law of Cosines. As in the previous problem, switch a and c and replace A with C in the formula given: $c^2 = a^2 + b^2 - 2ab \cos C$.

$$c^2 = 2900^2 + 3800^2 - 2(2900)(3800)\cos 110° = 30388123.96$$

$$c = \sqrt{30388123.96} = 5512.542422$$

To the *nearest meter*, the tunnel is 5513 meters long.

12.5 Draw a diagram.

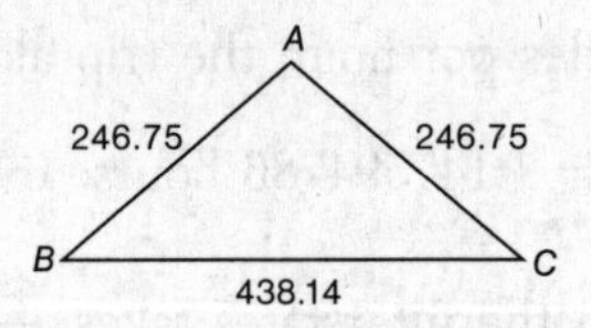

You know three sides of a triangle and need one angle; use the Law of Cosines. If you call the required angle A, then you can use the formula exactly as written on the reference sheet: $a^2 = b^2 + c^2 - 2bc \cos A$. With $b = c = 246.75$, and $a = 438.14$:

$$(438.14)^2 = (246.75)^2 + (246.75)^2 - 2(246.75)(246.75)\cos A$$
$$191966.6596 = 60885.5625 + 60885.5625 - 121771.125 \cos A$$
$$70195.5346 = -121771.125 \cos A$$
$$-0.5764546776 = \cos A$$
$$\cos^{-1}(-0.5764546776) = A$$
$$A = 125.2015679$$

To the *nearest tenth of a degree*, the angle between the walls is 125.2°.

12.6 $m\angle B = 180° - 30° - 50° = 100°$. Now you have a side, 13, and the angle opposite that side, 100°. You can use the Law of Sines.

$$\frac{AB}{\sin 50°} = \frac{13}{\sin 100°}$$

$$AB \sin 100° = 13 \sin 50°$$

$$AB = \frac{13 \sin 50°}{\sin 100°} = 10.1122 \approx 10.1$$

The correct choice is **(2)**.

12.7 Draw a rough diagram, do not worry about scale or accuracy for this.

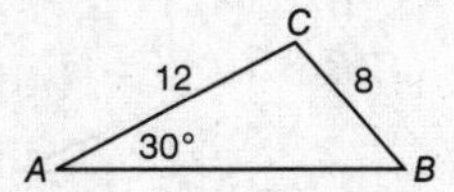

This is the ambiguous case. Use the Law of Sines to find m∠*B*.

$$\frac{8}{\sin 30^\circ} = \frac{12}{\sin B}$$

$$8 \sin B = 12 \sin 30^\circ$$

$$\sin B = \frac{12 \sin 30^\circ}{8} = 0.75$$

Sine is positive in Quadrants I and II, so there are two possible angles. Use $\sin^{-1}$ on your calculator to find one angle; the second is 180° minus that answer. If angle *B* is acute:

$$\begin{aligned} m\angle B &= \sin^{-1}(0.75) \\ &\approx 48.59^\circ \end{aligned}$$

If angle *B* is obtuse:

$$\begin{aligned} m\angle B &= 180^\circ - 48.59^\circ \\ &\approx 131.41^\circ \end{aligned}$$

Finally, find m∠*C* for each of the values found for m∠*B*. If m∠*B* = 48.59°, then m∠*C* = 180° – 30° – 48.59° = 101.41°. At least one triangle is possible. If m∠*B* = 131.41°, then m∠*C* = 180° – 30° – 131.41° = 18.59°. A second triangle is possible. Remember that there can never be more than two possible triangles in the ambiguous case.

The correct choice is **(2)**.

12.8 Draw a diagram (the labeling of the vertices is arbitrary). Proceed as in the previous problem. Use the Law of Sines.

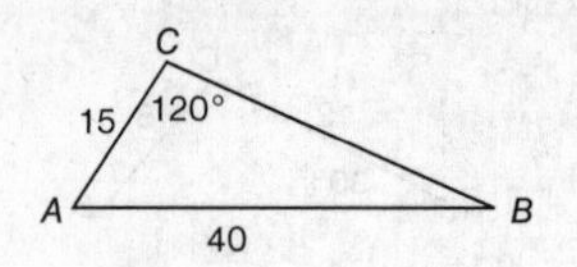

$$\frac{40}{\sin 120°} = \frac{15}{\sin B}$$

$$40 \sin B = 15 \sin 120°$$

$$\sin B = \frac{15 \sin 120°}{40} = 0.3247595$$

Two angles are possible.

Angle B Acute $\quad m\angle B = \sin^{-1}(0.3247595) \approx 18.95°$

Angle B Obtuse $\quad m\angle B = 180 - 18.95° \approx 161.05°$

Find $m\angle A$ for each of the values found for $m\angle B$.

Angle B Acute $\quad m\angle A = 180° - 120° - 18.95° = 41.05°$. At least one triangle is possible.

Angle B Obtuse $\quad m\angle A = 180° - 120° - 161.05° = -101.05°$. This is impossible, so there is *not* a second triangle.

The correct choice is **(1)**.

12.9 Draw a diagram (the labeling of the vertices is arbitrary).

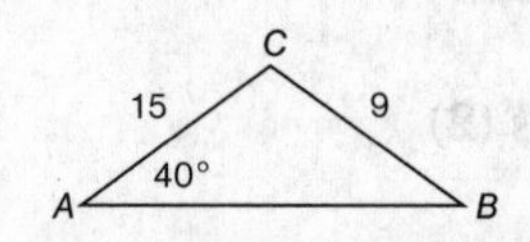

Use the Law of Sines.

$$\frac{9}{\sin 40°} = \frac{15}{\sin B}$$

$$9 \sin B = 15 \sin 40°$$

$$\sin B = \frac{15 \sin 40°}{9} = 1.071312683$$

Since no angle can have a sine greater than 1, no triangle is possible.

The correct choice is **(4)**.

12.10 Draw a diagram of the two forces. The resultant is drawn using the parallelogram rule, as shown in the second diagram.

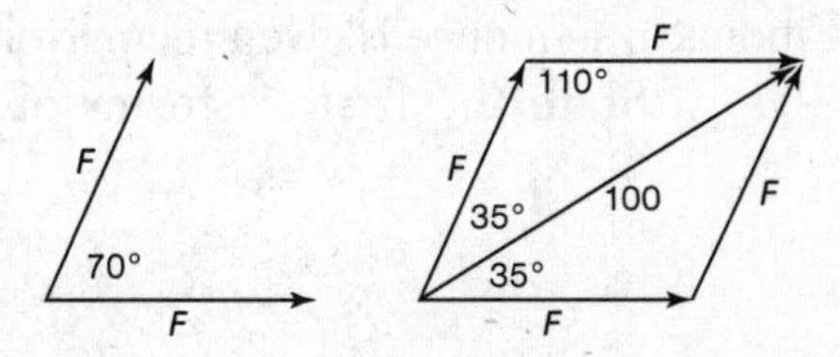

Because the forces are equal, the parallelogram is a rhombus. The resultant will bisect the angle, giving 35° between either force and the resultant. Use the Law of Sines.

$$\frac{F}{\sin 35°} = \frac{100}{\sin 110°}$$

$$F \sin 110° = 100 \sin 35°$$

$$F = \frac{100 \sin 35°}{\sin 110°} = 61.0387$$

To the *nearest hundredth*, the forces are both 61.04 newtons.

13. SEQUENCES AND SERIES

13.1 SIGMA NOTATION

The symbol Σ is the uppercase Greek letter sigma. It is used as a symbol for sum.

$$\sum_{k=\text{start}}^{\text{end}} f(k) = f(start) + f(start+1) + f(start+2) + \cdots + f(end)$$

13.2 SEQUENCE GIVEN BY A RECURSIVE RULE

A recursive rule gives the next terms in a sequence based on the previous terms. The beginning value(s) must be stated as part of the rule. a_{n+1} is the term after a_n, and a_n is the term after a_{n-1}.

EXAMPLE

The famous Fibonacci sequence is given recursively. $a_1 = 1$, $a_2 = 1$, and $a_n = a_{n-1} + a_{n-2}$. State the first six terms of the Fibonacci sequence.

Solution:

n	$a_n = a_{n-1} + a_{n-2}$	a_n
1	$a_1 = 1$, given	1
2	$a_2 = 1$, given	1
3	$a_3 = a_2 + a_1 = 1 + 1$	2
4	$a_4 = a_3 + a_2 = 2 + 1$	3
5	$a_5 = a_4 + a_3 = 3 + 2$	5
6	$a_6 = a_5 + a_4 = 5 + 3$	8

The first six terms of the Fibonacci sequence are {1, 1, 2, 3, 5, 8}.

13.3 ARITHMETIC SEQUENCE AND SERIES

An arithmetic sequence is a list of values where each consecutive pair differs by the same constant. In the arithmetic sequence {5, 8, 11, 14}, the first term is $a_1 = 5$, the common difference is $d = 3$, there are $n = 4$ terms, and the last term is $a_4 = 14$.

The explicit rule for an arithmetic sequence is $a_n = a_1 + d(n - 1)$, where a_n is the term in the nth position and d is the common difference, $d = a_2 - a_1$. To write this rule, fill in the values for a_1 and d from the sequence. To find a particular term, evaluate the rule at n. Note that the explicit rule for an arithmetic sequence is a linear equation with the slope equal to the common difference.

The recursive rule for an arithmetic sequence states the value for a_1 and then states that $a_n = a_{n-1} + d$. This rule says to start at a_1 and then repeatedly add d to get the next terms in the sequence.

An arithmetic series is the sum of an arithmetic sequence. The following formula is available during the exam on the reference sheet:

$$S_n = \frac{n(a_1 + a_n)}{2}$$

In this formula, S_n is the sum of the terms, a_1 is the first term, a_n is the last term, and n is the number of terms.

13.4 GEOMETRIC SEQUENCE AND SERIES

A geometric sequence is a list of values where each consecutive pair has the same ratio. In the geometric sequence {5, 10, 20, 40, 80, 160}, the first term is $a_1 = 5$, the common ratio is $r = 2$, there are $n = 6$ terms, and the last term is $a_6 = 160$.

The explicit rule for a geometric sequence is $a_n = a_1 r^{n-1}$, where a_n is the term in the nth position and r is the common ratio, $r = \frac{a_2}{a_1}$. To write this rule, fill in the values for a_1 and r from the sequence. To find a particular term, evaluate the rule at n. Note that the explicit rule for a geometric sequence is an exponential equation with the base equal to the common ratio.

The recursive rule for a geometric sequence states the value for a_1 and then $a_n = a_{n-1}r$. This rule says to start at a_1 and then repeatedly multiply by r to get the next terms in the sequence.

A geometric series is the sum of a geometric sequence. The following formula is available during the exam on the reference sheet:

$$S_n = \frac{a_1\left(1-r^n\right)}{1-r}$$

In this formula, S_n is the sum of the terms, a_1 is the first term, r is the common ratio, and n is the number of terms.

Practice Exercises

13.1 A ball is dropped from a height of 8 feet and allowed to bounce. Each time the ball bounces, it bounces back to half its previous height. The vertical distance the ball travels, d, is given by the formula $d = 8 + 16\sum_{k=1}^{n}\left(\frac{1}{2}\right)^k$, where n is the number of bounces. Based on this formula, what is the total vertical distance that the ball has traveled after 4 bounces?

(1) 8.9 ft
(2) 15.0 ft
(3) 22.0 ft
(4) 23.0 ft

13.2 Evaluate: $2\sum_{n=1}^{5}(2n-1)$

13.3 Which expression represents the sum of the sequence {3, 5, 7, 9, 11}?

(1) $\sum_{n=0}^{5}(2n+1)$
(2) $\sum_{n=1}^{5}3n$
(3) $\sum_{n=1}^{5}(3n+1)$
(4) $\sum_{n=1}^{5}(2n+1)$

13.4 Bill is trying the all-new Nutty Chocolate Diet. On the first day, he eats 400 chocolate-covered peanut candies. The next day he eats 380, the third day 360, and so on, each day eating 20 less than the previous day. Find a formula to tell how many candies Bill eats on the *n*th day of his diet.

13.5 Truman throws a ball hard against the gym floor. On its first bounce, the ball reaches a height of 18 feet. On each successive bounce, the ball rebounds to two-thirds the height of its previous bounce.

a. Find an expression for the height of the *n*th bounce.

b. Find the total distance, rounded to the *nearest hundredth of a foot*, the ball has traveled, up and down, from its first bounce until it hits the floor after its 20th bounce.

13.6 Each row in an auditorium has 2 more seats than the row in front of it. If the first row seats 20 people and there are 40 rows in the auditorium, write a formula for the number of seats in the *n*th row, and find the total number of seats in the auditorium.

13.7 An arithmetic sequence is defined by the formula $a_n = 3n + 5$. What is the common difference of the terms in this sequence?

(1) 5 (3) 3
(2) 2 (4) 8

13.8 What is the common ratio in the geometric sequence $\{15, -24, 38.4, -61.44\}$?

13.9 A geometric sequence is defined by the formula $a_n = 15(1.2)^{n-1}$. What is the common ratio of the terms in this sequence?

(1) 1.2 (3) 15
(2) 12.5 (4) 18

13.10 The first term of an arithmetic sequence is $2z$, and the common difference is $d = 7z$. What is the tenth term of the sequence?

13.11 Write the first four terms of the sequence defined recursively by $a_1 = 2209$ and the formula $a_n = \sqrt{a_{n-1}} + 2$.

Solutions

13.1 With $n = 4$, write out each term.

$$d = 8 + 16\sum_{k=1}^{4}\left(\frac{1}{2}\right)^k = 8+16\left(\left(\frac{1}{2}\right)^1+\left(\frac{1}{2}\right)^2+\left(\frac{1}{2}\right)^3+\left(\frac{1}{2}\right)^4\right)$$

$$= 8+16\left(\frac{15}{16}\right) = 23$$

The correct choice is **(4)**.

13.2

$$2\sum_{n=1}^{5}(2n-1) = 2\left[(2(1)-1)+(2(2)-1)+(2(3)-1)+(2(4)-1)+(2(5)-1)\right]$$

$$= 2(1+3+5+7+9) = 2(25) = 50$$

$$2\sum_{n=1}^{5}(2n-1) = 50$$

13.3

- Choice (1): $\sum_{n=0}^{5}(2n+1) = 1+3+5+7+9+11$ No. This has an extra term at the beginning. (You may be able to see the correct choice now.)

- Choice (2): $\sum_{n=1}^{5}3n = 3+6+9+12+15$ No

- Choice (3) $\sum_{n=1}^{5}(3n+1) = 4+7+10+13+16$ No

- Choice (4): $\sum_{n=1}^{5}(2n+1) = 3+5+7+9+11$ Yes

The correct choice is **(4)**.

13.4 Bill is eating 20 fewer candies each day than he did the day before. This is an example of an arithmetic sequence; consecutive terms always have the same difference. One way to find a formula for the number of candies Bill eats on day n is to write out the actual computations for each day and look for a pattern. In the second column of the table, the number of candies eaten on day 2 was 400 – 20, the number on day 3 was 400 – 20 – 20, and so forth.

Day	Number of M&Ms	
1	400	400 – 20(0)
2	400 – 20	400 – 20(1)
3	400 – 20 – 20	400 – 20(2)
4	400 – 20 – 20 – 20	400 – 20(3)
n		$400 - 20(n - 1)$

The pattern seems to be "start with 400 and subtract a certain number of 20s." The third column of the table shows this more compactly. Note that on any given day, the number of 20s subtracted is one less than the number of the day. In other words, on day 4, we subtract 3 20s. The formula is $a_n = 400 - 20(n - 1)$.

Memorizing the formula for the terms of an arithmetic sequence may be helpful: $a_n = a_1 + d(n - 1)$, where a_n is the nth term, a_1 is the first term, and d is the common difference of consecutive terms. In this problem, $a_1 = 400$ and $d = 380 - 400 = -20$. So $a_n = 400 - 20(n - 1)$.

The correct answer is $a_n = 400 - 20(n - 1)$.

13.5 a. This is an example of a geometric sequence. Consecutive terms always have the same ratio, in this case $\frac{2}{3}$. To find a formula for the height of the *n*th bounce, you can use a process similar to that in problem 13.4.

Bounce	Height	
1	18	18
2	$18\left(\frac{2}{3}\right)$	$18\left(\frac{2}{3}\right)^1$
3	$18\left(\frac{2}{3}\right)\left(\frac{2}{3}\right)$	$18\left(\frac{2}{3}\right)^2$
4	$18\left(\frac{2}{3}\right)\left(\frac{2}{3}\right)\left(\frac{2}{3}\right)$	$18\left(\frac{2}{3}\right)^3$
n		$18\left(\frac{2}{3}\right)^{n-1}$

Memorizing the formula for the terms of a geometric sequence may be helpful: $a_n = a_1 r^{n-1}$, where a_n is the *n*th term, a_1 is the first term, and r is the common ratio of consecutive terms. In this problem, $a_1 = 18$ and $r = \frac{2}{3}$.

The *n*th bounce is shown by the formula $a_n = 18\left(\frac{2}{3}\right)^{n-1}$.

b. The formula for the sum of the terms in a finite geometric sequence is given on the reference sheet: $S_n = \frac{a_1\left(1-r^n\right)}{1-r}$. However, on each bounce, the ball travels up and then back down the same distance. To find the total distance the ball travels, you need to *double* the sum of the heights. With $a_1 = 18$, $r = \frac{2}{3}$, and $n = 20$,

$$2S_{20} = (2)\frac{(18)\left(1-\left(\frac{2}{3}\right)^{20}\right)}{1-\left(\frac{2}{3}\right)} = 107.9675 \text{ feet}$$

The total distance traveled after 20 bounces is 107.97 feet.

13.6 This is an arithmetic sequence with first term $a_1 = 20$ and common difference $d = 2$.

The formula for the sum of the terms in an arithmetic sequence is given on the reference sheet: $S_n = \frac{n\left(a_1 + a_n\right)}{2}$, where n is the number of terms in the sequence and a_n is the value of the nth term. In this problem, $n = 40$. We need to know a_{40}, the number of seats in the 40th row, before we can use the formula.

Look for the pattern: $a_1 = 20$, $a_2 = 20 + 2$, $a_3 = 20 + 2 + 2$, and so on. This means that $a_{40} = 20 + 2(39) = 98$. This can also be found using the formula for the nth term of an arithmetic sequence, as shown in problem 13.4: $a_n = a_1 + d(n - 1)$. In this problem, $a_n = 20 + 2(n - 1)$. When $n = 40$, there are $a_{40} = 20 + 2(40 - 1) = 98$ seats in the last row and a total of $S_{40} = \frac{(40)(20+98)}{2} = 2360$.

The formula for the number of seats in the nth row is $a_n = 20 + 2(n - 1)$. The auditorium contains 2360 seats.

13.7 The common difference of an arithmetic sequence written in linear form is the slope. In this case, $d = 3$. If you do not remember that, you can evaluate the first two terms of the sequence and find their difference: $a_1 = 3(1) + 5 = 8$, $a_2 = 3(2) + 5 = 11$, and $d = 11 - 8 = 3$.

The correct choice is **(3)**.

13.8 The ratio of the first two terms is $r = \frac{-24}{15} = -1.6$. Note that it is important to calculate the ratio in the correct order, $r = \frac{a_2}{a_1}$, not $r = \frac{a_1}{a_2}$. Check the ratio of the second and third terms, which is $r = \frac{38.4}{-24} = -1.6$.

The common ratio is -1.6.

13.9 The common ratio of a geometric sequence written in exponential form is the base of the exponent. In this case, $r = 1.2$. If you do not remember that, you can evaluate the first two terms of the sequence and find their ratio: $a_1 = 15(1.2)^0 = 15$, $a_2 = 15(1.2)^1 = 18$, and $r = \frac{18}{15} = 1.2$.

The correct choice is **(1)**.

13.10 In an arithmetic sequence, each new term is found by adding the common difference to the term before it. So the second term will be $2z + 7z$, the third term will be $2z + 7z + 7z$, and so forth. The tenth term will be $2z$ with $7z$ added to it nine times: $a_{10} = 2z + 7z(9) = 65z$.

If you memorize the formula for the nth term of an arithmetic sequence, $a_n = a_1 + d(n - 1)$, then with $a_1 = 2z$ and $d = 7z$, you get $a_{10} = 2z + 7z(10 - 1) = 65z$.

The tenth term of the sequence is $65z$.

13.11 The first term is $a_1 = 2209$. The recursive formula $a_{n+1} = \sqrt{a_{n-}} + 2$ says each new term is found by taking the square root of the previous term and adding 2. Setting up a table is helpful.

n	$a_{n+1} = \sqrt{a_n} + 2$	a_n
1	given	2209
2	$a_2 = \sqrt{2209} + 2 = 49$	49
3	$a_3 = \sqrt{49} + 2 = 9$	9
4	$a_4 = \sqrt{9} + 2 = 5$	5

The first four terms of the sequence are {2209, 49, 9, 5}.

14. STATISTICS

14.1 TYPES OF STUDIES AND OUTCOMES

Data can be collected and analyzed in many ways. These include a survey (opinion), an observational study (measure a situation as is), and a controlled experiment (measure the effects of a change). Depending on the methods used to collect the data, the outcomes may be fair or biased. In a survey, for example, what is asked, who is asked, and how it is asked may all bias the outcome.

14.2 MEASURES OF CENTRAL TENDENCY AND STANDARD DEVIATION

The measures of central tendency are mean ($\overline{x}$), median, and mode. These are different ways of describing the center of the data. The median is more representative of the central value if outliers are present.

Measures of dispersion give an indication of the spread of the data. The range is the difference between the maximum and minimum values. The quartiles break the range into fourths. The interquartile range is the difference between the upper quartile (Q_3) and the lower quartile (Q_1). The interquartile range is less affected by outliers.

The sample standard deviation, s, and the population standard deviation, σ, can both be computed on a graphing calculator. Variance, σ^2, is the square of the standard deviation. A small standard deviation means the data are grouped closely around the mean; a large standard deviation means the data are more spread out.

14.3 NORMAL DISTRIBUTION

Data that are normally distributed form a bell-shaped curve. The area under the entire curve is 100%. The area between two σ values represents the percentage of scores that can be expected to fall within that interval. The following normal distribution curve is available during the exam on the reference sheet. The center of the curve is the mean, which is also the median score. The horizontal axis is marked in standard deviations from the mean; each interval in the diagram represents a width of 0.5 of a standard deviation.

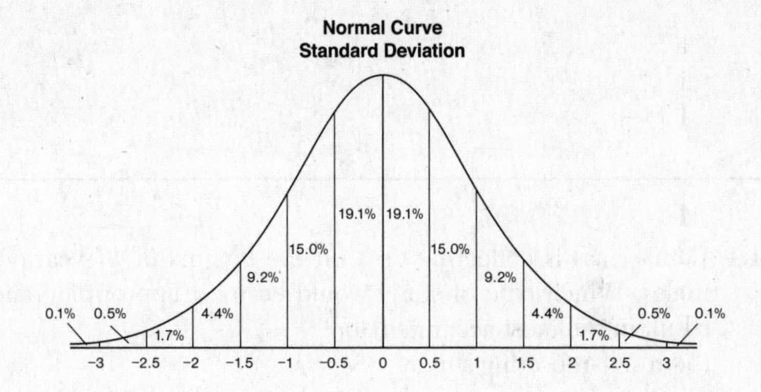

Practice Exercises

14.1 Lara's class is collecting data on the heights of 17-year-old males. Which type of study would be most appropriate and result in the most accurate data?
(1) a self-reporting survey
(2) measurements taken by a single trained individual
(3) measurements taken by friends of each person
(4) a controlled experiment

14.2 Which method of collecting data would most likely result in an unbiased random sample?
(1) selecting every third teenager leaving a movie theater to answer a survey about entertainment
(2) placing a survey in a local newspaper to determine how people voted in the 2008 presidential election
(3) selecting students by the last digit of their school ID number to participate in a survey about cafeteria food
(4) surveying honor students taking Algebra 2/Trigonometry to determine the average amount of time students in a school spend doing homework each night

14.3 Two social studies classes took the same current events exam that was scored on the basis of 100 points. Mr. Wong's class had a median score of 78 and a range of 4 points, while Ms. Rizzo's class had a median score of 78 and a range of 22 points. Explain how these classes could have the same median score while having very different ranges.

14.4 The term "snowstorms of note" applies to all snowfalls over 6 inches. The snowfall amounts for snowstorms of note in Utica, New York, over a four-year period are as follows:

7.1, 9.2, 8.0, 6.1, 14.4, 8.5, 6.1, 6.8, 7.7, 21.5, 6.7, 9.0, 8.4, 7.0, 11.5, 14.1, 9.5, 8.6

What are the mean and population standard deviation for these data, to the *nearest hundredth*?
(1) mean = 9.46; standard deviation = 3.74
(2) mean = 9.46; standard deviation = 3.85
(3) mean = 9.45; standard deviation = 3.74
(4) mean = 9.45; standard deviation = 3.85

14.5 The number of children of each of the first 41 United States presidents is given in the accompanying table.

Number of Children (x_i)	Number of Presidents (f_i)
0	6
1	2
2	8
3	6
4	7
5	3
6	5
7	1
8	1
10	1
15	1

For this population, what are the mean and the population standard deviation to the *nearest tenth*? How many of these presidents fall within one population standard deviation of the mean?

14.6 On a nationwide examination, the Adams School had a mean score of 875 and a standard deviation of 12. The Boswell School had a mean score of 855 and a standard deviation of 20. In which school were the scores more consistent? Explain how you arrived at your answer.

14.7 From 1984 to 1995, the winning scores for a golf tournament were 276, 279, 279, 277, 278, 278, 280, 282, 285, 272, 279, and 278. Using the standard deviation for the sample, find the percent of these winning scores that fall within one standard deviation of the mean.

14.8 The national mean for verbal scores on an exam was 428, and the standard deviation was 113. Approximately what percent of those taking this exam had verbal scores between 315 and 541?

(1) 68.2% (3) 38.2%
(2) 52.8% (4) 26.4%

14.9 The amount of orange soda that McDougal's restaurant sells in a day is normally distributed with a mean of 800 ounces and a standard deviation of 60 ounces. How many ounces of orange soda should the manager have on hand at the beginning of the day to be 97.7% sure she will *not* run out?

(1) 680 (3) 898
(2) 782 (4) 920

Solutions

14.1

- Choice (1): Many people may not know their own heights exactly. There is also a possibility that some might intentionally misrepresent (exaggerate, for example) their heights. This would probably *not* be the most accurate method.
- Choice (2): This eliminates both problems from choice (1) and seems like a good method. Since the individual is trained, one expects the measurements to be accurate.
- Choice (3): Many different individuals will be measuring, which could result in a wide range of measurement errors. This does not seem as accurate as choice (2).
- Choice (4): Since the class is not planning to compare heights under different circumstances, a controlled experiment is not appropriate.

The correct choice is **(2)**.

14.2

- Choice (1): Selecting every third person should get a random sample of people who went to that movie, but those people may not be representative of everyone in the target population. This is probably *not* a good way to get a representative sample.
- Choice (2): Not all the people read the local paper. In addition, you will get responses only from people who take the time to fill out and return the survey. This is called a self-selected sample and may not be representative of everyone in the target population.
- Choice (3): Assuming that the last digit of the school ID numbers is random, this should be a good way to get a representative sample.
- Choice (4): This sample is limited to honor students; it would not be representative of the whole student body. Even if you wanted data only about honor students, not all honor students take Algebra 2/Trigonometry. So this still would not be a good sample.

The correct choice is **(3)**.

14.3 The two statistics measure completely different things. The median tells the value of the middle score when all the scores are in numerical order. The range is one measure of the spread of the scores. In this problem, the two classes had the same middle score but Ms. Rizzo's scores were more spread out than Mr. Wong's. For example, Mr. Wong's scores might have ranged from 76 to 80 with a median of 78 while Ms. Rizzo's might have ranged from 70 to 92 with a median of 78.

Stating a specific example adds detail to your explanation and will help the grader understand your response.

14.4 Enter the data carefully into list L1 on your graphing calculator. Then use the 1-variable statistics function by typing STAT TRACE **1:1-Var Stats** ENTER ENTER.

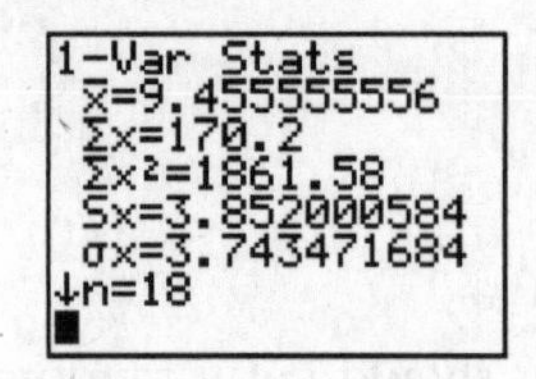

Rounded to the *nearest hundredth* (as in all the choices), the mean is $\overline{x} = 9.46$ and the population standard deviation is $\sigma = 3.74$. The correct choice is **(1)**.

14.5 Carefully enter the number of children into list L1 and the frequencies into list L2. Use the 1-variable statistics function (1-Var Stats L1, L2) on your calculator.

Note: You must key in L1, L2 (press 2nd 1, 2nd 2 ENTER) after 1-Var Stats so the calculator will use the frequencies in L2. If you do not, you will get the wrong answer. Check the number of scores, *n*; there are 41 presidents on the list.

Correct	**Wrong**
1-Var Stats	1-Var Stats
$\bar{x}$=3.634146341	$\bar{x}$=5.545454545
Σx=149	Σx=61
Σx²=893	Σx²=529
Sx=2.964423195	Sx=4.367233366
σx=2.928048527	σx=4.163993633
↓n=41	↓n=11

From the screen on the left, rounded to the *nearest tenth*, the mean is $\overline{x} = 3.6$ and the population standard deviation is $\sigma = 2.9$.

The interval of scores within one population standard deviation of the mean is $\overline{x} - \sigma$ to $\overline{x} + \sigma$. The interval is $3.6 - 2.9$ to $3.6 + 2.9$, or 0.7 to 6.5. This means you want to count all presidents having from one to six children. These presidents are shown in the box below.

Number of Children (x_i)	Number of Presidents (f_i)
0	6
1	2
2	8
3	6
4	7
5	3
6	5
7	1
8	1
10	1
15	1

The number of presidents within one population standard deviation of the mean is $2 + 8 + 6 + 7 + 3 + 5 = 31$.

Thirty-one presidents are within one population standard deviation of the mean for number of children.

U.S. history bonus question: Which president had 15 children?

Answer: John Tyler, term 1841–1845. Did you know?

14.6 The scores at Adams School had a smaller standard deviation than the scores at Boswell School. This means that on average, the scores at Adams tend to be closer to the mean than the scores at Boswell. So the scores at Adams show more consistency than the scores at Boswell.

14.7 Enter the scores into list L1 in your graphing calculator and use 1-variable statistics to find the mean and sample standard deviation.

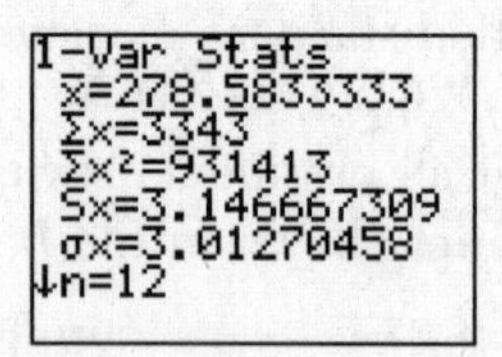

The mean score is $\overline{x} = 278.6$, and the sample standard deviation is $s = 3.1$.

The interval of scores within one sample standard deviation of the mean is $\overline{x} - s$ to $\overline{x} + s$. This interval is $278.6 - 3.1$ to $278.6 + 3.1$, or 275.5 to 281.7. You want all the scores from 276 to 281. There are 9 scores, out of a total of 12, within that interval.

The percent of winning scores that fall within one standard deviation of the mean is $\frac{9}{12}(100\%) = 75\%$.

14.8 For a national exam, it is reasonable to assume the scores follow a normal distribution. In this problem, 428 will be at the center of the curve, 428 – 113 = 315 will be one standard deviation below the mean, and 428 + 113 = 541 will be one standard deviation above the mean. To find the percent of scores between 315 and 541, add the percentages between –1 and 1 standard deviation on the normal curve given on the reference sheet, as shaded in the graph below.

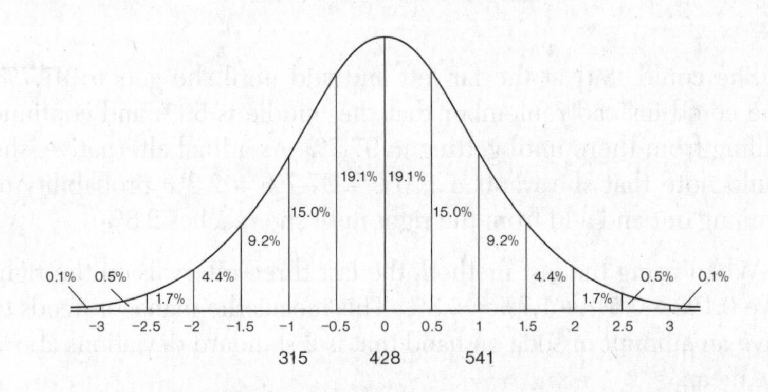

The percent of scores is 15.0% + 19.1% + 19.1% + 15.0% = 68.2%.

Memorizing the fact that in a normal distribution, 68.2% of the scores always lie within one standard deviation of the mean may be worthwhile.

The correct choice is **(1)**.

14.9 In order *not* to run out of orange soda, sales for the day must be *less than* what is on hand at the beginning of the day. The manager must find the value on the normal distribution curve where the sum of the percentages to the left is 97.7, as shaded in the diagram below.

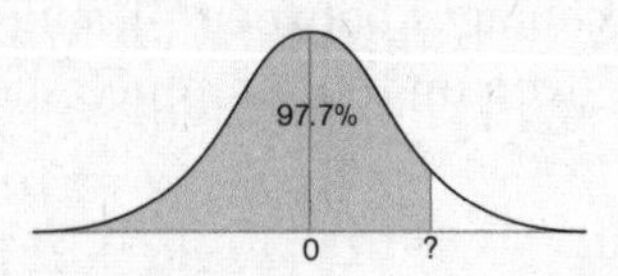

She could start at the far left and add until she gets to 97.7%. She could instead remember that the middle is 50% and continue adding from there until getting to 97.7%. As a final alternative, she could note that she wants a 100% – 97.7% = 2.3% probability of running out and add from the right until she reaches 2.3%.

When using the last method, the last three intervals on the right give 0.1% + 0.5% + 1.7% = 2.3%. This means the manager needs to have an amount of soda on hand that is 2 standard deviations above the mean.

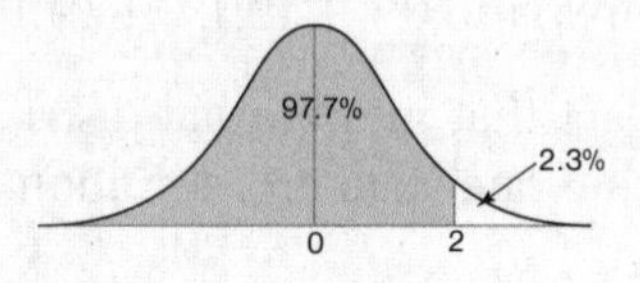

Two standard deviations above the mean is 800 + 2(60) = 920 ounces.

The correct choice is **(4)**.

15. REGRESSIONS

15.1 TYPES OF REGRESSIONS

Modeling a set of data with a line or curve is called regression analysis. The specific regression model, whether linear, exponential, logarithmic, or power, depends on the data. The data are usually represented in a table and then plotted as a scatter plot. The regression feature of a graphing calculator will find the equation for each type of regression model. The calculator will give the coefficients for the model, but you should write the equation in standard form on the exam. The regression equation can be used to make predictions both inside the domain (interpolation) and outside the domain (extrapolation). Follow all rounding instructions carefully.

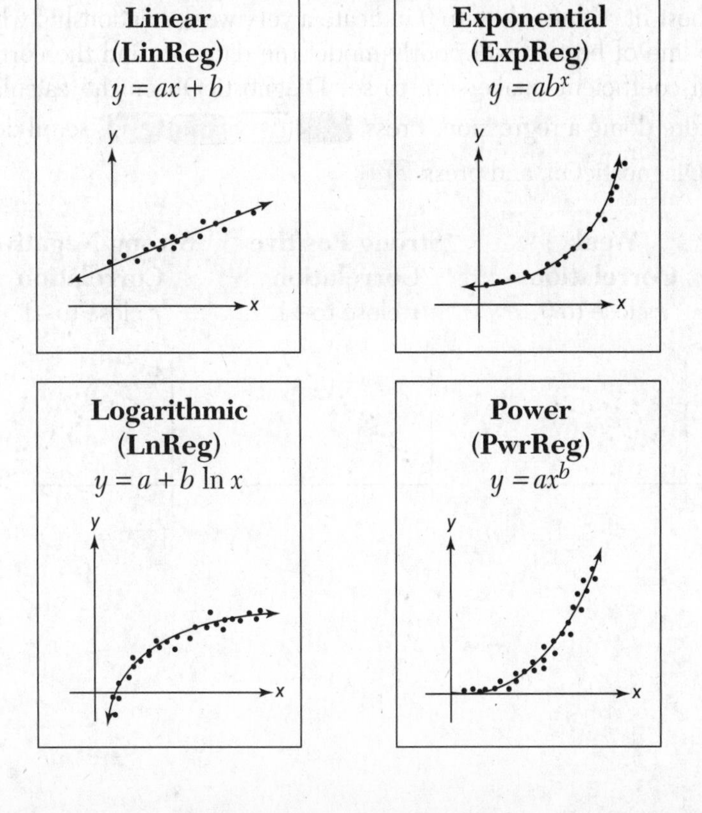

The linear model is, of course, a line. The exponential, logarithmic, and power models may have similar shapes. The exponential model can have a y-intercept but no root. The logarithmic model can have a root but no y-intercept. The power model has data in only Quadrant I.

15.2 LINEAR CORRELATION COEFFICIENT

The linear correlation coefficient, r, is a number from –1 to +1 that represents the magnitude and direction of the linear relationship, if any, between two sets of data. The sign of the correlation coefficient matches the sign of the slope of the line of best fit for the data. Positive values of r mean the y-values are increasing as x-values increase, and negative values of r mean the y-values are decreasing as x-values increase. Values of r close to +1 indicate a strong linear relationship, which may be well modeled by the line of best fit. Values close to 0 indicate a very weak relationship where the line of best fit will poorly model the data. To find the correlation coefficient, make sure to set DiagnosticOn on the calculator before doing a regression. Press 2nd 0 ALPHA x^{-1}, scroll down to DiagnosticOn, and press ENTER.

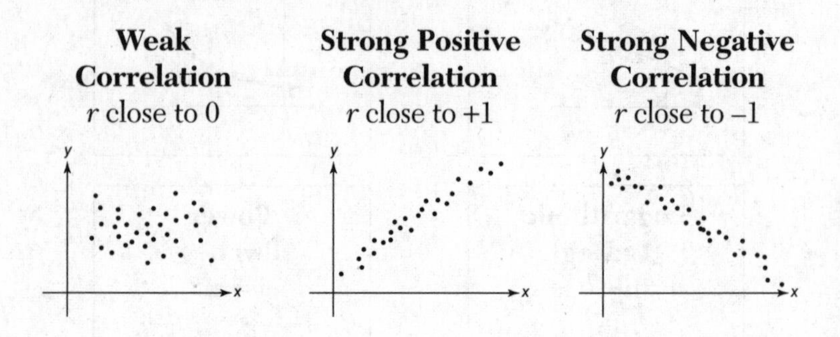

Practice Exercises

15.1 The number of newly reported crime cases in a county in New York State is shown in the accompanying table.

Year (x)	New Cases (y)
1999	440
2000	457
2001	369
2002	351

a. Write the linear regression equation that represents this set of data. (Let $x = 0$ represent 1999.)

b. Using this equation, find the projected number of new cases for 2009, rounded to the *nearest whole number*.

15.2 A box containing 1,000 coins is shaken, and the coins are emptied onto a table. Only the coins that land heads up are returned to the box, and then the process is repeated. The accompanying table shows the number of trials and the number of coins returned to the box after each trial.

Trial	0	1	3	4	6
Coins Returned	1,000	610	220	132	45

a. Write an exponential regression equation, rounding the calculated values to the *nearest ten-thousandth*.

b. Use the equation to predict how many coins will be returned to the box after the eighth trial.

15.3 The accompanying table shows the number of new cases reported by the Nassau and Suffolk County Police Crime Stoppers program for the years 2000 through 2002.

Year (x)	New Cases (y)
2000	457
2001	369
2002	353

a. If $x = 1$ represents the year 2000 and y represents the number of new cases, find the equation of best fit using a power regression, rounding all values to the *nearest thousandth*.

b. Using this equation, find the estimated number of new cases, to the *nearest whole number*, for the year 2007.

15.4 The accompanying table shows wind speed and the corresponding wind chill factor when the air temperature is 10°F.

Wind Speed (mi/h) x	Wind Chill Factor (°F) y
4	3
5	1
12	–5
16	–7
22	–10
31	–12

a. Write the logarithmic regression equation for this set of data, rounding coefficients to the *nearest ten-thousandth*.

b. Using this equation, find the wind chill factor, to the *nearest degree*, when the wind speed is 50 miles per hour.

c. Based on your equation, if the wind chill factor is 0°, what is the wind speed, to the *nearest mile per hour*?

15.5 A linear regression equation of best fit between a student's attendance and the degree of success in school is $h = 0.5x + 68.5$. The correlation coefficient, r, for these data is

(1) $0 < r < 1$ (3) $r = 0$
(2) $-1 < r < 0$ (4) $r = -1$

15.6 Which scatter diagram shows the strongest positive correlation?

(1) (3)

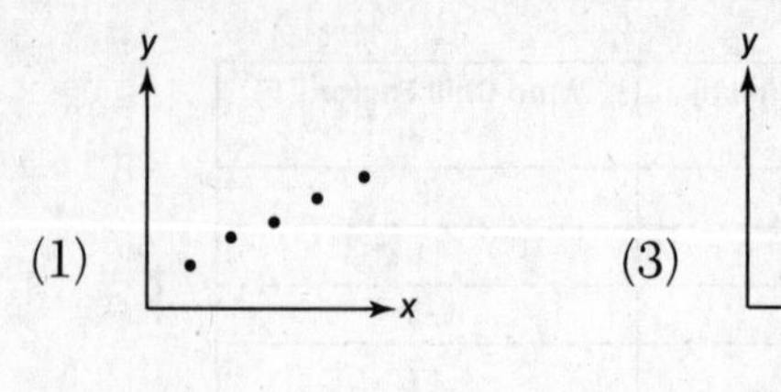

(2) (4)

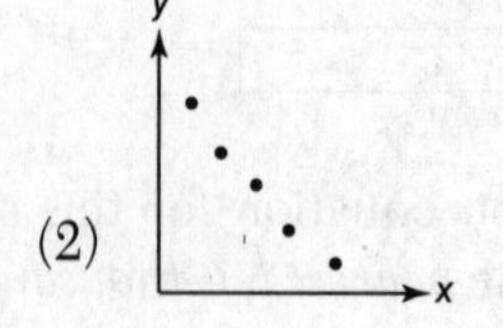

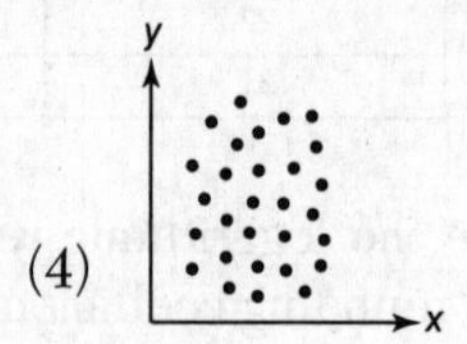

Solutions

15.1 a. The problem says to let $x = 0$ represent 1999. This means the given data will have $x = 0, 1, 2,$ and 3. If you did not get the correct regression equation, check your x-values first. Enter these x-values into list L1 on your graphing calculator and the corresponding y-values into list L2. Use the linear regression function on your calculator.

```
LinReg
 y=ax+b
 a=-35.5
 b=457.5
 r²=.7742282292
 r=-.8799023975
```

Note that you will not get r^2 and r unless you have previously set DiagnosticOn. However, r^2 and r are not important in this problem.

The linear regression equation is $y = -35.5x + 457.5$. You must write it in this form on your paper. Do *not* write $y = ax + b$, $a = -35.5$, $b = 457.5$.

b. The year 2009 corresponds to $x = 2009 - 1999 = 10$. Substitute $x = 10$ into the regression equation: $y = -35.5(10) + 457.5 = 102.5$. You must write the substitution on your paper to get full credit.

To the *nearest whole number*, the projected number of new cases in 2009 is 103.

15.2 a. Enter the trial number, x, into list L1 on your graphing calculator and the number of coins returned, y, into list L2. Use the exponential regression function on your calculator.

The exponential regression equation, rounded to the *nearest ten-thousandth*, is $y = 1018.2839(0.5969)^x$.

b. When $x = 8$, $y = 1018.2839(0.5969)^8 = 16.409$.

After the eighth trial, the model predicts 16 coins will be returned.

15.3 a. Enter the year number as $x = 1$, 2, and 3 into list L1 on your graphing calculator and the number of number of new cases, y, into list L2. (If you listed 0, 1, and 2 in L1, read the problem again.) Use the power regression function on your calculator.

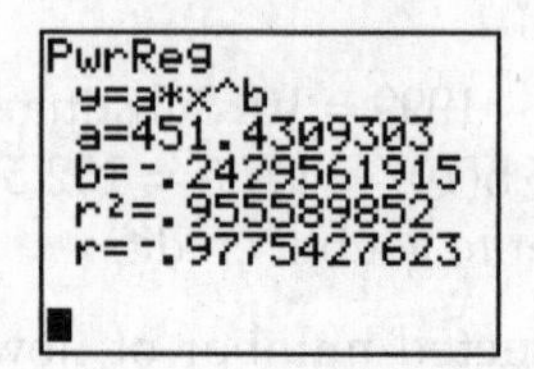

The power regression equation, rounded to the *nearest thousandth*, is $y = 451.431x^{-0.243}$.

b. The year 2007 corresponds to $x = 8$, giving $y = 451.431(8)^{-0.243}$ $= 272.358$.

The estimated number of new cases in 2007 is 272.

15.4 a. Enter the wind speeds, x, into list L1 on your graphing calculator and the wind chill factor, y, into list L2. Use the logarithmic regression function on your calculator.

The logarithmic regression equation, rounded to the *nearest ten-thousandth*, is $y = 13.0134 - 7.3135 \ln x$.

b. Evaluate the equation at $x = 50$ to find the wind chill factor.

$$y = 13.0134 - 7.3135 \ln(50) = -15.597$$

At 50 miles per hour, the wind chill factor to the *nearest degree* is –16°.

c. To find the wind speed that makes a wind chill factor 0°, solve $0 = 13.0134 - 7.3135 \ln x$.

$$7.3173 \ln x = 13.0134$$

$$\ln x = \frac{13.0134}{7.3135} = 1.779366924$$

$$x = e^{1.779366924} = 5.9261$$

To the nearest mile per hour, the wind speed that produces a wind chill factor of 0° is 6 mph.

15.5 Since the slope is positive, the correlation coefficient must also be positive.

The correct choice is **(1)**.

15.6 To have a positive correlation, the data must increase from left to right. As x increases, y tends to also increase. This eliminates choice (2), which clearly decreases, and choice (4), which shows no real trend. The stronger the correlation, the more closely the data points should come to forming a straight line. Choice (1) shows stronger correlation than choice (3).

The correct choice is **(1)**.

16. PROBABILITY

16.1 PERMUTATIONS AND COMBINATIONS

A permutation counts the number of possible arrangements of r items picked from n possibilities. Remember that permutations count all the different orders. When selecting three letters from among A, B, C, D, and E, ABD and DAB are two different permutations. The formula to find the number of permutations is

$${}_nP_r = \frac{n!}{(n-r)!}.$$

A combination counts the number of groups of r items chosen from n possibilities. Combinations do *not* count different orders within the same group. When selecting three letters from among A, B, C, D, and E, ABD and DAB are the same combination. The formula to find the number of combinations is ${}_nC_r = \frac{n!}{(n-r)!r!}$.

Graphing calculators have both these functions in the  PRB menu.

16.2 PROBABILITY

The theoretical probability of an event when all the outcomes are equally likely is the ratio $\frac{\text{number of favorable outcomes}}{\text{number of possible outcomes}}$. Combinations and permutations can be used to determine the number of favorable outcomes and the total number of possible outcomes in a theoretical probability. Empirical (or experimental) probability uses actual data to estimate probabilities and can be found with the ratio $\frac{\text{number of successes}}{\text{number of trials}}$.

16.3 GEOMETRIC PROBABILITIES

Probabilities based on geometric figures are the ratio of areas. Know common area formulas:

- Circle: $A = \pi r^2$
- Rectangle: $A = bh$
- Triangle: $A = \frac{1}{2}bh$

16.4 BINOMIAL PROBABILITIES AND THE NORMAL DISTRIBUTION

If only two outcomes are possible for an experiment having a probability of success p for each trial, the probability of exactly r successes in n trials is given by ${}_nC_r p^r(1-p)^{n-r}$. Remembering this in words may help:

$$_{\text{number of trials}}C_{\text{number of successes}}\left(\text{probability of success}\right)^{\text{number of successes}}\left(\text{probability of failure}\right)^{\text{number of failures}}$$

To compute the probability of at least (more than or equal to) k successes, sum the probabilities for $k \le r \le n$. Similarly, to compute the probability of at most (less than or equal to) k successes, sum the probabilities for $0 \le r \le k$.

For a large number of trials, a graph of the probabilities for $0 \le r \le n$ will approximate the normal distribution curve with mean $= np$ and standard deviation $= \sqrt{np(1-p)}$.

Practice Exercises

16.1 A basketball team has 12 players, and the coach needs to choose 5 to put into a game. How many different possible ways can the coach choose a center, a point guard, a shooting guard, a small forward, and a power forward if each person has an equal chance of being selected?

(1) ${}_{12}P_5$ (3) ${}_{12}C_5$
(2) ${}_5P_{12}$ (4) ${}_5C_{12}$

16.2 At the next Olympics, the United States can enter 4 athletes in the diving competition. How many different teams of 4 divers can be selected from a group of 9 divers?

(1) 36 (3) 3,024
(2) 126 (4) 6,561

16.3 A committee of 5 members is to be randomly selected from a group of 9 freshmen and 7 sophomores. Which expression represents the number of different committees of 3 freshmen and 2 sophomores that can be chosen?

(1) ${}_9C_3 + {}_7C_2$ (3) ${}_{16}C_3 \cdot {}_{16}C_2$
(2) ${}_9C_3 \cdot {}_7C_2$ (4) ${}_9P_3 \cdot {}_7P_2$

16.4 Three roses will be selected for a flower vase. The florist has 4 red roses, 3 white roses, 5 yellow roses, 6 orange roses, and 2 pink roses from which to choose.

a. How many different 3-rose selections can be formed from the 20 roses?

b. What is the probability that 3 roses selected at random will contain 1 red rose, 1 white rose, and 1 pink rose?

c. What is the probability that 3 roses selected at random will *not* contain an orange rose?

16.5 The accompanying diagram shows a square dartboard. The side of the dartboard measures 30 inches. The square shaded region at the center has a side that measures 10 inches. If darts thrown at the board are equally likely to land anywhere on the board, what is the theoretical probability that a dart does *not* land in the shaded region?

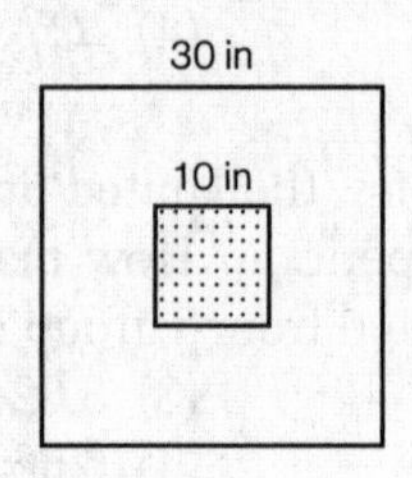

16.6 The amount of time that a teenager plays video games in any given week is normally distributed. If a teenager plays video games an average of 15 hours per week with a standard deviation of 3 hours, what is the probability of the teenager playing video games between 15 and 18 hours a week?

16.7 If the probability that the Islanders will beat the Rangers in a game is $\frac{2}{5}$, which expression represents the probability that the Islanders will win exactly 4 out of 7 games in a series against the Rangers?

(1) $\left(\frac{2}{5}\right)^4\left(\frac{3}{5}\right)^3$

(2) ${}_5C_2\left(\frac{4}{7}\right)^2\left(\frac{3}{7}\right)^3$

(3) ${}_7C_4\left(\frac{2}{5}\right)^4\left(\frac{2}{5}\right)^3$

(4) ${}_7C_4\left(\frac{2}{5}\right)^4\left(\frac{3}{5}\right)^3$

16.8 Team *A* and team *B* are playing in a league. They will play each other five times. If the probability that team *A* wins a game is $\frac{1}{3}$, what is the probability that team *A* will win at least 3 of the 5 games?

16.9 Students in Ms. Nazzeer's mathematics class tossed a six-sided number cube whose faces are numbered 1 to 6. The results are recorded in the table below.

Result	Frequency
1	3
2	6
3	4
4	6
5	4
6	7

Based on these data, what is the empirical probability of tossing exactly two 4s in five tosses of the number cube?

(1) $\frac{2}{5}$

(2) $\frac{10}{7776}$

(3) $\frac{1}{36}$

(4) $\frac{640}{3125}$

Solutions

16.1 Since the coach is selecting players for specific positions, the selection order matters. Use permutations, ${}_{12}P_5$.

The correct choice is **(1)**.

16.2 Since the divers are not being selected for specific positions on the team, the selection order does not matter. Use combinations, ${}_9C_4 = 126$.

The correct choice is **(2)**.

16.3 When selecting committees, the order of selection is not important. Use combinations. There are ${}_9C_3$ different groups of 3 freshmen that can be chosen from 9 freshmen. There are ${}_7C_2$ groups of 2 sophomores that can be chosen from 7 sophomores. To find the total number of committees that have 3 freshmen and 2 sophomores, multiply those two numbers: ${}_9C_3 \cdot {}_7C_2$.

The correct choice is **(2)**.

16.4 a. The order in which the three roses are chosen does not matter. Since the order does not change the final selection, use combinations. The number of ways to choose 3 roses from a total of 20 is ${}_{20}C_3 = 1140$.

b. Choose each color with a separate combination. The fraction below shows the combinations in the order given in the problem: red, white, yellow, orange, and pink. Multiply the results together to find the total number of successes. Divide by the total number of rose combinations found in part a.

$$\frac{{}_4C_1 \cdot {}_3C_1 \cdot {}_5C_0 \cdot {}_6C_0 \cdot {}_2C_1}{{}_{20}C_3} = \frac{4 \cdot 3 \cdot 1 \cdot 1 \cdot 2}{1140} = \frac{24}{1140} \text{ or } \frac{2}{95}$$

Note that reducing the answer is optional. Do not take the chance of making a mistake.

Tip: Write a combination for each color even if 0 are chosen. This will help you see that the total of all colors should add to 20 and the total chosen should add to 3, matching the values in the denominator's combination.

c. Model this problem after the previous solution. If you choose no orange roses, you still have 20 – 6 = 14 roses from which to choose. You cannot break out the colors further because no criteria are given for any specific colors.

$$\frac{{}_6C_0 \cdot {}_{14}C_3}{{}_{20}C_3} = \frac{1 \cdot 364}{1140} = \frac{364}{1140} \text{ or } \frac{91}{285}$$

Again, there is no need to reduce the answer.

16.5 For a geometric probability, use the ratio of areas. The probability that the dart does not land on the shaded region is the same as the probability that the dart lands on the unshaded region.

$$\frac{\text{area of unshaded region}}{\text{total area of dartboard}} = \frac{30^2 - 10^2}{30^2} = \frac{900 - 100}{900} = \frac{800}{900} \text{ or } \frac{8}{9}$$

The probability that the dart does not land in the shaded region is $\frac{8}{9}$.

16.6 In this problem, 15 hours is the mean and 18 hours is 1 standard deviation above the mean. To find the probability the teenager plays between 15 and 18 hours of video games, add the percentages between 0 and 1 standard deviation shown on the normal curve standard deviation given on the reference sheet and shaded in the graph below.

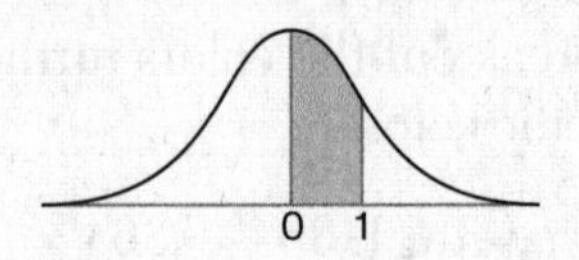

$$P = 19.1\% + 15.0\% = 34.1\%$$

The probability the teenager plays between 15 and 18 hours of video games is 34.1%

Note that probabilities can be expressed as percents as well as fractions. Using the percent form is easier in this problem. The equivalent fraction answer is $\frac{341}{1000}$. Why do the extra work?

16.7 This is an example of the binomial distribution. Use the formula

$${}_{\text{number of trials}}C_{\text{number of successes}}(\text{probability of success})^{\text{number of successes}}(\text{probability of failure})^{\text{number of failures}}$$

In the given problem, there are 7 trials (games), 4 successes (Islanders win), and 7 – 4 = 3 failures (Islanders do not win). The probability of success is $\frac{2}{5}$, and the probability of failure is

$1 - \frac{2}{5} = \frac{3}{5}$.

$$P(\text{Islanders win exactly 4 of 7 games}) = {}_7C_4\left(\frac{2}{5}\right)^4\left(\frac{3}{5}\right)^3$$

The correct choice is **(4)**.

16.8 This is another example of a binomial distribution. Use the same formula as in the previous problem. For team A to win at least 3 games out of 5, they could win 3 games, 4 games, or all 5 games. The probability is found by adding the three cases:

$$P(\geq 3) = P(3) + P(4) + P(5)$$

$$= {}_5C_3\left(\frac{1}{3}\right)^3\left(\frac{2}{3}\right)^2 + {}_5C_4\left(\frac{1}{3}\right)^4\left(\frac{2}{3}\right)^1 + {}_5C_5\left(\frac{1}{3}\right)^5\left(\frac{2}{3}\right)^0$$

$$= \frac{40}{243} + \frac{10}{243} + \frac{1}{243}$$

$$= \frac{51}{243} \text{ or } \frac{17}{81}$$

Tip: You can enter the entire formula in your calculator and use MATH **1:Frac** to get a reduced fraction answer or to get the decimal form. If you write out the intermediate values, use only fractions to avoid rounding errors. If you write the intermediate fraction results in unreduced form, you will always have a common denominator, which will make adding them easier. The final probability does not need to be simplified.

16.9 The empirical probability of getting a 4 on one roll of the number cube is

$$P = \frac{\text{\# of times got a 4}}{\text{total \# of rolls}} = \frac{6}{(3+6+4+6+4+7)} = \frac{6}{30} = \frac{1}{5}$$

Using the binomial distribution, if the same number cube is rolled five times, the probability of getting a 4 exactly twice is

$${}_5C_2\left(\frac{1}{5}\right)^2\left(\frac{4}{5}\right)^3 = \frac{640}{3125}$$

The correct choice is **(4)**.

Glossary of Terms

abscissa The x-coordinate of a point in the coordinate plane. The abscissa of the point (2, 3) is 2.

absolute value The absolute value of a number x, denoted by $|x|$, is its distance from zero on the number line. Thus, $|x|$ always represents a nonnegative number.

acute angle An angle whose degree measure is less than 90° and greater than 0°.

ambiguous case The situation in which the measures of two sides and a nonincluded angle of a triangle are given. These measures may determine one triangle, two triangles, or no triangles.

amplitude The amplitude of a sine or a cosine function of the form $y = a \sin bx$ or $y = a \cos bx$ is $|a|$, which is half the distance from the maximum value to the minimum value.

angle The union of two rays that have the same endpoint.

angle of depression An angle measured downward from the horizontal to a line of sight below the horizontal.

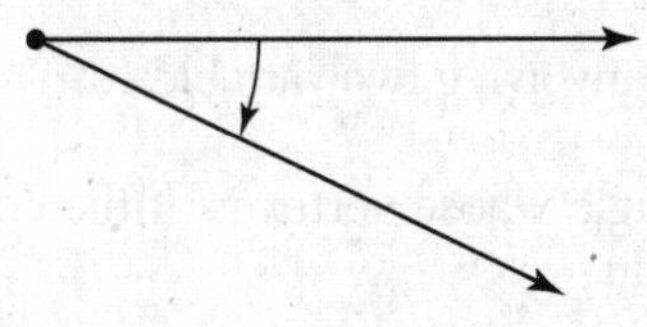

angle of elevation An angle measured upward from the horizontal to a line of sight above the horizontal.

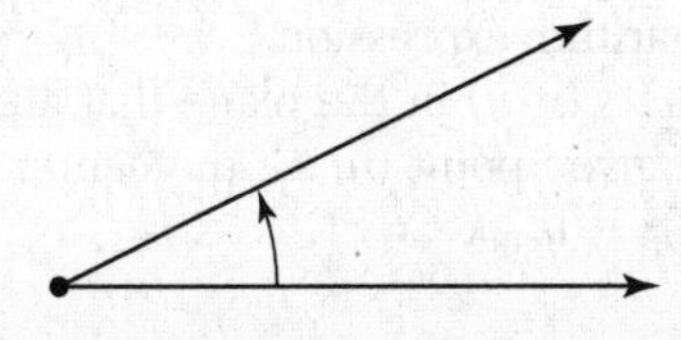

antilogarithm The number whose logarithm is given.

arccos x The angle A such that $\cos A = x$ and $0° \le A \le 180°$; same as $\cos^{-1} x$.

arc length The distance along part of the circumference of a circle.

arcsin x The angle A such that $\sin A = x$ and $-90° \le A \le 90°$; same as $\sin^{-1} x$.

arctan x The angle A such that $\tan A = x$ and $-90° < A < 90°$; same as $\tan^{-1} x$.

argument The input to a function.

arithmetic sequence A set of numbers in which the difference between each term and the preceding term is a constant.

arithmetic series The sum of an arithmetic sequence.

asymptote A line that is the limiting value of a graph. The graph of $y = 2^x$ has the negative x-axis as an asymptote. The line $x = \frac{\pi}{2}$ is one of the asymptotes of the graph of $y = \tan x$.

Bernoulli experiment A probability experiment in which each trial has exactly two possible outcomes, success and failure, and the probability of success is the same for all trials. The probability of k successes out of n trials is given by the expression ${}_nC_k p^k(1-p)^{n-k}$.

biased sample A sample that does not fairly represent the population from which it is drawn.

bimodal Having two modes.

binomial A polynomial with two unlike terms, as in $2x + y$.

binomial theorem A formula that tells how to expand a binomial of the form $(a + b)^n$, where n is a positive integer, without performing repeated multiplications. The formula is given on the reference sheet.

bivariate data Data involving two variables.

central angle An angle whose vertex is at the center of a circle and whose sides are radii.

change of base formula The formula, $\log_b x = \frac{\log x}{\log b}$ or $\log_b x = \frac{\ln x}{\ln b}$, used to change from a logarithm with base b to an equivalent common or natural logarithm expression.

circle The set of points (x, y) in the plane that are a fixed distance r, the radius, from a given point (h, k), the center. Thus, an equation of a circle is $(x - h)^2 + (y - k)^2 = r^2$.

cofunctions A pair of trigonometric functions that have equal values when their arguments are complementary angles. The three pairs of cofunctions are sine and cosine, tangent and cotangent, and secant and cosecant.

combination A selection of objects in which the order of the individual objects is not considered. For example, in selecting a committee of three students from {Alice, Bob, Carol, Dave, Emma}, the selections Alice-Bob-Emma and Bob-Emma-Alice represent the same combination.

common difference The difference between any two successive terms of an arithmetic sequence.

common logarithm A logarithm whose base is 10, written as $\log x$.

complementary angles Two angles whose degree measures add up to 90° or whose radian measures add up to $\frac{\pi}{2}$.

completing the square A process used to change an expression in standard form $ax^2 + bx + c$ into a perfect square binomial $a(x+h)^2 + k$.

complex fraction A fraction with fractions in the numerator, the denominator or both, as in $\dfrac{\frac{2}{x}+1}{\frac{3}{x+1}}$.

complex number A number that can be written in the form $a + bi$, where $i = \sqrt{-1}$ and a and b are real numbers. The set of real numbers is a subset of the set of complex numbers.

conjugate pair The sum and difference of the same two terms, as in $a + b$ and $a - b$.

constant A quantity that is fixed in value. In the equation $y = x + 3$, x and y are variables and 3 is a constant.

constant function A function where each member of the domain is mapped to the same element in the range. The graph of a constant function is a horizontal line.

controlled experiment An experiment that compares the results from an experimental sample to a control sample.

coordinate The real number that corresponds to the position of a point on the number line.

coordinate plane The plane determined by a horizontal number line and a vertical number line intersecting at their zero points, called the origin.

correlation coefficient A number from –1 to +1, denoted by r, that represents the magnitude and direction of the linear relationship, if any, between two sets of data. If a set of data points is closely clustered about a line, then $|r| \approx 1$. The sign of r depends on the sign of the slope of the line about which the data points are clustered.

cosecant The reciprocal of the sine function.

cosine ratio In a right triangle, the ratio of the length of the leg adjacent to a given acute angle to the length of the hypotenuse. If an angle θ is in standard position, $\cos \theta = \frac{x}{r}$, where $P(x, y)$ is any point on the terminal side of angle θ and $r = \sqrt{x^2 + y^2}$.

cotangent The reciprocal of the tangent function.

coterminal angles Angles in standard position whose terminal sides coincide.

degree A unit of angle measurement defined as $\frac{1}{360}$ of one complete rotation of a ray about its vertex.

degree of a monomial The sum of the exponents of a monomial's variable factors. For example, the degree of $3x^4$ is 4 and the degree of $-5xy^2$ is 3 since 1 (the exponent of x) plus 2 (the exponent of y) equals 3.

degree of a polynomial The greatest degree of a polynomial's monomial terms. For example, the degree of $x^2 - 4x + 5$ is 2.

dependent variable The output of a function. For a function of the form $y = f(x)$, y is the dependent variable because the value of y depends on the value of the input x.

difference of two perfect squares A binomial of the form $a^2 - b^2$. This can be factored into $(a - b)(a + b)$.

dilation A transformation of points in the coordinate plane that changes lengths by a certain factor. The change can be horizontal only, vertical only, or a combination of both.

direct variation A relationship in which the ratio of two variables is a constant, $\frac{y_1}{x_1} = \frac{y_2}{x_2}$. A direct variation has the equation $y = kx$, where k is the constant of variation.

discriminant In the quadratic formula $x = \frac{-b \pm \sqrt{b^2 - 4ac}}{2a}$, the discriminant is the quantity underneath the radical sign, $b^2 - 4ac$. If the discriminant is positive, the two roots are real. If the discriminant is 0, the two roots are equal. If the discriminant is negative, the two roots are imaginary.

domain The set of values of the independent variable for which a given function is defined; the set of first coordinates in the ordered pairs of a relation.

double root A root of an equation that occurs twice. If $(x - r)^2$ is a factor of $f(x)$, then r is a double root of $f(x) = 0$.

e The base of the natural logarithm. *e* is an irrational number that is approximately 2.71828.

empirical probability A probability calculated from experimental results rather than by analyzing the theoretical outcomes, also called experimental probability.

event A subset of the sample space of a probability experiment. When selecting a letter from the word NUMBER, the sample space is {N, U, M, B, E, R}. The event described by "get a vowel" would be {U, E}.

exact value The value of an expression that has not been rounded. The exact value is frequently written as a fraction or radical.

experimental probability See *empirical probability*.

explicit formula The function rule used to find the nth term of a sequence without necessarily knowing the preceding terms.

exponent In x^n, the number n is the exponent and indicates the number of times the base x is used as a factor in a product. For example, $x^3 = x \cdot x \cdot x$.

exponential equation An equation in which the variable appears in an exponent, as in $2^{x+1} = 16$.

exponential function A function of the form $y = b^x$, where b is a positive number other than 1.

exponential regression model See *regression model*.

extraneous root A solution of a derived equation that is not a solution to the original equation.

extrapolation The process of estimating a y-value from a table, graph, or equation using a value of x that falls outside the range of observed x-values.

factor A number or an algebraic expression that is being multiplied in a product. A number or an algebraic expression is a factor of a given product if it divides that product with no remainder.

factorial *n* Denoted by $n!$ and defined for any positive integer n as the product of consecutive integers from n to 1. For example, $5! = 5 \cdot 4 \cdot 3 \cdot 2 \cdot 1 = 120$.

factoring The process by which a number or polynomial is written as the product of two or more factors.

factoring completely Factoring a number or polynomial into factors that cannot be factored further.

fractional exponent An exponent that is a rational number. The denominator of the exponent is the index of the equivalent radical.

frequency (of a data set) A number that represents how often each value occurs in the data set.

frequency (of a periodic function, specifically angular frequency) The number of complete cycles that occur in the interval from 0 to 2π.

function A relation in which no two ordered pairs have the same first member and different second members.

geometric sequence A set of numbers in which the ratio between consecutive terms is a constant.

geometric series The sum of a geometric sequence.

greatest common factor (GCF) The GCF of two or more monomials is the monomial with the greatest coefficient and the variable factors of the greatest degree that are common to all the given monomials. For example, the GCF of $8a^2b^2$ and $20ab^3$ is $4ab^2$.

growth factor The base of an exponential function. For example, in $y = 2^x$, 2 is the growth factor and the values double each time x increases by 1.

half-life The time required for an amount of a substance to decrease by half.

horizontal line test If no horizontal line intersects the graph of a function in more than one point, then the graph represents a one-to-one function. If a function is one-to-one, then it has an inverse function.

identity An equation that is true for all possible replacements of the variable, as in $\sin^2 A + \cos^2 A = 1$.

imaginary number A number of the form bi, where b is a real number and $i = \sqrt{-1}$.

independent variable The input of a function. For a function of the form $y = f(x)$, x is the independent variable.

index (of a radical) The number k in the expression $\sqrt[k]{x}$ that tells what root of x is to be taken. In a square root radical, the index is omitted and is understood to be 2.
index (of a series) The variable in the sigma notation that indicates the position of the term. In the series $\sum_{k=1}^{3} 2^k = 2^1 + 2^2 + 2^3$, k is the index.
integer A number from the set {. . . , –3, –2, –1, 0, 1, 2, 3, . . .}.
interpolation The process of estimating a y-value from a table, graph, or equation using a value of x that falls within the range of observed x-values.
interquartile range The difference between the first and third quartiles of a set of data; a measure of dispersion that is resistant to outliers.
inverse function The function obtained by interchanging the first and second members (x and y) in each ordered pair of a one-to-one function. The graphs of a function and its inverse are symmetric about the line $y = x$.
inverse relation The relation obtained by interchanging the first and second members (x and y) of each ordered pair of a relation.
inverse variation A relationship in which the product of two variables is a constant, $x_1y_1 = x_2y_2$. If y varies inversely with x, then $xy = k$ or, equivalently, $y = \frac{k}{x}$, where $k \neq 0$.
irrational number A number that cannot be expressed as the ratio of two integers.

Law of Cosines A relationship among the cosine of an angle of a triangle and the lengths of the three sides of the triangle. It is a generalization of the Pythagorean Theorem that is not restricted to right triangles. The formula is on the reference sheet.
Law of Sines A relationship among the sides of a triangle and the sines of the angles opposite these sides. The formula is on the reference sheet.
least squares regression A statistical calculation that finds an equation of a line or curve that best fits a set of measurements by minimizing the sum of the squares of the vertical distances between the plotted measurements and the fitted line or curve. A graphing calculator has a built-in regression feature that performs the required calculations. See also *regression model*.

linear equation An equation in which all the terms have degree 1. The graph of a linear equation is a straight line.

linear function A function defined by a linear equation.

linear regression model See *regression model*.

logarithmic function The function $y = \log_b x$, which is the inverse of the exponential function $y = b^x$, where b is positive and not equal to 1.

logarithmic regression model See *regression model.*

logarithm of x An exponent that represents the power to which a given base must be raised to produce a positive number x. For example, $\log_2 8 = 3$ because $2^3 = 8$.

mean A measure of the center of a set of n data values found by dividing the sum of the data values by n; also called average.

measure of central tendency A statistic that measures the center of a data set. The most common examples are the mean, median, and mode.

measure of dispersion A statistic that measures the spread of the data values about the center. Common examples are range, quartiles, interquartile range, standard deviation (both sample and population), and variance (both sample and population.)

median The middle value when a set of numbers is arranged in size order. If the set has an even number of values, the median is the average of the two middle values.

minute A unit of degree measure equal to $\frac{1}{60}$ of a degree.

mode The data value that occurs most frequently in a given set of data.

monomial A number, a variable, or a product of numbers and variables.

natural logarithm A logarithm whose base is e, written as $\ln x$.

negative angle An angle in standard position whose terminal side rotates in a clockwise direction.

negative exponent An exponent that is a negative number. For $x \neq 0$, $x^{-a} = \dfrac{1}{x^a}$.

normal curve A bell-shaped curve that describes a distribution of data values in which the mean is the most likely value and the likelihood of a value decreases the farther the value is from the mean.

nth term An arbitrary term in a sequence. The value of n indicates the position of the term. It is frequently used as the last term of a finite sequence.

one-to-one function A function in which no two ordered pairs have different x-values paired with the same y-value.
onto function A function, f, that maps set A to set B and in which the range of f is equal to set B.
ordinate The y-coordinate of a point in the coordinate plane. The ordinate of the point (2, 3) is 3.
origin The zero point on a number line; the point in the coordinate plane where the two axes intersect.
outlier A data value far from the majority of the data values. Outliers greatly affect the mean and the range of a data set but have little or no effect on the median and interquartile range.

parabola The U-shaped graph of a quadratic equation in two variables in which either x or y is squared but not both. A parabola in which x is squared has a vertical axis of symmetry. A parabola in which y is squared has a horizontal axis of symmetry.
perfect square A rational number whose square root is also rational.
percentile The score below which a certain percentage of the scores in a distribution occur. For example, if a test score of 78 is the 60th percentile, then 60% of the scores are less than 78.
period The length of one complete cycle of a periodic function. The period of a sine or cosine function in the form $y = a \sin bx$ or $y = a \cos bx$ is $\frac{2\pi}{|b|}$. The period of a tangent function in the form $y = \tan bx$ is $\frac{\pi}{|b|}$.
permutation A selection of objects in which the order of the individual objects is considered. For example, in selecting three students from {Aisha, Ben, Chandra, Daniel, Emma}, the selections Aisha-Ben-Emma and Ben-Emma-Aisha are two different permutations.
phase shift (of a periodic function) The horizontal translation of a periodic graph.
polynomial A monomial or the sum of two or more monomials.
positive angle An angle in standard position whose terminal side rotates in a counterclockwise direction.

power regression model See *regression model.*

probability (of an event) A number from 0 to 1 that represents the likelihood that an event will occur. Probability 0 means an event will not occur; probability 1 means the event is sure to occur.

Pythagorean Theorem In a right triangle, the square of the length of the hypotenuse is equal to the sum of the squares of the lengths of the legs.

quadrant One of the four regions into which the coordinate plane is divided by the axes.

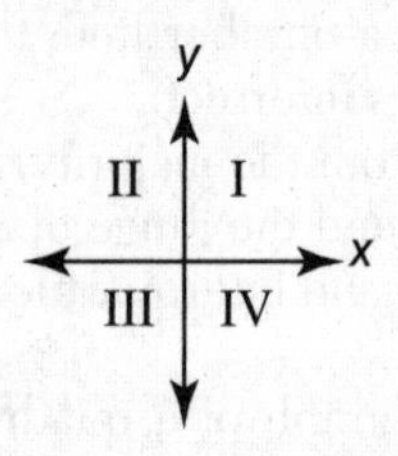

quadrantal angle An angle in standard position whose terminal side coincides with a coordinate axis.

quadratic equation An equation that can be put into the standard form $ax^2 + bx + c = 0, a \neq 0$.

quadratic formula The roots of the quadratic equation $ax^2 + bx + c = 0, a \neq 0$, are given by the formula $x = \frac{-b \pm \sqrt{b^2 - 4ac}}{2a}$.

quadratic function A function that has the form $y = ax^2 + bx + c$, $a \neq 0$.

quartiles Values that divide a data set into four equal-size groups when the data set is arranged in size order. The second quartile, Q_2, is the median. The first (or lower) quartile, Q_1, is the median of the lower half of the data set. The third (or upper) quartile, Q_3, is the median of the upper half of the data set.

radian The measure of a central angle of a circle that intercepts an arc whose length equals the radius of the circle.

radical equation An equation in which the variable appears as part of a radicand, such as $\sqrt{x} + 3 = 6$.

radical form An expression containing a radical symbol such, as $2\sqrt{5}$ or $\sqrt[3]{7x^2}$.

radicand The expression that appears underneath a radical sign.

random sample A sample in which all members of the population have an equal chance of being selected.

range (of a function or relation) The set of all possible values of the dependent variable of a function; the set of second coordinates in the ordered pairs that compose a relation.

range (in statistics) The difference between the greatest and the smallest data values.

rationalize a denominator An algebraic technique to rewrite a fraction with a complex or irrational denominator as a fraction with a rational denominator by multiplying the numerator and denominator by the conjugate of the denominator.

rational expression A quotient of two polynomials, such as $\frac{x^2-9}{x+2}$, $x \neq -2$.

rational number A number that can be written in the form $\frac{a}{b}$, where a and b are integers with $b \neq 0$. Decimals in which a set of digits endlessly repeat, like 0.2500 . . . $\left(=\frac{1}{4}\right)$ and 0.3333 . . . $\left(=\frac{1}{3}\right)$, represent rational numbers.

real number A number that is a member of the set that consists of all rational and irrational numbers.

recursive rule A formula that computes the next term in a sequence based on the preceding term(s). A starting value for the sequence must be given.

reference angle When an angle is placed in standard position, the acute angle formed by the terminal side and the x-axis.

reflection over x-axis A transformation of the coordinate plane where the y-value of every point in the plane is negated. The effect is to flip figures in the plane over the x-axis.

reflection over y-axis A transformation of the coordinate plane where the x-value of every point in the plane is negated. The effect is to flip figures in the plane over the y-axis.

regression model An equation of the line or curve that is fitted to a set of data by a statistical calculation. A linear regression model has the form $y = ax + b$, an exponential regression model has the form $y = ab^x$, a logarithmic regression model has the form $y = a + b \ln x$, and a power regression model has the form $y = ax^b$. The regression feature of a graphing calculator allows you to choose the type of regression model and then calculates the constants a and b for the regression model selected. See also *least squares regression.*

relation A set of ordered pairs.

replacement set The set of values that a variable can have; another name for the domain of a variable.

restricted domain A smaller domain for a function based on meeting certain criteria. For example, the domain of a function may be restricted to create a one-to-one function that will have an inverse function.

resultant vector The vector obtained by adding two other vectors. Graphically, this is represented as the diagonal of a parallelogram with the original vectors as adjacent sides of the parallelogram.

root of an equation A solution to $f(x) = 0$. An equation may have more than one root. Graphically, a root of an equation is an x-value where the graph intersects the x-axis. The value of the root checks in the original equation.

sample space The set of all possible outcomes for an experiment.

scatter plot A graph that results from plotting bivariate data in the coordinate plane.

secant The reciprocal of the cosine function.

sigma (σ) The lowercase Greek letter σ represents the population standard deviation.

sigma (Σ) The uppercase Greek letter Σ represents the successive summation of terms, as in $\sum_{k=1}^{3} 2^k = 2^1 + 2^2 + 2^3$.

sine ratio In a right triangle, the ratio of the length of the leg opposite a given acute angle to the length of the hypotenuse. If an angle θ is in standard position, then $\sin \theta = \frac{y}{r}$, where $P(x, y)$ is any point on the terminal side of angle θ and $r = \sqrt{x^2 + y^2}$.

standard deviation A statistic that measures how spread out numerical data are from the mean.

standard position An angle whose vertex is at the origin and whose initial side coincides with the positive x-axis.

survey A method of collecting data by asking people questions using an interview or a questionnaire.

tangent ratio In a right triangle, the ratio of the length of the leg opposite a given acute angle to the length of the leg adjacent to that angle. If an angle θ is in standard position, then $\tan \theta = \frac{y}{x}$, where $P(x, y)$ is any point on the terminal side of angle θ.

terminal side The side of an angle in standard position that rotates about the origin while the initial side of the angle remains fixed on the positive x-axis.

theoretical probability The probability of an event determined from counting principles or geometric considerations rather than from experimentation.

transformation A function that maps each point in the coordinate plane onto its image point in the coordinate plane.

translation A transformation in which every point in the plane moves the same distance in the same direction.

trinomial A polynomial with three unlike terms, as in $x^2 - 3x + 7$.

undefined An expression that does not have a mathematical value. Division by zero and the logarithm of zero are always undefined. The square root or logarithm of a negative number is undefined unless you are working with complex numbers.

unit circle A circle centered at the origin with a radius of 1.

variance A measure of dispersion equal to the square of the standard deviation.

vector A quantity that has both magnitude and direction. Graphically, a vector is represented as a directed line segment.

vertical line test If no vertical line intersects the graph of a relation in more than one point, the graph represents a function.

zero of a function Any value of the variable for which the function evaluates to 0. Each x-intercept of the graph of a function, if any, represents a zero of the function. A zero is another name for a root.

Regents Examinations, Answers, and Self-Analysis Charts

Important Note

The first two tests that follow are *not* actual Regents Examinations. They are practice tests that have the same format, credit distribution, and topic average as an actual Regents examination in Algebra 2/Trigonometry.

An Official Test Sampler from the New York State Education Department begins on page 345.

The last two exams in the book (starting on page 391) were actually administered in June and August 2010. These are the first actual tests available for your review, so they should be very helpful for you as you prepare for the Regents.

Examination Sample Test 1

Algebra 2/Trigonometry

REFERENCE SHEET

Area of a Triangle

$K = \frac{1}{2}ab \sin C$

Law of Cosines

$a^2 = b^2 + c^2 - 2bc \cos A$

Functions of the Sum of Two Angles

$\sin(A + B) = \sin A \cos B + \cos A \sin B$

$\cos(A + B) = \cos A \cos B - \sin A \sin B$

$\tan(A + B) = \dfrac{\tan A + \tan B}{1 - \tan A \tan B}$

Functions of the Double Angle

$\sin 2A = 2 \sin A \cos A$

$\cos 2A = \cos^2 A - \sin^2 A$

$\cos 2A = 2 \cos^2 A - 1$

$\cos 2A = 1 - 2 \sin^2 A$

$\tan 2A = \dfrac{2 \tan A}{1 - \tan^2 A}$

Functions of the Difference of Two Angles

$\sin(A - B) = \sin A \cos B - \cos A \sin B$

$\cos(A - B) = \cos A \cos B + \sin A \sin B$

$\tan(A - B) = \dfrac{\tan A - \tan B}{1 + \tan A \tan B}$

Functions of the Half Angle

$$\sin\frac{1}{2}A = \pm\sqrt{\frac{1-\cos A}{2}}$$

$$\cos\frac{1}{2}A = \pm\sqrt{\frac{1+\cos A}{2}}$$

$$\tan\frac{1}{2}A = \pm\sqrt{\frac{1-\cos A}{1+\cos A}}$$

Law of Sines

$$\frac{a}{\sin A} = \frac{b}{\sin B} = \frac{c}{\sin C}$$

Sum of a Finite Arithmetic Sequence

$$S_n = \frac{n(a_1 + a_n)}{2}$$

Sum of a Finite Geometric Sequence

$$S_n = \frac{a_1(1-r^n)}{1-r}$$

Binomial Theorem

$$(a+b)^n = {}_nC_0a^nb^0 + {}_nC_1a^{n-1}b^1 + {}_nC_2a^{n-2}b^2 + \cdots + {}_nC_na^0b^n$$

$$(a+b)^n = \sum_{r=0}^{n} {}_nC_r a^{n-r}b^r$$

Normal Curve
Standard Deviation

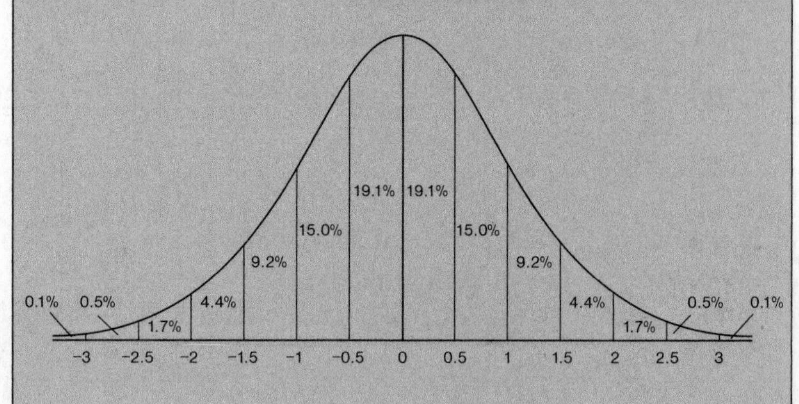

PART I

Answer all 27 questions in this part. Each correct answer will receive 2 credits. No partial credit will be allowed. For each question, write in the space provided the numeral preceding the word or expression that best completes the statement or answers the question. [54 credits]

1 The number of degrees equal to $\frac{5}{9}\pi$ radians is

(1) 45 (3) 100
(2) 90 (4) 900

1 _____

2 Find all values of x that satisfy the equation $|x - 1| = 5 - 2x$

(1) {2,4} (3) No solution
(2) {2} (4) {4}

2 _____

3 Rebecca is conducting a statistical study for psychology class. She asks a group of students to do a timed arithmetic test. Then she has a similar group of students take the timed arithmetic test after they eat a peppermint candy. She compares the scores to see if eating peppermint improves math skills. What type of study is this?

(1) an opinion poll
(2) an observational study
(3) a survey
(4) a controlled experiment

3 _____

4 Which of the following is the solution to the system of equations?

$$\frac{y}{x-2} = x - 2$$

$$y = 3x - 6$$

(1) (5, 9)
(2) (2, 0)
(3) (5, 9) and (2, 0)
(4) $\left(\frac{7}{3}, 1\right)$

4 ____

5 From his summer earnings, Nate is saving $500 for a car. If he puts his money into a certificate of deposit at the bank for five years, the bank will pay 3.25% interest compounded continuously. The formula for the value, V, of the account with interest compounded continuously is $V = Pe^{rt}$ where P is the principal, which is the original amount of money; r is the annual interest rate; and t is the number of years invested. How much will the account be worth at the end of 5 years?

(1) $220.86
(2) $586.71
(3) $588.22
(4) $2539.21

5 ____

6 What is the value of x in the equation $81^{x+2} = 27^{5x+4}$?

(1) $-\frac{2}{11}$
(2) $-\frac{3}{2}$
(3) $\frac{4}{11}$
(4) $-\frac{4}{11}$

6 ____

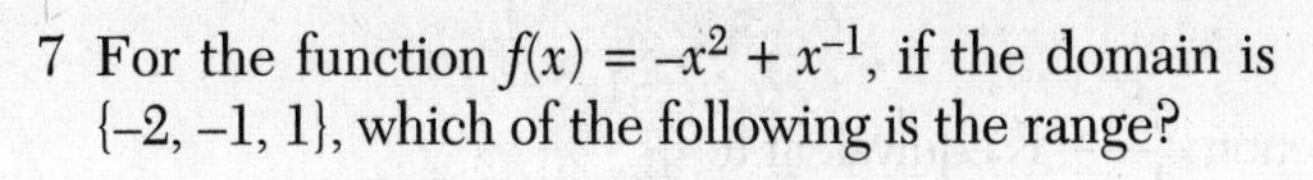

7 For the function $f(x) = -x^2 + x^{-1}$, if the domain is $\{-2, -1, 1\}$, which of the following is the range?

(1) $\left\{\frac{-9}{2}, -2, 0\right\}$ (3) $\left\{0, 2, \frac{7}{2}\right\}$

(2) $\{-2, 0, 0\}$ (4) $\{0, 2, 6\}$ 7 ____

8 Which of the following is the exact value of sec 30°?

(1) $\frac{999}{1000}$ (3) $\frac{23}{20}$

(2) 2 (4) $\frac{2\sqrt{3}}{3}$ 8 ____

9 The minimum stopping distance for a moving car is directly proportional to the square of its velocity. Suppose a certain car has a stopping distance of 43 feet when traveling at 30 miles per hour. To the nearest foot, what will be the stopping distance for the same car traveling at 55 miles per hour?

(1) 58 (3) 145
(2) 79 (4) 168 9 ____

10 There are 50 rows of seats in a theater. The first row has 19 seats and each row has two more seats than the row in front of it. If 117 seats are in the last row, how many seats are in the theater?

(1) 1050 (3) 4900
(2) 3400 (4) 6800 10 ____

11 The fraction $\dfrac{\frac{x}{y}+x}{\frac{1}{y}+1}$ is equivalent to

(1) $\dfrac{2xy}{1+y}$ (3) x

(2) $\dfrac{x^2y}{1+y}$ (4) $2x$ 11 ____

12 The volume of a soap bubble is represented by the equation $V = 0.094\sqrt{A^3}$, where A represents the surface area of the bubble. Which expression is also equivalent to V?

(1) $0.094A^{\frac{3}{2}}$ (3) $0.094A^6$

(2) $0.094A^{\frac{2}{3}}$ (4) $(0.094A^3)^{\frac{1}{2}}$ 12 ____

13 What is the amplitude of the function shown in the accompanying graph?

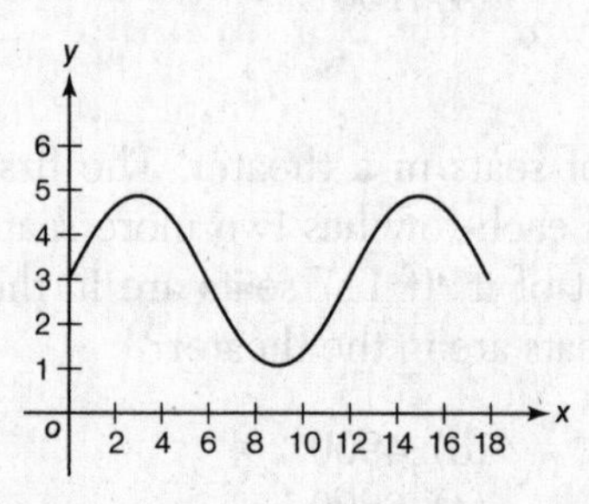

(1) 5 (3) 3

(2) 2 (4) 12 13 ____

14 What is the solution set of the equation $\frac{x}{x-4}-\frac{1}{x+3}=\frac{28}{x^2-x-12}$?

(1) { } (3) {–6}
(2) {4,–6} (4) {4} 14 _____

15 Factored completely, the expression $2x^3+4x^2-3x-6$ is equivalent to

(1) $2x(x+2)-3(x+2)$
(2) $2x(x+2)-3(x-2)$
(3) $(x+2)(2x^2-3)$
(4) It cannot be factored. 15 _____

16 Which of the following inequalities is graphed below?

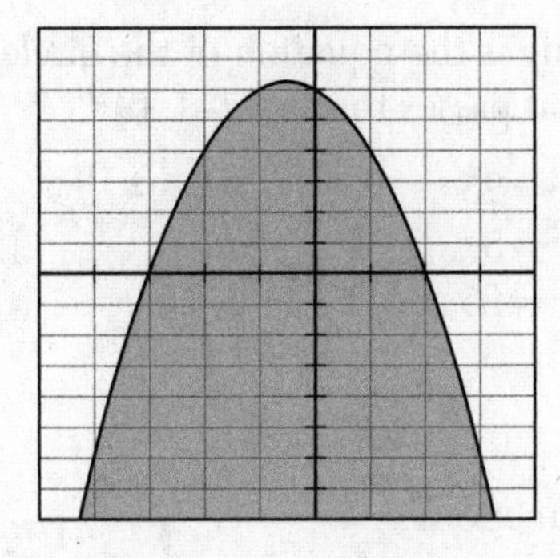

(1) $y>-x^2-x+6$ (3) $y<-x^2-x+6$
(2) $y\geq -x^2-x+6$ (4) $y\leq -x^2-x+6$ 16 _____

17 Yvonne throws darts at the target shown. The target consists of three concentric circles of radii 1, 3, and 5. Yvonne never misses entirely, but her shots hit the target randomly. What is the probability a thrown dart hits the shaded region?

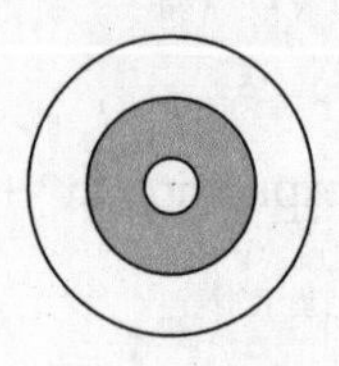

(1) $\frac{2}{25}$ (3) $\frac{9}{25}$

(2) $\frac{8}{25}$ (4) $\frac{3}{5}$ 17 ______

18 Which of the following is the equation of the circle with center (2, –4) that passes through (–1, 0)?

(1) $(x-2)^2+(y+4)^2=25$
(2) $(x-2)^2+(y+4)^2=5$
(3) $(x+2)^2+(y-4)^2=25$
(4) $(x+2)^2+(y-4)^2=5$ 18 ______

19 Which relation is a function?

(1) $x=4$ (3) $y=\sin x$
(2) $x=y^2+1$ (4) $x^2+y^2=16$ 19 ______

20 Jonathan's teacher asked him to express the sum $\frac{2}{3}+\frac{3}{4}+\frac{4}{5}+\frac{5}{6}+\frac{6}{7}$ using sigma notation. Jonathan proposed four possible answers. Which of these four answers is *not* correct?

(1) $\sum_{k=3}^{7}\frac{k-1}{k}$

(2) $\sum_{k=1}^{5}\frac{k}{k+1}$

(3) $\sum_{k=1}^{5}\frac{k+1}{k+2}$

(4) $\sum_{k=2}^{6}\frac{k}{k+1}$

20 ______

21 Battery lifetime is normally distributed for large samples. The mean lifetime is 500 days, and the standard deviation is 61 days. Approximately what percent of batteries have lifetimes *longer than* 561 days?

(1) 16%

(2) 34%

(3) 68%

(4) 84%

21 ______

22 The graph of the function, $f(x)$, has been translated to the left 3 units and down 5 units.

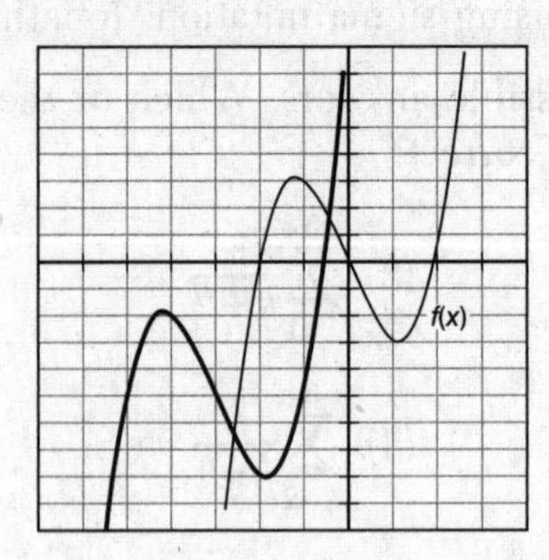

Which of the following represents the new function?

(1) $f(x+3)+5$
(2) $f(x-3)+5$
(3) $f(x+3)-5$
(4) $f(x-3)-5$

22 ______

23 The expression $\frac{2+i}{3+i}$ is equivalent to

(1) $\frac{6+5i}{8}$
(2) $\frac{6+i}{8}$
(3) $\frac{7}{10}-\frac{1}{2}i$
(4) $\frac{7}{10}+\frac{1}{10}i$

23 ______

24 If $\log x = a$, $\log y = b$, and $\log z = c$, then $\log \frac{x^2y}{\sqrt{z}}$ is equivalent to

(1) $2a+b+\frac{1}{2}c$
(2) $2ab-\frac{1}{2}c$
(3) $a^2+b-\frac{1}{2}c$
(4) $2a+b-\frac{1}{2}c$

24 ______

25 How many different 8-letter arrangements can be formed from the word PARALLEL?

(1) 1680 (3) 6720
(2) 3360 (4) 40,320 25 ______

26 Classical mathematics uses the term Golden Ratio for the ratio $\frac{1+\sqrt{5}}{2}$. The Golden Ratio was used by many famous artists to determine the dimensions of their paintings. If the ratio of the length to the width of a painting is $(1+\sqrt{5})$ to 2, find the length, in feet, of a painting that has a width of 14 feet.

(1) $14+14\sqrt{5}$ (3) $7+\sqrt{35}$

(2) $7+7\sqrt{5}$ (4) $\frac{1+\sqrt{5}}{28}$ 26 ______

27 What is the value of $\cot(\tan^{-1}\frac{3}{4})$?

(1) $\frac{3}{4}$ (3) $\frac{\sqrt{7}}{4}$

(2) $\frac{4}{3}$ (4) $\frac{4}{\sqrt{7}}$ 27 ______

PART II

Answer all 8 questions in this part. Each correct answer will receive 2 credits. Clearly indicate the necessary steps, including appropriate formula substitutions, diagrams, graphs, charts, etc. For all questions in this part, a correct numerical answer with no work shown will receive only 1 credit. [16 credits]

28 Using the fact that $15° = 45° - 30°$, determine the exact value of $\sin 15°$ in simplest radical form.

29 Determine the domain and range for $f(x) = \sqrt{2x-1}$.

[Use of the grid is optional.]

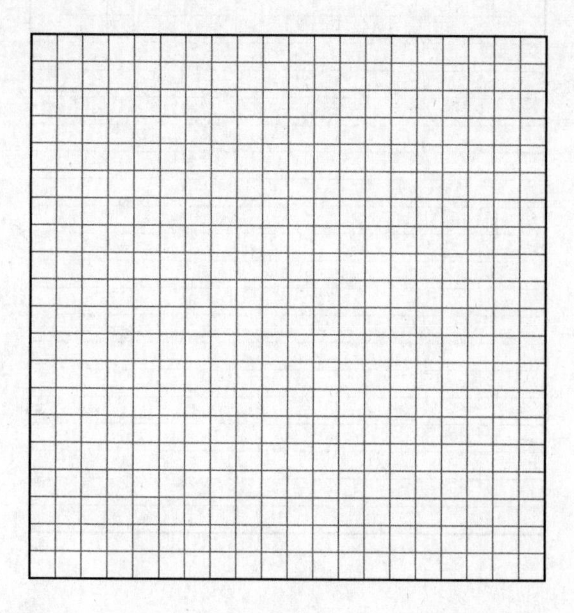

30 On the grid below, graph $y = \sin^{-1} x$ over the domain for which it is defined.

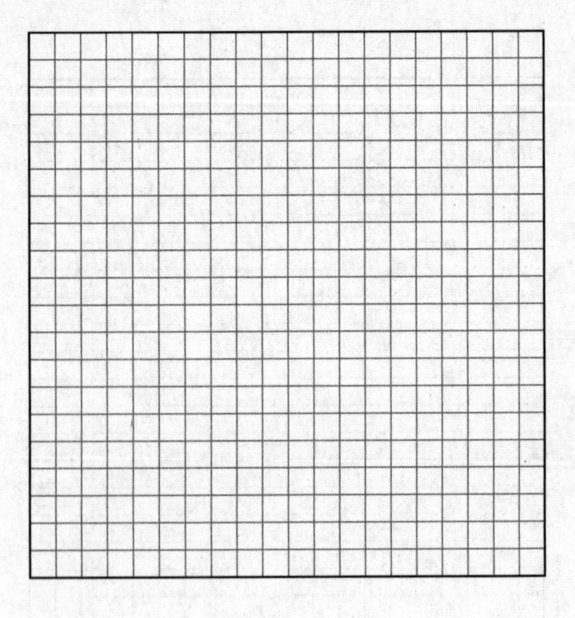

31 Juan has been told to write a quadratic equation where the sum of the roots is equal to –3 and the product of the roots is equal to –9. He writes $2x^2 + 6x - 18 = 0$ as his answer. His classmate, Tasha, wrote $x^2 + 3x - 9 = 0$ for her answer. Who should get credit for a correct response? Justify your answer.

32 A wrecking ball suspended from a chain is a type of pendulum. The relationship between the rate of speed of the ball R, the mass of the ball m, the length of the chain L, and the force F, is $R = 2\pi\sqrt{\frac{mL}{F}}$. Determine the force, F, to the *nearest hundredth*, when $L = 12$, $m = 50$, and $R = 0.6$.

33 Find, to the *nearest tenth of a degree*, all values of θ in the interval $0° \leq \theta < 360°$ that satisfy the equation $3 \sin \theta - 1 = 0$.

34 Factor completely: $6x^3 + 10x^2 - 24x$

35 A ski lift begins at ground level 0.75 mile from the base of a mountain whose face has a 50° angle of elevation, as shown in the accompanying diagram. The ski lift ascends in a straight line at an angle of 20°. Find the length of the ski lift from the beginning of the ski lift to the top of the mountain, to the *nearest hundredth of a mile*.

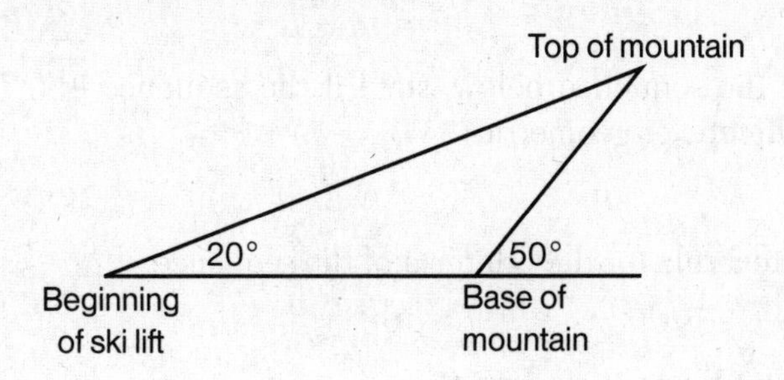

PART III

Answer all 3 questions in this part. Each correct answer will receive 4 credits. Clearly indicate the necessary steps, including appropriate formula substitutions, diagrams, graphs, charts, etc. For all questions in this part, a correct numerical answer with no work shown will receive only 1 credit. [12 credits]

36 For the sequence below, state if the sequence is arithmetic or geometric.

Write a rule for the nth term of the sequence.

Find the value of the 12th term.

{4, 12, 36, 108…}

37 A weight hangs from a spring in the physics lab next to a vertical meterstick. At rest, the weight is at the 0 cm mark on the meterstick. Roscoe pulls the weight down a certain amount and releases it. With a stopwatch, Roscoe observes the weight's position at various times and records the data in a table.

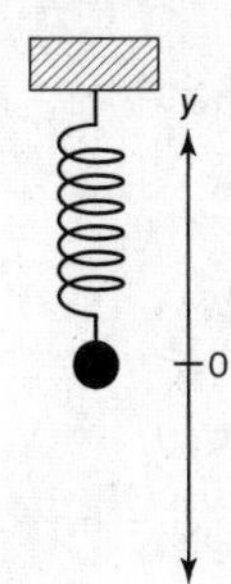

Time (s):	0.0	0.5	1.0	1.5	2.0	2.5	3.0
Position (cm):	−8	0	8	0	−8	0	8

Write an equation for the weight's position, y, as a function of time, t, in the form $y = a \sin(b(t + c))$.

[Use of the grid is optional.]

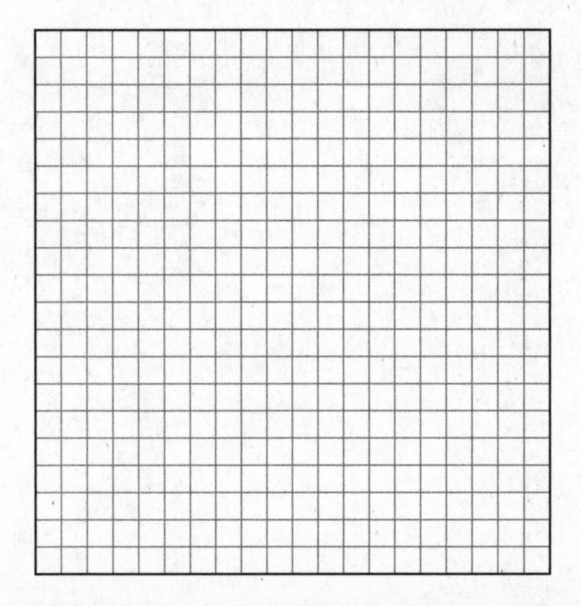

38 Two tow trucks try to pull a car out of a ditch. One tow truck applies a force of 1,500 pounds, while the other truck applies a force of 2,000 pounds. The resultant force is 3,000 pounds. Find the angle between the two applied forces, rounded to the *nearest degree.*

PART IV

Answer the question in this part. The correct answer will receive 6 credits. Clearly indicate the necessary steps, including appropriate formula substitutions, diagrams, graphs, charts, etc. For the question in this part, a correct numerical answer with no work shown will receive only 1 credit. [6 credits]

39 The accompanying table shows the average salary of baseball players starting in 1996. Using the data in the table, state the linear regression equation with the coefficients rounded to the *nearest hundredth*.

Baseball Players' Salaries	
Number of Years Since 1990	**Average Salary (thousands of dollars)**
6	1263
7	1472
8	1516
9	1864
10	2067
11	2378
12	2385

State the exponential regression equation with the coefficient and base rounded to the *nearest hundredth*.

Using your written regression equations, by how much does the exponential model exceed the average salary in 2010 predicted by the linear model, to the *nearest thousand dollars*?

Answers Sample Test 1 Algebra 2/Trigonometry

Answer Key

PART I

1. (3)
2. (2)
3. (4)
4. (1)
5. (3)
6. (4)
7. (1)
8. (4)
9. (3)
10. (2)
11. (3)
12. (1)
13. (2)
14. (3)
15. (3)
16. (4)
17. (2)
18. (1)
19. (3)
20. (2)
21. (1)
22. (3)
23. (4)
24. (4)
25. (2)
26. (2)
27. (2)

PART II

28. $\frac{\sqrt{6}-\sqrt{2}}{4}$

29. domain: $x \geq \frac{1}{2}$; range: $y \geq 0$

30. *See the accompanying detailed solution.*

31. Both students should get credit for a correct response.

32. 65797.36

33. {19.5°, 160.5°}

34. $2x(3x-4)(x+3)$

35. 1.15 miles

PART III

36. geometric; $a_n = 4(3)^{n-1}$; $a_{12} = 708588$

37. $y = 8\sin(\pi(t - 0.5))$

38. 63°

PART IV

39. $y = 204.61x + 7.82$; $y = 648.98(1.12)^x$; \$2,160,000

Answers Explained

PART I

1. To convert radians to degrees, multiply by $\frac{180°}{\pi}$:

$$\frac{5}{9}\pi \cdot \frac{180°}{\pi} = \frac{900°}{9} = 100°$$

CALC Check

Set the calculator to the mode you want for the answer. Insert the expression for the angle in parentheses. Put the symbol for the original unit (found in the ANGLE menu) after the expression and select ENTER.

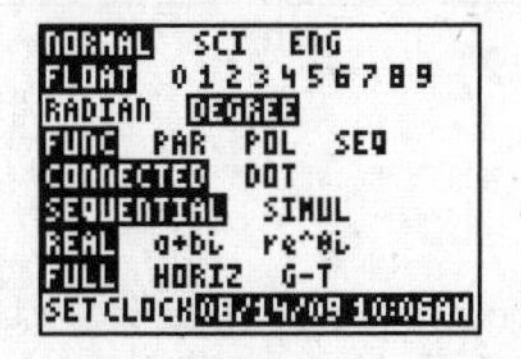

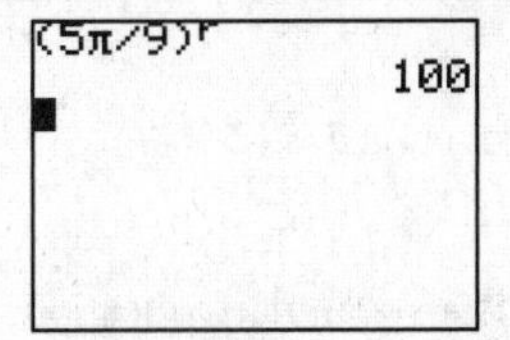

The correct choice is **(3)**.

2. To solve an absolute value equation algebraically:

Write two separate equations to eliminate the absolute value. Solve both.

$$\begin{aligned} x - 1 &= 5 - 2x &\quad \text{or} \quad -(x-1) &= 5 - 2x \\ & & -x + 1 &= 5 - 2x \\ 3x - 1 &= 5 &\quad \text{or} \quad 1 &= 5 - x \\ 3x &= 6 &\quad \text{or} \quad -4 &= -x \\ x &= 2 &\quad \text{or} \quad x &= 4 \end{aligned}$$

Check both candidate solutions, $x = 2$ and $x = 4$, in the original problem.

Check $x = 2$: $|(2) - 1| \stackrel{?}{=} 5 - 2(2)$

$1 = 1$ ✔

Check $x = 4$: $|(4) - 1| \stackrel{?}{=} 5 - 2(4)$

$3 = -3$ No; reject.

Only $x = 2$ checks.

This problem can be solved graphically. The graph shows one solution at $x = 2$ only.

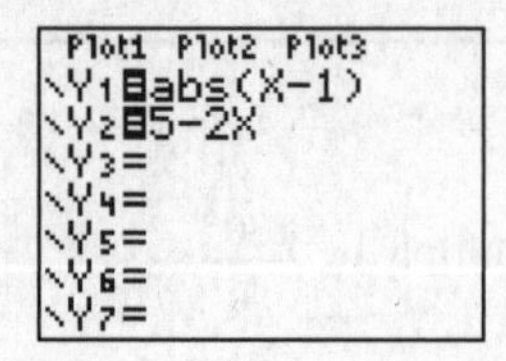

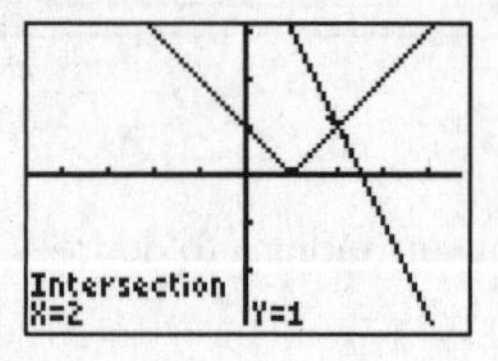

CALC Check

Plot1 Plot2 Plot3
\Y1=abs(X-1)
\Y2=5-2X
\Y3=
\Y4=
\Y5=
\Y6=
\Y7=

TABLE SETUP
TblStart=0
ΔTbl=1
Indpnt: Auto Ask
Depend: Auto Ask

X	Y1	Y2
2	1	1
4	3	-3

X=

The correct choice is **(2)**.

3. This is a vocabulary question about types of statistical studies. Choice (4) is the best answer. A controlled experiment compares the results from two groups; the experimental condition is changed for only one of the groups.

WRONG CHOICES EXPLAINED:

(1), (3) An opinion poll and a survey are the same. An interviewer or a questionnaire asks people questions. An opinion poll and a survey are not applicable.

(2) In an observational study, an outside observer collects measurements without controlling any of the conditions in the study. An observational study is not applicable.

The correct choice is **(4)**.

4. This system can be solved by substitution. Substitute $3x - 6$ in place of y in the first equation.

$$\frac{y}{x-2} = x - 2 \text{ and } y = 3x - 6$$

$$\frac{3x-6}{x-2} = x - 2 \text{ or } \frac{3x-6}{x-2} = \frac{x-2}{1}$$

Cross multiply.

$$\begin{aligned} 3x - 6 &= (x-2)(x-2) \\ &= x^2 - 2x - 2x + 4 \\ &= x^2 - 4x + 4 \end{aligned}$$

This equation is quadratic; rearrange it in standard form.

$$\begin{aligned} 3x - 6 &= x^2 - 4x + 4 \\ -3x + 6 &\quad\;\; - 3x + 6 \\ \hline 0 &= x^2 - 7x + 10 \end{aligned}$$

Factor and solve.

$$0 = (x-2)(x-5)$$
$$x = 2 \text{ or } x = 5$$

For each x-value, solve for y using one of the original equations.

When $x = 2$, $y = 3(2) - 6 = 0$
When $x = 5$, $y = 3(5) - 6 = 9$

Finally, check the candidate solutions in *both* of the original equations.

The candidate solutions are (2, 0) and (5, 9).

Check (2, 0): $\frac{0}{2-2} \stackrel{?}{=} 2 - 2$ or

$\frac{0}{0} \stackrel{?}{=} 0$. The left side is undefined, *not* 0, so this does not check.

Check (5, 9): $\frac{9}{5-2} \stackrel{?}{=} 5 - 2$

or $3 = 3$ ✔

and $9 = 3(5) - 6$ or $9 = 9$ ✔

Only (5, 9) checks.

Alternatively, after the initial substitution, if one notices the left side factors and simplifies, the problem can be done more easily.

$$\frac{3x-6}{x-2}=x-2$$

$$\frac{3\cancel{(x-2)}}{\cancel{x-2}}=x-2$$

$$3=x-2$$

It is still a good idea to check the solution in the original equations.

$x=5$; $y=3(5)-6=9$ and the solution is $(5, 9)$.

CALC Check

The original equation $\frac{y}{x-2}=x-2$ cannot be entered into the calculator.

Solving for $y=(x-2)(x-2)$ produces an equation equivalent everywhere EXCEPT at $x=2$, which is undefined in the original equation. $x=2$ seems to check in the calculator. This is why all answers must be checked in the *original* equations.

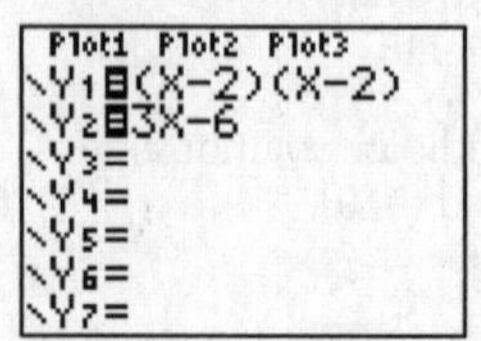

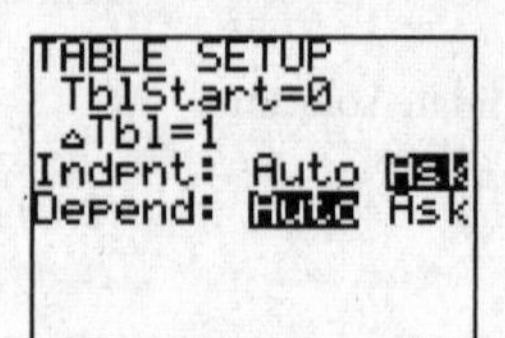

X	Y1	Y2
5	9	9
2	0	0
2.3333	.11111	1

X=

The correct choice is **(1)**.

5. This is an exponential evaluation problem. The principal P is given as \$500 and the number of years is $t=5$. Before it is used in the formula, the interest rate must be converted to a decimal:

$$r=3.25\%=\frac{3.25}{100}=0.0325.$$

The value after five years is given by the formula

$$V=\$500e^{0.0325(5)}=\$588.22.$$

The correct choice is **(3)**.

6. Since 81 and 27 can both be written as powers of 3, this equation can be rewritten to have common bases.

$$81 = 3^4;\ 27 = 3^3$$
$$81^{x+2} = 27^{5x+4}$$
$$(3^4)^{x+2} = (3^3)^{5x+4}$$

When a power is raised to a power, multiply the exponents.

$$3^{4(x+2)} = 3^{3(5x+4)}$$

If two powers of the same base are equal, the exponents must be equal.

$$4(x+2) = 3(5x+4)$$
$$4x + 8 = 15x + 12$$
$$-4 = 11x$$

Solve for x.

$$x = -\frac{4}{11}$$

CALC Check

This problem can be done on the calculator by checking all four choices. Even though the table shows decimals for the x-values, the fraction answer choices were keyed in. If you use your calculator to evaluate functions, make sure you understand the calculator notation for scientific notation. For example, in the table at $X = -1.5$, $Y_2 = 9.8E-6$ which is $9.8 \times 10^{-6} = 0.0000098$.

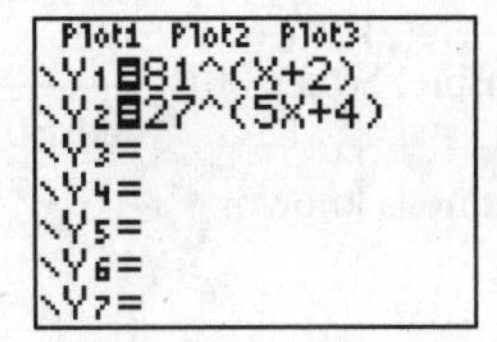

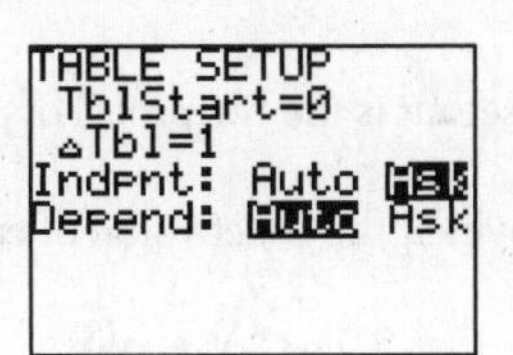

X	Y1	Y2
-.1818	2951	26559
-1.5	9	9.8E-6
.36364	32431	2.13E8
-.3636	1327.3	1327.3

X=

The correct choice is **(4)**.

7. The domain of a function gives all the possible inputs, or x-values. The range is all the possible outputs, or values of $f(x)$, that can come from those inputs. Evaluate the function for each value in the domain:

- When $x = -2$: $f(-2) = -(-2)^2 + (-2)^{-1} = -4 + \frac{1}{-2} = -4 - \frac{1}{2} = \frac{-9}{2}$.
- When $x = -1$: $f(-1) = -(-1)^2 + (-1)^{-1} = -1 + \frac{1}{-1} = -1 - 1 = -2$
- When $x = 1$: $f(1) = -(1)^2 + (1)^{-1} = -1 + \frac{1}{1} = -1 + 1 = 0$

The range includes the three values $\frac{-9}{2}$, -2, and 0.

CALC Check

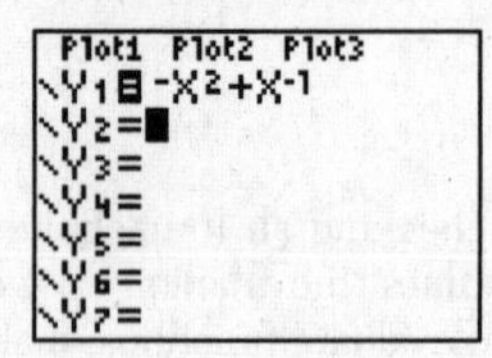

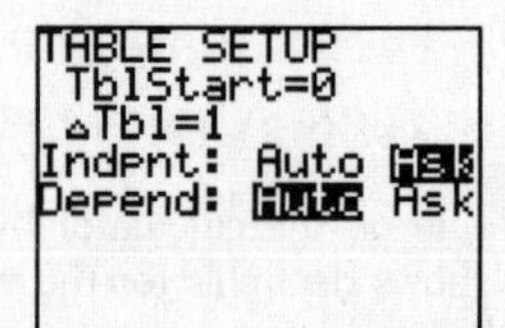

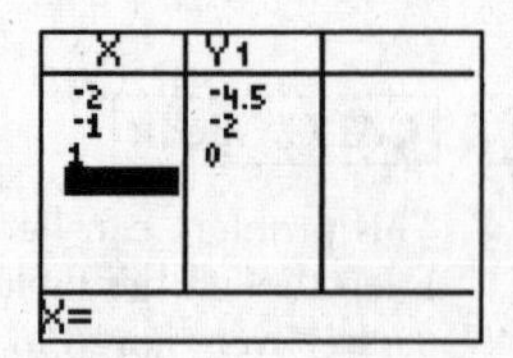

The correct choice is **(1)**.

8. The definition of secant is the reciprocal of cosine. So $\sec 30° = \frac{1}{\cos 30°}$.

30° is a special angle for which the exact value of cosine is known:

$$\cos 30° = \frac{\sqrt{3}}{2}.$$

$$\sec 30° = \frac{1}{\frac{\sqrt{3}}{2}} = \frac{2}{\sqrt{3}} = \frac{2}{\sqrt{3}} \cdot \frac{\sqrt{3}}{\sqrt{3}} = \frac{2\sqrt{3}}{3}$$

CALC Check

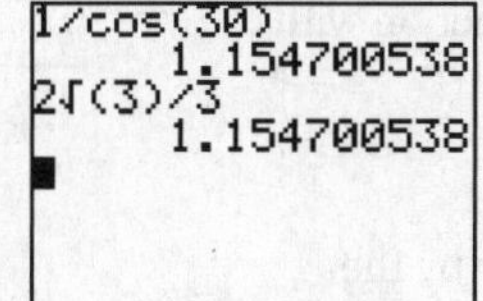

The correct choice is **(4)**.

9. Direct variation means d and v^2 always have the same ratio. Set up and solve a proportion:

$$\frac{\text{distance}}{\text{velocity}^2}: \frac{43}{30^2} = \frac{d}{55^2}$$

$$900d = 130075$$

$$d = \frac{130075}{900} = 144.53$$

$$\approx 145 \text{ to the nearest foot}$$

The correct choice is **(3)**.

10. Since each row has two more seats than the one in front of it, the number of seats in each row form an arithmetic sequence. The formula for the sum of a finite arithmetic sequence is on the reference sheet, $S_n = \frac{n(a_1 + a_n)}{2}$, where n is the number of terms (rows), 50; a_1 is the first term, 19; and a_n is the last term, 117. The number of seats in the theater is

$$S_{40} = \frac{50(19+117)}{2} = 3400$$

The correct choice is **(2)**.

11. To simplify a complex fraction, rewrite both the numerator and the denominator as a single fraction with a common denominator.

$$\frac{\frac{x}{y}+x}{\frac{1}{y}+1}=\frac{\left(\frac{x}{y}+x\cdot\frac{y}{y}\right)}{\left(\frac{1}{y}+1\cdot\frac{y}{y}\right)}=\frac{\frac{x+xy}{y}}{\frac{1+y}{y}}$$

Multiply the numerator by the reciprocal of the denominator.

$$=\frac{x+xy}{y}\cdot\frac{y}{1+y}$$

Factor where possible (here, only the left numerator factors), and then divide out common factors.

$$=\frac{x\cancel{(1+y)}}{(1+\cancel{y})}\cdot\frac{\cancel{y}}{\cancel{1+y}}=x$$

Alternatively, you can multiply the numerator and denominator by the least common denominator for both. Here the LCD $= y$, so multiply the numerator and denominator by y.

$$\frac{\frac{x}{y}+x}{\frac{1}{y}+1}=\frac{y\left(\frac{x}{y}+x\right)}{y\left(\frac{1}{y}+1\right)}=\frac{x+xy}{1+y}$$

Then factor and reduce.

$$=\frac{x\cancel{(1+y)}}{\cancel{(1+y)}}=x$$

The correct choice is **(3)**.

12. The expression $\sqrt{A^3}$ can be expressed as the rational exponent $A^{\frac{n}{d}}$ where the numerator is the whole-number exponent of A, $n = 3$, and the denominator is the index of the radical on A, $d = 2$ (square root). So $\sqrt{A^3} = A^{\frac{3}{2}}$ and $V = 0.094\sqrt{A^3} = 0.094A^{\frac{3}{2}}$, which is choice (1).

WRONG CHOICES EXPLAINED:

(2) This is not equivalent to V, $0.094A^{\frac{2}{3}} = 0.094\sqrt[3]{A^2}$.

(3) $0.094A^6$ does not have a fractional exponent, so this is not a radical.

(4) The parentheses make this different from choice (1),

$$(0.094A^3)^{\frac{1}{2}} = \sqrt{0.094A^3}\,.$$

The correct choice is **(1)**.

13. The amplitude of a sine or cosine graph is half the distance between the minimum value and the maximum value of the function. In this case, the minimum value appears to be 1 and the maximum value appears to be 5.

The amplitude is $\frac{1}{2}(5-1)=2$, which is choice (2).

WRONG CHOICES EXPLAINED:

(1) This is the maximum value of the function.

(3) This is the vertical shift of the function.

(4) This is the period of the function.

The correct choice is **(2)**.

14. To solve a rational equation, first factor the denominators.

$$\frac{x}{x-4}-\frac{1}{x+3}=\frac{28}{x^2-x-12}$$

$$\frac{x}{x-4}-\frac{1}{x+3}=\frac{28}{(x-4)(x+3)}$$

Next get common denominators for all the terms in the equation.

$$\frac{x(x+3)}{(x-4)(x+3)}-\frac{(1)(x-4)}{(x+3)(x-4)}=\frac{28}{(x-4)(x+3)}$$

Then equate the numerators (this is equivalent to multiplying both sides by $(x-4)(x+3)$). Solve.

$$x(x+3)-(1)(x-4)=28$$
$$x^2+3x-x+4=28$$
$$x^2+2x-24=0$$
$$(x+6)(x-4)=0$$
$$x=-6 \text{ or } x=4$$

Finally, check the candidate solutions in the original equation. Use a calculator if necessary.

Check $x=-6$:

$$\frac{-6}{(-6)-4}-\frac{1}{(-6)+3}\stackrel{?}{=}\frac{28}{(-6)^2-(-6)-12}$$

$$\frac{28}{30}=\frac{28}{30} \checkmark$$

Check $x = 4$:

$$\frac{4}{4-4} - \frac{1}{4+3} \stackrel{?}{=} \frac{28}{4^2 - 4 - 12} \quad \text{NO}$$

Note that the first term and the last are undefined. Thus 4 is not a valid solution to the equation.

Only $x = -6$ works. The solution set is $\{-6\}$.

Note: For multiple-choice problems asking for the solution to an equation, it is often quicker and more accurate simply to check all the offered solutions in the original equation. In this problem, the only two candidates were -6 and 4. As shown above, only $x = -6$ works.

CALC Check

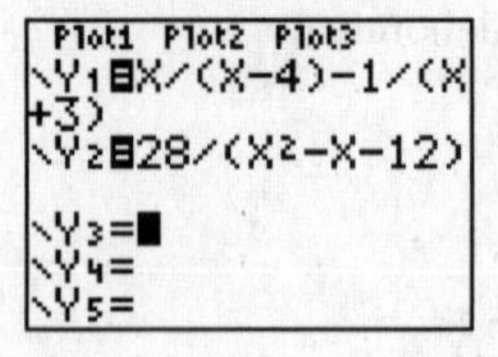

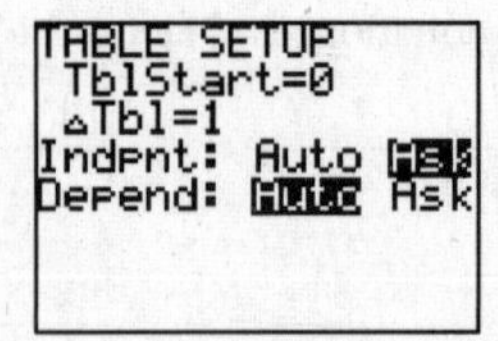

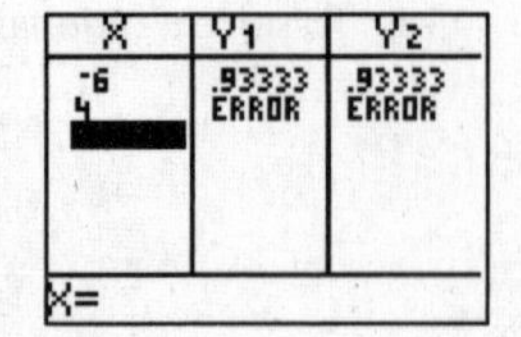

The correct choice is **(3)**.

15. There are four terms so try factor by grouping.

$$\begin{aligned} 2x^3 + 4x^2 - 3x - 6 &= 2x^2(x + 2) - 3(x + 2) \\ &= (x + 2)(2x^2 - 3) \end{aligned}$$

Choice (1) is a partial answer; the factoring is not complete
Choice (2) has a sign error in factoring the second binomial.
Choice (4) is a choice for students who do not recognize factor by grouping.

The correct choice is **(3)**.

16. Examine the choices given. All four have the same equation for the parabola, so the only question is whether the inequality is $>$, $\geq$, $<$, or $\leq$. Since the shaded area is below the graph of the parabola, the answer must include y being less than the parabola. Since the parabola is graphed as a solid curve instead of a dashed curve, the answer includes y being equal to the parabola. The correct inequality is $y \leq -x^2 - x + 6$.

The correct choice is **(4)**.

17. The probability of hitting the shaded region is

$$P = \frac{\text{Area of shaded region}}{\text{Total area of target}}.$$

The area of the shaded region is found by subtracting the area of the center circle, radius 1, from the area of the middle circle, radius 3:

$$A_{\text{middle ring}} = \pi(3)^2 - \pi(1)^2 = 8\pi.$$

The area of the target is $A_{\text{target}} = \pi(5)^2 = 25\pi$. So the probability of hitting the shaded region is $P = \frac{8\pi}{25\pi} = \frac{8}{25}$.

The correct choice is **(2)**.

18. The choices are all given in center-radius form, and the center is given in the problem as (2, –4). The length of the radius is the distance from the center to a point on the circle. This can be found with the distance formula.

Center: $(h, k) = (2, -4)$ (Given)

Point on circle: $(-1, 0)$ (Given)

Radius: $r = \sqrt{(x_P - h)^2 + (y_P - k)^2} = \sqrt{((-1) - 2)^2 + (0 - (-4))^2} = 5$

The center and radius are then substituted into the equation of a circle.

$$(x - h)^2 + (y - k)^2 = r^2$$
$$(x - 2)^2 + (y - (-4))^2 = 5^2 \text{ or}$$
$$(x - 2)^2 + (y + 4)^2 = 25$$

Remember that the signs inside the parentheses in the equation of a circle are *opposite* to the signs of the center's coordinates. Also remember that in the equation of a circle, the radius is *squared*.

The correct choice is **(1)**.

19. In a function, each input (x-value) has only one output (y-value). On a graph, any vertical line will intersect the graph of a function in at most one point (Vertical Line Test.)

- Choice (1): $x = 4$ is a vertical line. When $x = 4$, y can be infinitely many different values. This is *not* a function.
- Choice (2): This is quadratic in y. When a value is chosen for x, the resulting quadratic equation will usually give two values for y. For example, when $x = 5$, $y = -2$, or $y = 2$. This equation might also be recognized as a sideways parabola; many vertical lines will intersect it twice. This is *not* a function.
- Choice (3): Sine is a trigonometric function. For each x-value, $\sin x$ gives a unique output. Any vertical line will intersect the sine graph once. This is a function.
- Choice (4): Again, this relation is quadratic in y. Most x-values will lead to two y-values. For example, when $x = \sqrt{7}$, $y = -3$ or $y = 3$. This should be recognized as the equation of a circle; some vertical lines will intersect a circle twice. It is *not* a function.

The correct choice is **(3)**.

20. Check each choice:

- Choice (1):

$$\sum_{k=3}^{7} \frac{k-1}{k} = \frac{3-1}{3} + \frac{4-1}{4} + \frac{5-1}{5} + \frac{6-1}{6} + \frac{7-1}{7} = \frac{2}{3} + \frac{3}{4} + \frac{4}{5} + \frac{5}{6} + \frac{6}{7}$$

This is correct.

- Choice (2):

$$\sum_{k=1}^{5} \frac{k}{k+1} = \frac{1}{1+1} + \frac{2}{2+1} + \frac{3}{3+1} + \frac{4}{4+1} + \frac{5}{5+1} = \frac{1}{2} + \frac{2}{3} + \frac{3}{4} + \frac{4}{5} + \frac{5}{6}$$

This is *not* correct.

- Choice (3):

$$\sum_{k=1}^{5} \frac{k+1}{k+2} = \frac{1+1}{1+2} + \frac{2+1}{2+2} + \frac{3+1}{3+2} + \frac{4+1}{4+2} + \frac{5+1}{5+2} = \frac{2}{3} + \frac{3}{4} + \frac{4}{5} + \frac{5}{6} + \frac{6}{7}$$

This is correct.

- Choice (4):

$$\sum_{k=2}^{6}\frac{k}{k+1}=\frac{2}{2+1}+\frac{3}{3+1}+\frac{4}{4+1}+\frac{5}{5+1}+\frac{6}{6+1}=\frac{2}{3}+\frac{3}{4}+\frac{4}{5}+\frac{5}{6}+\frac{6}{7}$$

This is correct.

The question asked which answer is *not* correct.

The correct choice is **(2)**.

21. Since 561 – 500 = 61 is one standard deviation, a battery that lasts 561 days would be one standard deviation above the average. To find the probability a battery lasts longer than that, add all the percentages to the right of 1 on the normal curve given on the reference sheet, shaded in the graph below.

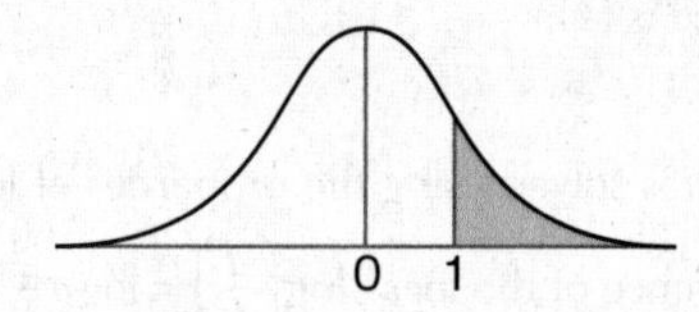

This gives a probability of 9.2% + 4.4% + 1.7% + 0.5% + 0.1% = 15.9% or approximately 16%.

The correct choice is **(1)**.

22. The transformation $f(x + h)$ translates a function to the left by h units; $f(x - h)$ translates it to the right by h units. The transformation $f(x) + k$ translates a function up by k units; $f(x) - k$ translates it down by k units. The problem asks for a function translated left 3 units and down 5 units, so the appropriate transformation is $f(x + 3) - 5$.

The correct choice is **(3)**.

23. To rationalize a complex denominator, multiply by the conjugate of the denominator:

$$\frac{2+i}{3+i}=\frac{(2+i)}{(3+i)}\cdot\frac{(3-i)}{(3-i)}=\frac{6-2i+3i-i^2}{9-i^2}=\frac{6+i-(-1)}{9-(-1)}=\frac{7+i}{10}=\frac{7}{10}+\frac{1}{10}i$$

CALC Check

(2+i)/(3+i)
.7+.1i
Ans▸Frac
7/10+1/10i

The correct choice is **(4)**.

24. This problem is solved using the properties of logarithms. The log of a quotient is the difference of the logs, $\log\left(\frac{a}{b}\right)=\log a-\log b$:

$$\log\frac{x^2y}{\sqrt{z}}=\log x^2y-\log\sqrt{z}$$

The log of a product is the sum of the logs, $\log(ab)=\log a+\log b$:

$$\log x^2y-\log\sqrt{z}=\log x^2+\log y-\log\sqrt{z}$$

The log of a power is the product of the exponent and the log of the base, $\log a^n=n\log a$. Note that $\sqrt{z}=z^{\frac{1}{2}}$:

$$\log x^2+\log y-\log\sqrt{z}=2\log x+\log y-\frac{1}{2}\log z$$

Now substitute $\log x=a$, $\log y=b$, and $\log z=c$:

$$2\log x+\log y-\frac{1}{2}\log z=2a+b-\frac{1}{2}c$$

The correct choice is **(4)**.

25. An arrangement of letters is a permutation because the order of the letters matters. There are 8 letters, so there are ${}_8P_8$ orders. However, two letters are repeated, A and L. The number of permutations for each of these letters must be divided from the total:

$$\frac{{}_8P_8}{{}_3P_3 \cdot {}_2P_2} = \frac{8!}{3! \cdot 2!} = \frac{40320}{6 \cdot 2} = \frac{40320}{12} = 3360$$

The correct choice is **(2)**.

26. Set up and solve a proportion:

$$\frac{\text{Length}}{\text{Width}} = \frac{1+\sqrt{5}}{2} = \frac{x}{14}$$

$$2x = 14\left(1+\sqrt{5}\right)$$

$$x = \frac{14\left(1+\sqrt{5}\right)}{2} = 7\left(1+\sqrt{5}\right) = 7 + 7\sqrt{5}$$

The correct choice is **(2)**.

27. $\tan^{-1}\left(\frac{3}{4}\right)$ means "the angle whose tan is $\frac{3}{4}$." The problem doesn't actually ask for that angle; it asks for the cotangent of that angle. Since cotangent is the reciprocal of tangent, the problem is asking "what is the reciprocal of the tangent of the angle whose tangent is $\frac{3}{4}$." In other words, what is the reciprocal of $\frac{3}{4}$? The answer is $\frac{4}{3}$.

CALC Check

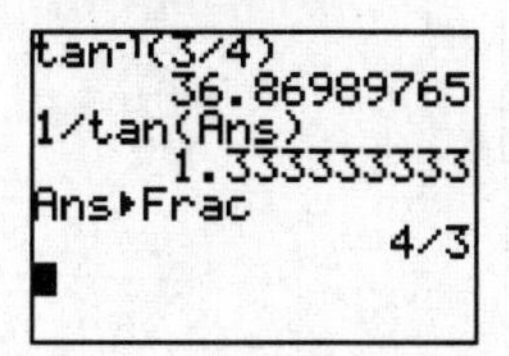

The correct choice is **(2)**.

PART II

28. The problem says to use the fact that 15° = 45° – 30°. Use the formula shown on the reference sheet for the sine of the difference of two angles. The formula is $\sin(A - B) = \sin A \cos B - \cos A \sin B$. In this problem, $A = 45°$ and $B = 30°$; 45° and 30° are special angles for which the exact values are known:

$$\sin 45° = \cos 45° = \frac{\sqrt{2}}{2}$$

$$\sin 30° = \frac{1}{2} \text{ and } \cos 30° = \frac{\sqrt{3}}{2}$$

$$\sin 15° = \sin(45° - 30°)$$

$$= \sin 45° \cos 30° - \cos 45° \sin 30°$$

$$= \left(\frac{\sqrt{2}}{2}\right)\left(\frac{\sqrt{3}}{2}\right) - \left(\frac{\sqrt{2}}{2}\right)\left(\frac{1}{2}\right)$$

$$= \frac{\sqrt{6}}{4} - \frac{\sqrt{2}}{4} \text{ or } \frac{\sqrt{6} - \sqrt{2}}{4}$$

CALC Check

If the question specifies you must find the exact value or express the answer in simplest radical form, do NOT write the decimal you get from your calculator. You can, however, compare your radical answer to the decimal on the calculator to check your answer.

```
sin(15)
          .2588190451
(√(6)-√(2))/4
          .2588190451
```

The exact value of sin 15° is $\frac{\sqrt{6} - \sqrt{2}}{4}$.

29. Unless a problem states otherwise, assume that both the domain and the range consist of real values. For a radical to produce a real value, the radicand must be greater than or equal to zero. For $f(x) = \sqrt{2x-1}$ to be real:

$$2x - 1 \geq 0$$

$$2x \geq 0$$

$$x \geq \frac{1}{2}$$

The domain of the function $f(x) = \sqrt{2x-1}$ is $x \geq \frac{1}{2}$.

The range is most easily read from a graph. Be sure you evaluate $f\left(\frac{1}{2}\right) = 0$ to know that the function does equal 0 at $x = \frac{1}{2}$. The range is $y \geq 0$.

CALC Check

Using ZoomDecimal gives an accurate graph for interpreting both domain and range. Be sure to sketch and label this curve on the grid provided.

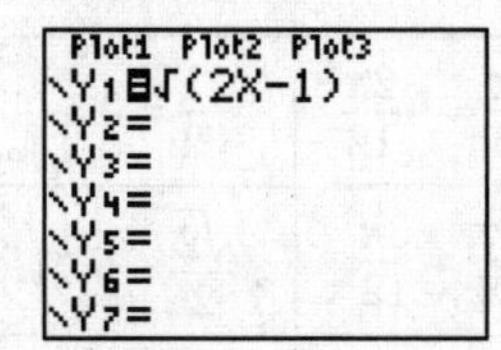

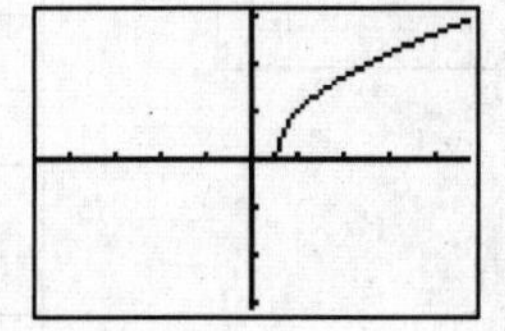

CALC Caution

If you used ZoomStandard, you might think the y-values never equal 0.

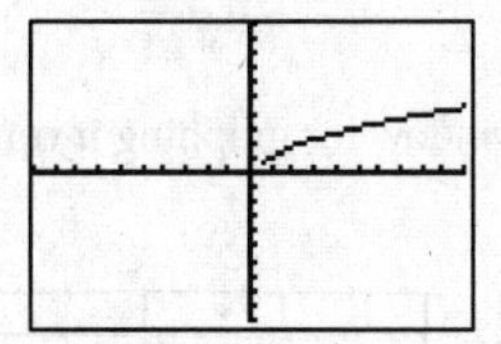

The domain of $f(x)$ is $x \geq \frac{1}{2}$, and the range of $f(x)$ is $y \geq 0$.

30. Remember that the range for $\sin^{-1}x$ is $-\frac{\pi}{2} \le x \le \frac{\pi}{2}$. Make a table of values for $y = \sin x$ with that domain. It is convenient to rewrite all the x-values with the same denominator.

Since $\sin^{-1}x$ is the inverse of $\sin x$, to graph $\sin^{-1}x$, reverse the coordinates of all the points in the table for $\sin x$: $(\frac{\pi}{2}, 1)$ in the table is graphed as $(1, \frac{\pi}{2})$, etc.

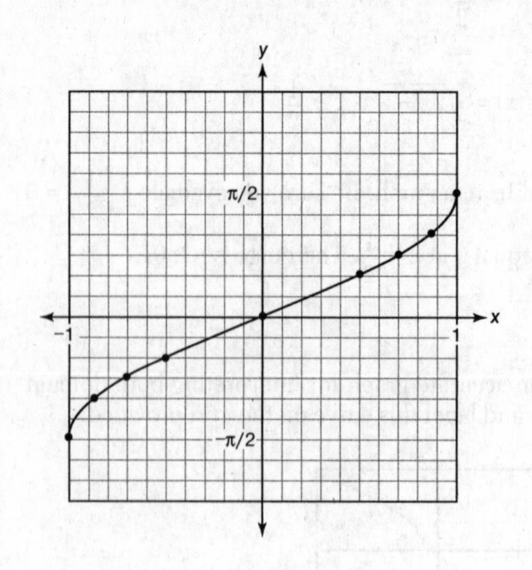

x	$\sin x$
$-\frac{\pi}{2} = -\frac{6\pi}{12}$	-1
$-\frac{\pi}{3} = -\frac{4\pi}{12}$	$-\frac{\sqrt{3}}{2} \approx -0.87$
$-\frac{\pi}{4} = -\frac{3\pi}{12}$	$-\frac{\sqrt{2}}{2} \approx -0.71$
$-\frac{\pi}{6} = -\frac{2\pi}{12}$	$-\frac{1}{2} = -0.5$
0	0
$\frac{\pi}{6} = -\frac{2\pi}{12}$	$\frac{1}{2} = 0.5$
$\frac{\pi}{4} = \frac{3\pi}{12}$	$\frac{\sqrt{2}}{2} \approx 0.71$
$\frac{\pi}{3} = \frac{4\pi}{12}$	$\frac{\sqrt{3}}{2} \approx 0.87$
$\frac{\pi}{2} = \frac{6\pi}{12}$	1

CALC Check

Select Radian Mode. ZoomDecimal is a better window for graphing inverse trigonometric functions than ZoomTrig.

```
Plot1 Plot2 Plot3
\Y1=sin⁻¹(X)
\Y2=
\Y3=
\Y4=
\Y5=
\Y6=
\Y7=
```

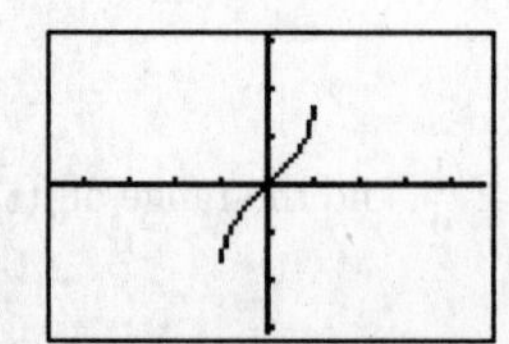

X	Y1
-1.5	ERROR
-1	-1.571
-.5	-.5236
0	0
.5	.5236
1	1.5708
1.5	ERROR

X=1.5

31. For any quadratic equation in standard form, $ax^2 + bx + c = 0$, the product of the roots is $-\frac{b}{a}$ and the sum of the roots is $\frac{c}{a}$. In Juan's equation, the product is $-\frac{6}{2} = -3$, and the sum is $\frac{-18}{2} = -9$, both as required. For Tasha's equation, the product is $-\frac{3}{1} = -3$, and the sum is $\frac{-9}{1} = -9$, again as required.

An alternate way to do the problem is to check that Tasha's equation has the required product and sum of its roots. Then observe that Juan's equation is the same as Tasha's except for a factor of 2. So Juan's equation will have the same roots as Tasha's.

Both students should get credit for a correct response because both equations satisfy the requirements of the problem.

32. This problem can be done by solving for F first and then substituting in values.

$$R = 2\pi\sqrt{\frac{mL}{F}}$$

Isolate the radical.

$$\frac{R}{2\pi} = \sqrt{\frac{mL}{F}}$$

Square both sides.

$$\left(\frac{R}{2\pi}\right)^2 = \frac{mL}{F}$$

Multiply both sides by F.

$$F\left(\frac{R}{2\pi}\right)^2 = mL$$

Divide to solve for F.

$$F = \frac{mL}{\left(\frac{R}{2\pi}\right)^2}$$

Substitute and evaluate. When using a calculator, put 2π in its own set of parentheses.

$$F = \frac{(50)(12)}{\left(\frac{0.6}{2\pi}\right)^2} = 65797.36267$$

OR

Substitute the values first, and then solve for F. Be careful to place the numbers accurately.

$$R = 2\pi\sqrt{\frac{mL}{F}}$$

$$0.6 = 2\pi\sqrt{\frac{(50)(12)}{F}}$$

Isolate the radical. Remember, when using a calculator to divide, use parentheses around 2π.

$$0.0954929659 = \sqrt{\frac{600}{F}}$$

Square both sides.

$$0.0091189065 = \frac{600}{F}$$

Multiply both sides by F.

$$0.0091189065F = 600$$

Divide to solve for F.

$$F = \frac{600}{0.0091189065} = 65797.36267$$

Note: If the problem is done the second way, do *not* round any intermediate results.

CALC Caution

Always use parentheses in a fraction whenever either the numerator or the denominator contains an operation. The values below are clearly not equal.

```
.6/2π
        .9424777961
.6/(2π)
        .0954929659
```

To the *nearest hundredth*, the force $F = 65797.36$.

33. First isolate the trigonometric function. Add 1 and then divide by 3.

$$3 \sin \theta - 1 = 0$$

$$3 \sin \theta = 1$$

$$\sin \theta = \frac{1}{3}$$

Use inverse sine to find one angle that satisfies the equation. Make sure the calculator is set to degree mode.

$$\theta = \sin^{-1}\frac{1}{3} = 19.47122°$$

Since $\sin \theta > 0$, there is a second solution in Quadrant II that has the same reference angle as the first angle, 19.47122°. Using a diagram is optional but may be helpful.

$$180° - 19.47° = 160.53°$$

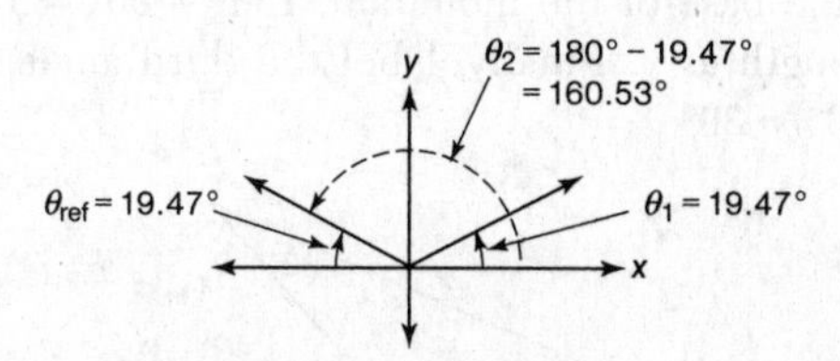

The problem does not specify which solution method must be used. You can solve this problem graphically. Since no grid is provided, you must accurately sketch and label the scale on the graph on your exam. To get the graphs below, in Window, set Xmin = 0, Xmax = 360 and select ZoomFit. Use CALC Zero to find the two roots.

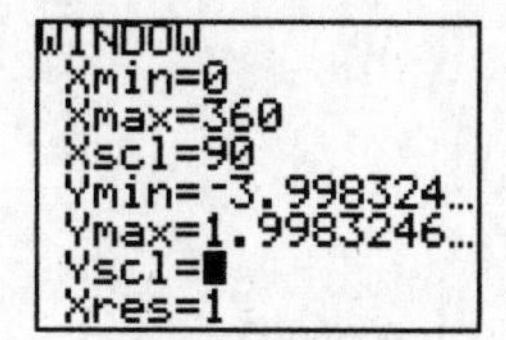

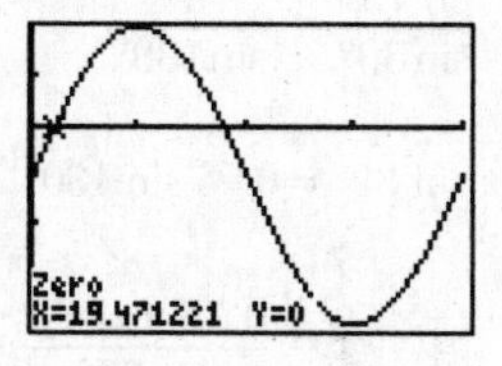

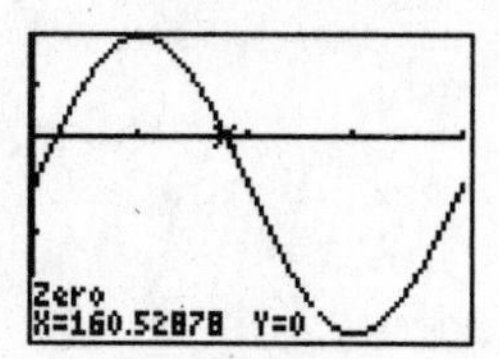

To the nearest *tenth of a degree*, the solutions are {19.5°, 160.5°}.

34. To factor completely, first factor out the greatest common factor, if any. In this problem, all three terms can be divided by $2x$.

$$6x^3 + 10x^2 - 24x = 2x(3x^2 + 5x - 12)$$

Then factor the remaining polynomial.

$$= 2x(3x - 4)(x + 3)$$

Check your factoring (especially the binomial factors) by multiplying.

$$= 2x(3x - 4)(x + 3)$$
$$= 2x(3x^2 + 9x - 4x - 12)$$
$$= 2x(3x^2 + 5x - 12)$$
$$= 6x^3 + 10x^2 - 24x$$

When completely factored, $6x^3 + 10x^2 - 24x$ is equal to $2x(3x - 4)(x + 3)$.

35. First label the known side of the triangle and the interior angle of the triangle at the base of the mountain, $180° - 50° = 130°$. Then label the unknown length as x. Finally, label the third angle in the triangle, $180° - (130° + 20°) = 30°$.

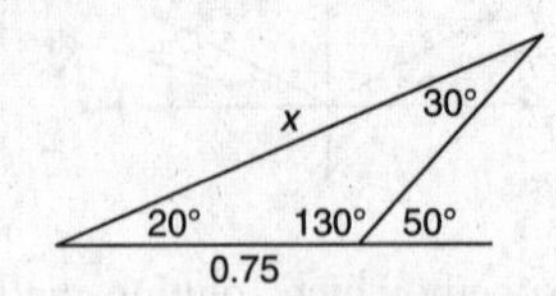

Since you know both the length of a side and the angle opposite it, 0.75 and 30°, you can use the Law of Sines shown on the reference sheet. Make sure the calculator is set to degree mode.

$$\frac{0.75}{\sin 30°} = \frac{x}{\sin 130°}$$

$$x \sin 30° = 0.75 \sin 130°$$

$$x = \frac{0.75 \sin 130°}{\sin 30°} = 1.149066665$$

To the *nearest hundredth* of a mile, the length of the ski lift is 1.15 miles.

PART III

36. If the sequence is arithmetic, each pair of consecutive terms must have the same difference. Since $12 - 4 = 8$ and $36 - 12 = 24 \neq 8$, the sequence is not arithmetic.

If the sequence is geometric, each pair of consecutive terms must have the same ratio: $\frac{12}{4} = 3$, $\frac{36}{12} = 3$, and $\frac{108}{36} = 3$. The sequence is geometric with common ratio $r = 3$.

The sequence can be rewritten as follows: $\{4, 4 \cdot 3^1, 4 \cdot 3^2, 4 \cdot 3^3, \ldots\}$ or $\{4, 4 \cdot 31, 4 \cdot 32, 4 \cdot 33, \ldots\}$. Notice that the fourth term has only three factors of 3. The fifth term will have only 4 factors of 3. In general, the nth term will have $n - 1$ factors of 3. So the nth term can be written as $a_n = 4(3)^{n-1}$. It is a good idea to memorize the explicit rule for a geometric sequence, $a_n = a_1(r)^{n-1}$.

Using $n = 12$, the 12th term is $a_{12} = 4(3)^{12-1} = 4(3)^{11} = 708588$.

If you cannot find the rule for the nth term, you can still get partial credit for finding the 12th term—just keep multiplying by 3; be sure you write down each term.

CALC Check

This shows how to repeatedly multiply by 3 without typing it over and over, which reduces errors. Press Enter after each term to get the next term. Count carefully.

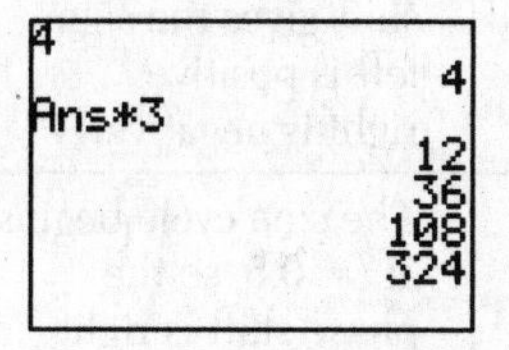

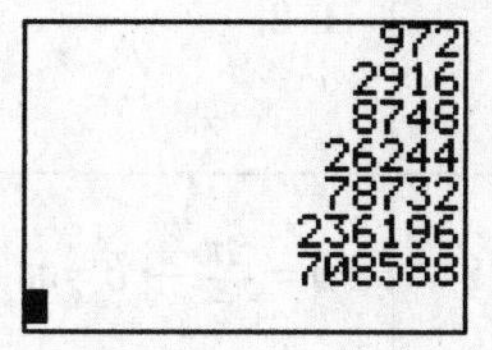

The sequence is geometric. A rule for the nth term is $a_n = 4(3)^{n-1}$, and the 12th term is 708588.

37. Making a sketch of the graph is helpful.

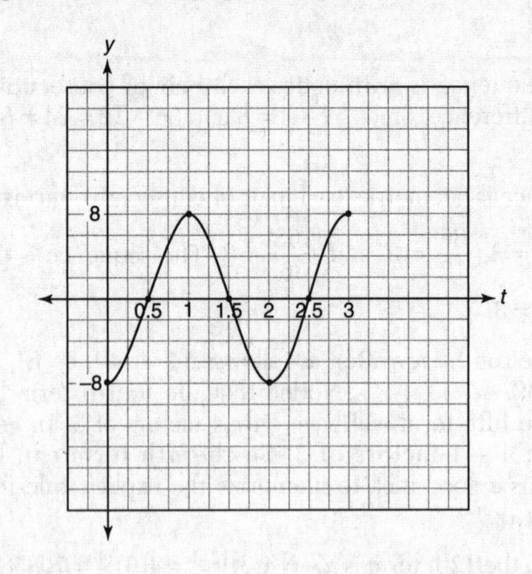

To write the equation, $y = a\sin(b(t + c))$, find the values of a, b, and c.

$\|a\|$ = Amplitude	b = Frequency	c = Phase Shift
The amplitude is half the distance from the minimum value to the maximum value.	$b = \dfrac{2\pi}{\text{period}}$ The period, the length of one complete cycle, is 2.	The phase shift is the t-value at the beginning of the sine cycle. The direction of the shift gives the sign: left is positive; right is negative.
$a = \dfrac{8-(-8)}{2} = \dfrac{16}{2} \rightarrow a = 8$	$b = \dfrac{2\pi}{2} \rightarrow b = \pi$	The sine cycle begins at $t = 0.5$, so the phase shift is right 0.5 units. $c = -0.5$

One equation of the weight's position is $y = 8\sin(\pi(t - 0.5))$.

Note: This problem has multiple correct answers. As an example, $y = -8\sin(\pi(t - 1.5))$ also works.

CALC Check

Be sure you are in radian mode. Change the Window values to match the data in the given table of values.

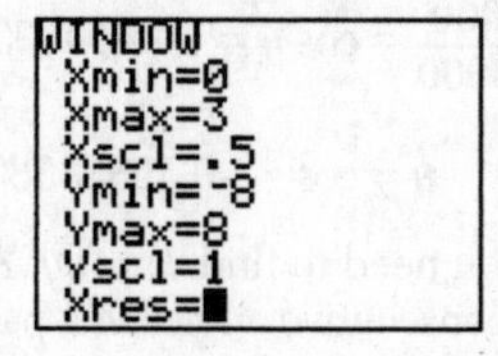

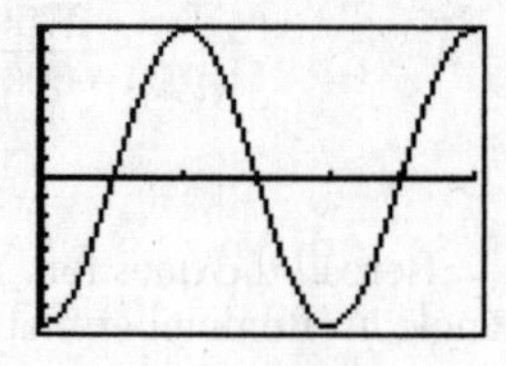

The equation of the weight's position is $y = 8\sin(\pi(t - 0.5))$.

38. First draw a diagram. It is convenient to let one of the tow trucks pull along the horizontal. Your initial diagram should look something like this, where x is the angle to be found.

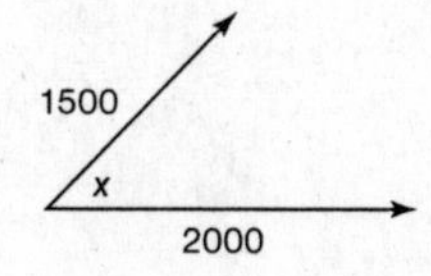

Now use the parallelogram rule to add the vectors and show the resultant force on your diagram.

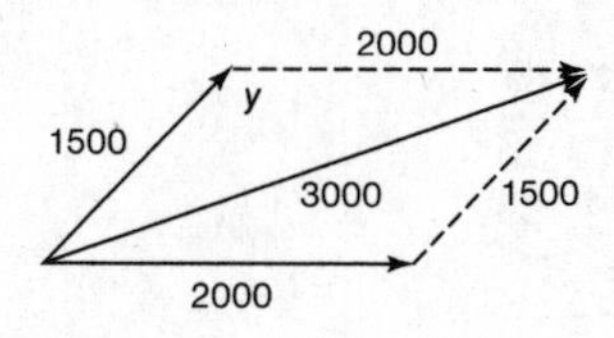

Since angle x is not part of either triangle, find the measure of angle y instead. Since you have all three sides, use the Law of Cosines, shown on the reference sheet. *Be careful solving for y.* Order of operations requires you to subtract 6250000 first and then divide by –6000000.

$$3000^2 = 2000^2 + 1500^2 - 2(2000)(1500)\cos y$$

$$9000000 = 6250000 - 6000000\cos y$$

$$2750000 = -6000000 \cos y$$

$$\frac{2750000}{-6000000} = \cos y = -\,0.454833333$$

$$y = \cos^{-1}(-0.45833333) = 117.2796°$$

Reread the question. You need to find x, not y. Angle x is the supplement of angle y. (Remember that consecutive angles in a parallelogram sum to 180°.)

$$x = 180° - y$$

$$= 180° - 117.2796°$$

$$= 62.7204°$$

To the nearest degree, the angle between the tow trucks measures 63°.

PART IV

39. Enter the data into lists L1 and L2 on your calculator with number of years since 1990 as x in L1 and average salary as y in L2. Use STAT CALC to find both regression equations, rounding the coefficients according to the directions. The linear regression equation is $y = 204.61x + 7.82$. The exponential regression equation is $y = 648.98(1.12)^x$.

The year 2010 corresponds to $x = 2010 - 1990 = 20$. Showing written work is needed to receive full credit. Be sure to show the evaluation by substituting 20 for x and using the rounded coefficients. Using the linear model, the average salary in 2010 is

$$y = 204.61(20) + 7.82 = 4100.02$$

Using the exponential model, the average salary in 2010 is

$$y = 648.98(1.12)^{20} \approx 6260.25$$

The exponential model exceeds the linear prediction by $6260.25 - 4100.02 = 2160.23$ thousand dollars. Rounded to the nearest thousand dollars, this is \$2,160,000.

The linear regression equation is $y = 204.61x + 7.82$. The exponential regression equation is $y = 648.98(1.12)x$. Rounded to the nearest thousand dollars, the exponential model exceeds the linear prediction by \$2,160,000.

Topic	Question Numbers	Number of Points	Your Points	Your Percentage
1. Exponents and Radicals: operations, equivalent expressions, simplifying, rationalizing	12, 26, 32	2 + 2 + 2 = 6		
2. Complex Numbers: operations, powers of i, rationalizing	23	2		
3. Quadratics and Higher-Order Polynomials: operations, factoring, binomial expansion, formula, quadratic equations and inequalities, nature of the roots, sum and product of the roots, quadratic-linear systems, completing the square, higher-order polynomials	4, 15, 16, 31, 34	2 + 2 + 2 + 2 + 2 = 10		
4. Rationals: operations, equations	11, 14	2 + 2 = 4		
5. Absolute Value: equations and inequalities	2	2		
6. Direct and Inverse Variation	9	2		
7. Circles: center-radius equation, completing the square	18	2		
8. Functions: relations, functions, domain, range, one-to-one, onto, inverses, compositions, transformations of functions	7, 19, 22, 29	2 + 2 + 2 + 2 = 8		
9. Exponential and Logarithmic Functions: equations, common bases, logarithm rules, e, ln	5, 6, 24	2 + 2 = 2 = 6		
10. Trigonometric Functions: radian measure, arc length, cofunctions, unit circle, inverses	1, 8, 27	2 + 2 + 2 = 6		
11. Trigonometric Graphs: graphs of basic functions and inverse functions, domain, range, restricted domain, transformations of graphs, amplitude, frequency, period, phase shift	13, 30, 37	2 + 2 + 4 = 8		
12. Trigonometric Identities, Formulas, and Equations	28, 33	2 + 2 = 4		
13. Trigonometry Laws and Applications: area of a triangle, sine and cosine laws, ambiguous cases	35, 38	2 + 4 = 6		
14. Sequences and Series: sigma notation, arithmetic, geometric	10, 20, 36	2 + 2 + 4 = 8		

Topic	Question Numbers	Number of Points	Your Points	Your Percentage
15. Statistics: studies, central tendency, dispersion including standard deviation, normal distribution	3, 21	2 + 2 = 4		
16. Regressions: linear, exponential, logarithmic, power, correlation coefficient	39	6		
17. Probability: permutation, combination, geometric, binomial	17, 25	2 + 2 = 4		

HOW TO CONVERT YOUR RAW SCORE TO YOUR ALGEBRA 2/TRIGONOMETRY REGENTS EXAMINATION SCORE

Below is a sample conversion chart that can be used to determine your final score on Sample Test 1. To estimate your final exam score, locate in the column labeled "Raw Score" the total number of points you scored out of a possible 88 points. Since partial credit is allowed in Parts II, III, and IV of the test, you may need to approximate the credit you would receive for a solution that is not completely correct. Then locate in the adjacent column to the right the scaled score that corresponds to your raw score. The scaled score is your final sample Regents Examination score.

*__Important Note:__ These are inexact estimates, since no official conversion charts for Algebra 2/Trigonometry exist as of press time.

Raw Score	Scaled Score	Raw Score	Scaled Score	Raw Score	Scaled Score
88	**100**	58	**75**	28	**46**
87	**99**	57	**74**	27	**45**
86	**99**	56	**74**	26	**43**
85	**97**	55	**73**	25	**42**
84	**96**	54	**72**	24	**41**
83	**95**	53	**71**	23	**39**
82	**94**	52	**70**	22	**38**
81	**93**	51	**69**	21	**37**
80	**92**	50	**69**	20	**35**
79	**92**	49	**68**	19	**34**
78	**91**	48	**67**	18	**32**
77	**90**	47	**66**	17	**31**
76	**89**	46	**65**	16	**29**
75	**89**	45	**64**	15	**28**
74	**88**	44	**63**	14	**26**
73	**87**	43	**62**	13	**24**
72	**86**	42	**61**	12	**23**
71	**86**	41	**60**	11	**21**
70	**85**	40	**59**	10	**19**
69	**84**	39	**58**	9	**17**
68	**83**	38	**57**	8	**15**
67	**83**	37	**56**	7	**14**
66	**82**	36	**55**	6	**12**
65	**81**	35	**54**	5	**10**
64	**80**	34	**53**	4	**8**
63	**79**	33	**52**	3	**6**
62	**79**	32	**51**	2	**4**
61	**78**	31	**49**	1	**2**
60	**77**	30	**48**	0	**0**
59	**76**	29	**47**		

Examination Sample Test 2

Algebra 2/Trigonometry

REFERENCE SHEET

Area of a Triangle

$$K = \frac{1}{2}ab \sin C$$

Law of Cosines

$$a^2 = b^2 + c^2 - 2bc \cos A$$

Functions of the Sum of Two Angles

$$\sin(A + B) = \sin A \cos B + \cos A \sin B$$

$$\cos(A + B) = \cos A \cos B - \sin A \sin B$$

$$\tan(A + B) = \frac{\tan A + \tan B}{1 - \tan A \tan B}$$

Functions of the Double Angle

$$\sin 2A = 2 \sin A \cos A$$

$$\cos 2A = \cos^2 A - \sin^2 A$$

$$\cos 2A = 2 \cos^2 A - 1$$

$$\cos 2A = 1 - 2 \sin^2 A$$

$$\tan 2A = \frac{2 \tan A}{1 - \tan^2 A}$$

Functions of the Difference of Two Angles

$$\sin(A - B) = \sin A \cos B - \cos A \sin B$$

$$\cos(A - B) = \cos A \cos B + \sin A \sin B$$

$$\tan(A - B) = \frac{\tan A - \tan B}{1 + \tan A \tan B}$$

Functions of the Half Angle

$$\sin\frac{1}{2}A = \pm\sqrt{\frac{1-\cos A}{2}}$$

$$\cos\frac{1}{2}A = \pm\sqrt{\frac{1+\cos A}{2}}$$

$$\tan\frac{1}{2}A = \pm\sqrt{\frac{1-\cos A}{1+\cos A}}$$

Law of Sines

$$\frac{a}{\sin A} = \frac{b}{\sin B} = \frac{c}{\sin C}$$

Sum of a Finite Arithmetic Sequence

$$S_n = \frac{n(a_1 + a_n)}{2}$$

Sum of a Finite Geometric Sequence

$$S_n = \frac{a_1(1-r^n)}{1-r}$$

Binomial Theorem

$$(a+b)^n = {}_nC_0 a^n b^0 + {}_nC_1 a^{n-1} b^1 + {}_nC_2 a^{n-2} b^2 + \cdots + {}_nC_n a^0 b^n$$

$$(a+b)^n = \sum_{r=0}^{n} {}_nC_r a^{n-r} b^r$$

Normal Curve
Standard Deviation

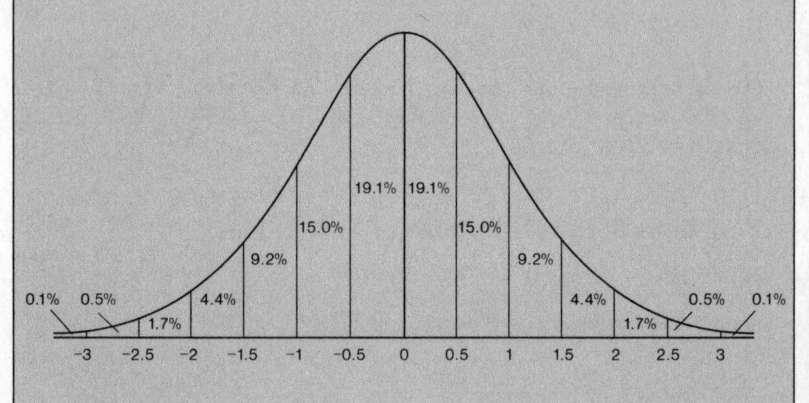

PART I

Answer all 27 questions in this part. Each correct answer will receive 2 credits. No partial credit will be allowed. For each question, write in the space provided the numeral preceding the word or expression that best completes the statement or answers the question. [54 credits]

1 What is 235°, expressed in radian measure?

(1) 235π (3) $\frac{36\pi}{47}$

(2) $\frac{\pi}{235}$ (4) $\frac{47\pi}{36}$ 1 ____

2 According to Boyle's Law, the pressure, p, of a compressed gas is inversely proportional to the volume, v. If a pressure of 20 pounds per square inch exists when the volume of the gas is 500 cubic inches, what is the pressure when the gas is compressed to 400 cubic inches?

(1) 16 lb/in^2 (3) 40 lb/in^2
(2) 25 lb/in^2 (4) 50 lb/in^2 2 ____

3 Which graph represents data used in a linear regression that produces a correlation coefficient closest to –1?

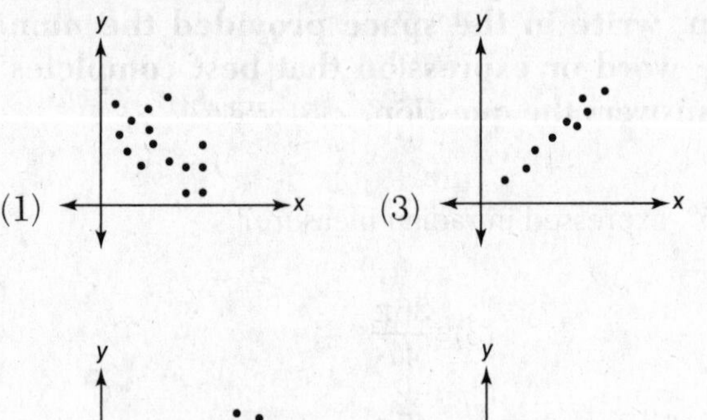

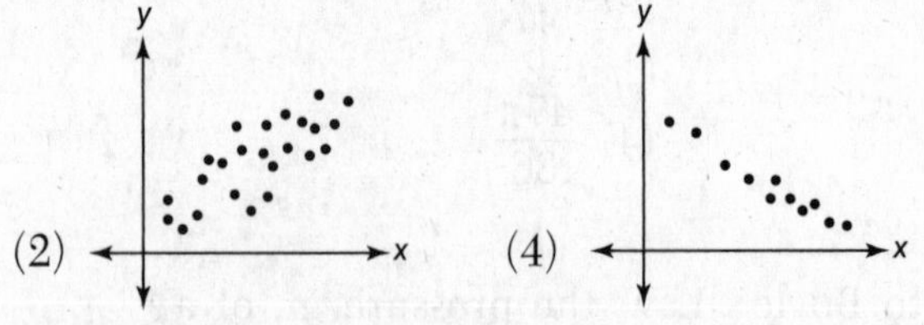

3 _____

4 The local school district is doing a survey to determine the support among the taxpayers for a new building project. Which of the following sampling methods would give the best estimate of support for the project?

(1) Put the survey in the local newspaper.
(2) Send the survey home with all the students who attend the school.
(3) Survey all taxpayers having a 3 as the last digit of their taxpayer ID number.
(4) Survey every fifth person who comes to the school concert.

4 _____

5 Which of the following functions is both one-to-one and onto?

(1) 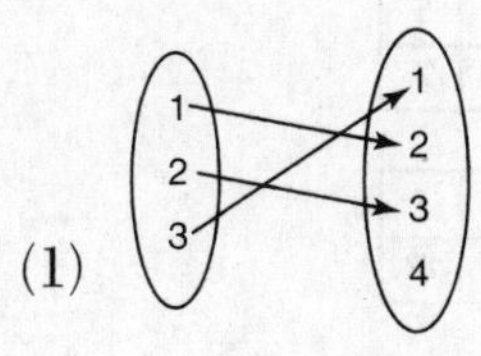

(3) 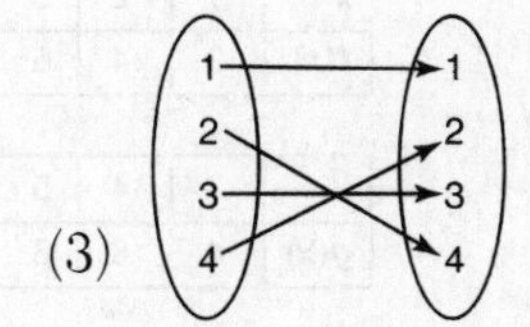

(2) 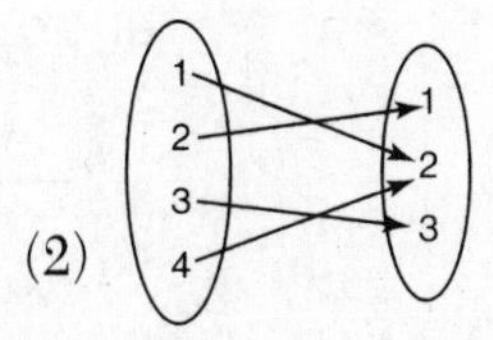

(4) 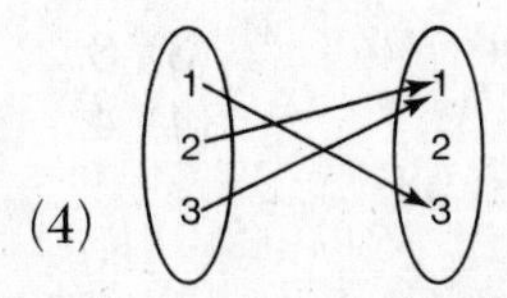

5 ____

6 The expression $\dfrac{4}{5-\sqrt{13}}$ is equivalent to

(1) $\dfrac{5+\sqrt{13}}{3}$ (3) $\dfrac{2(5+\sqrt{13})}{19}$

(2) $\dfrac{5-\sqrt{13}}{3}$ (4) $\dfrac{2(5-\sqrt{13})}{19}$

6 ____

7 Mrs. Donahue made up a game to help her class learn about imaginary numbers. The winner will be the student whose expression is equivalent to $-i$. Which expression will win the game?

(1) i^{46} (3) i^{48}

(2) i^{47} (4) i^{49}

7 ____

8 The accompanying tables define functions $f(x)$ and $g(x)$.

x	1	2	3	4	5
$f(x)$	3	4	5	6	7

x	3	4	5	6	7
$g(x)$	4	6	8	10	12

What is $(g \circ f)(3)$?

(1) 6
(2) 2
(3) 8
(4) 4

8 ______

9 What is the third term in the expansion of $(\cos x + 3)^5$?

(1) $90 \cos^2 x$
(2) $270 \cos^2 x$
(3) $60 \cos^3 x$
(4) $90 \cos^3 x$

9 ______

10 What is the common ratio in the geometric sequence $\left\{18, -6, 2, -\frac{2}{3}\right\}$?

(1) 3
(2) -3
(3) $\frac{1}{3}$
(4) $-\frac{1}{3}$

10 ______

11 The solution set of $|3x + 2| < 1$ contains

(1) only negative real numbers
(2) only positive real numbers
(3) both positive and negative real numbers
(4) no real numbers

11 ______

12 Which expression is equivalent to $\left(\sqrt{a^2 b^{\frac{1}{2}}}\right)^{-1}$, where a and b are both positive values?

(1) $a^{-2}b^{-\frac{1}{2}}$
(2) $-ab^{\frac{1}{4}}$
(3) $-ab^2$
(4) $\dfrac{1}{ab^{\frac{1}{4}}}$

12 ______

13 An architect commissions a contractor to produce a triangular window. The architect describes the window as $\triangle ABC$, where $m\angle A = 50°$, $AB = 12$ inches, and $BC = 10$ inches. How many distinct triangles can the contractor construct using these dimensions?

(1) 1
(2) 2
(3) more than 2
(4) 0

13 ______

14 Which of the following regressions best models the data shown in the scatter plot?

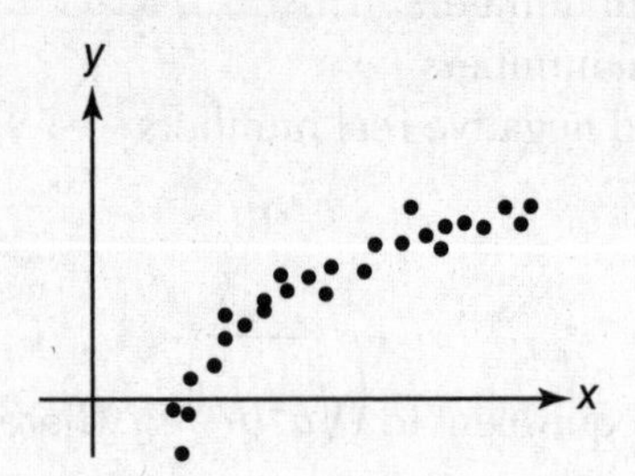

(1) logarithmic (3) linear
(2) exponential (4) power 14 _____

15 The graph of $f(x) = ax^4 - bx - 2$ is shown below.

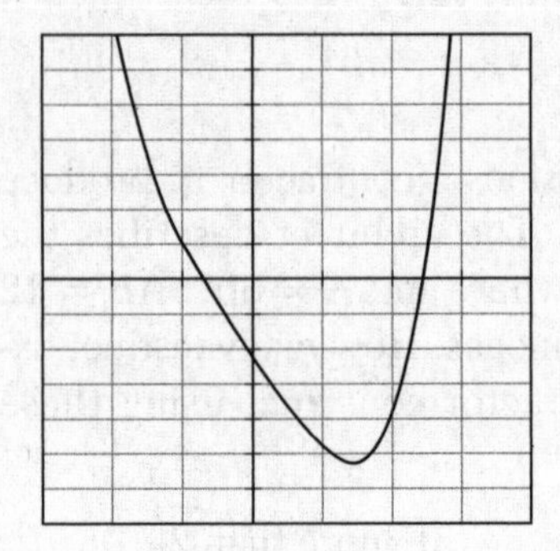

Estimate all the roots to the equation $ax^4 - bx - 2 = 0$.

(1) {–0.6, 2.5} (3) {1.5}
(2) {–2} (4) {–2, –0.6, 2.5} 15 _____

16 The roots of the equation $5x^2 + 1 = 4x$ are

(1) $2 \pm 3i$
(2) $2 \pm i$
(3) $\frac{2}{5} \pm \frac{3}{5}i$
(4) $\frac{2}{5} \pm \frac{1}{5}i$

16 _____

17 What is the product of $5 + \sqrt{-36}$ and $1 - \sqrt{-49}$, expressed in simplest $a + bi$ form?

(1) $-37 + 41i$
(2) $5 - 71i$
(3) $47 + 41i$
(4) $47 - 29i$

17 _____

18 In the equation $\log_x 4 + \log_x 9 = 2$, x is equal to

(1) $\sqrt{13}$
(2) 6
(3) 6.5
(4) 18

18 _____

19 The graph of a function is shown below.

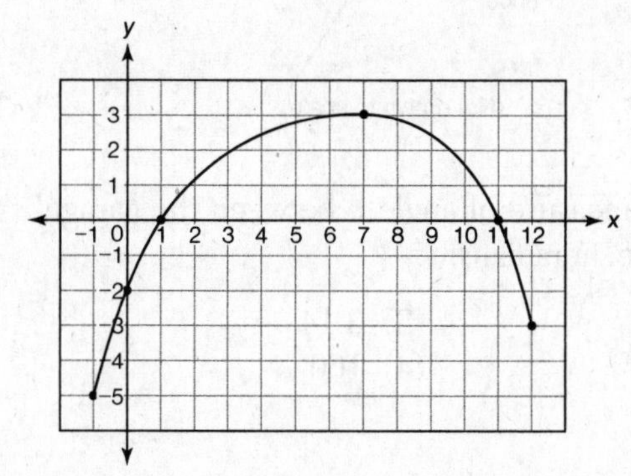

Which of the following is the domain of the function?

(1) [1, 11]
(2) [−1, 12]
(3) [−5, 3]
(4) [−3, 3]

19 _____

20 If $\tan A = \frac{-24}{7}$ and A is in Quadrant II, find the value of $\sin\left(\frac{1}{2}A\right)$.

(1) $\frac{3}{5}$ (3) $\frac{4}{5}$

(2) $-\frac{3}{5}$ (4) $-\frac{4}{5}$ 20 ______

21 A dog has a 20-foot leash attached to the corner where a garage and a fence meet, as shown in the accompanying diagram. When the dog pulls the leash tight and walks from the fence to the garage, the arc the leash makes is 55.8 feet.

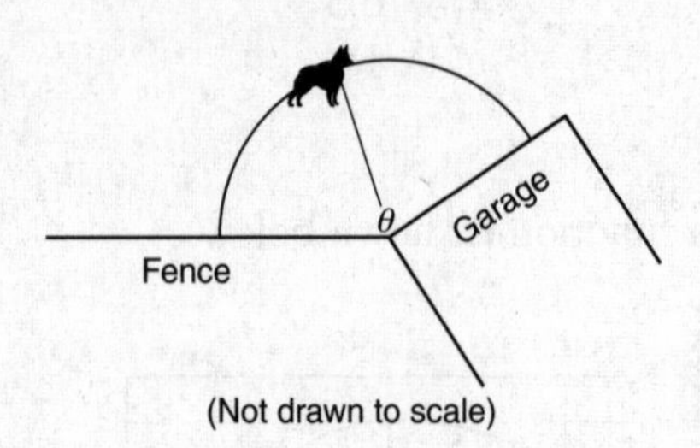

(Not drawn to scale)

What is the measure of angle θ between the garage and the fence, in radians?

(1) 0.36 (3) 3.14
(2) 2.79 (4) 160 21 ______

22 If θ is an angle in standard position and $P(-3,4)$ is a point on the terminal side of θ, what is the value of $\sin \theta$?

(1) $\frac{3}{5}$ (3) $\frac{4}{5}$

(2) $-\frac{3}{5}$ (4) $-\frac{4}{5}$ 22 ______

23 If $\sin(6A + B)^\circ = \cos(9A - B)^\circ$, then the value of A could be equal to

(1) 6 (3) 54
(2) 36 (4) 1.5 23 ______

24 A function $y = f(x)$ is shown in the graph below.

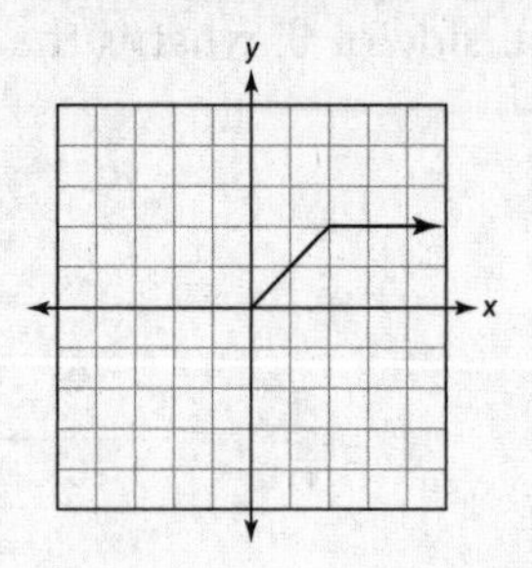

Which graph below is the graph of $y = f(-x)$?

(1) 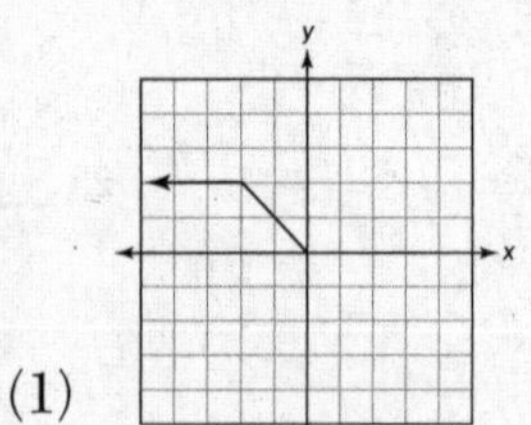

(3) 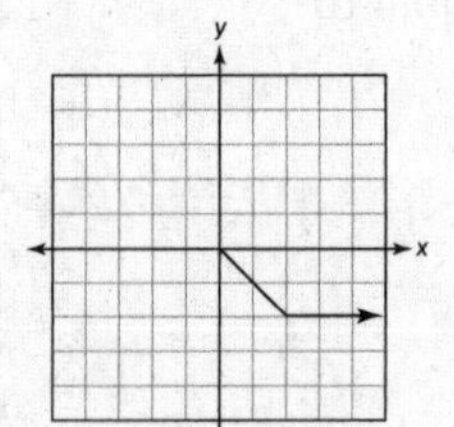

(2) 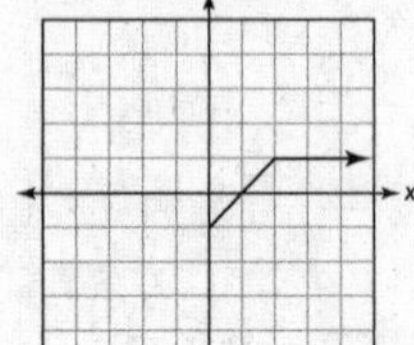

(4) 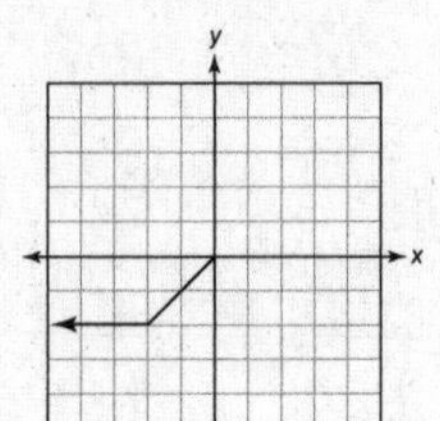

24 ______

25 Which of the following shows the graph of $y = \sec x$?

(1) 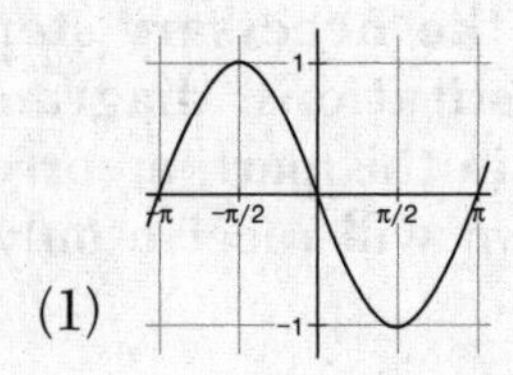

(3) 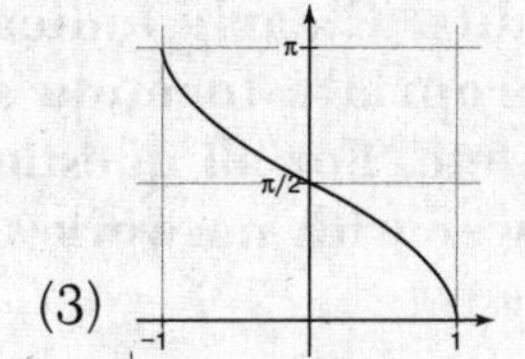

(2) 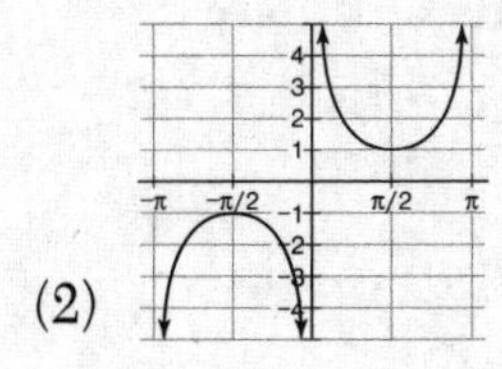

(4) 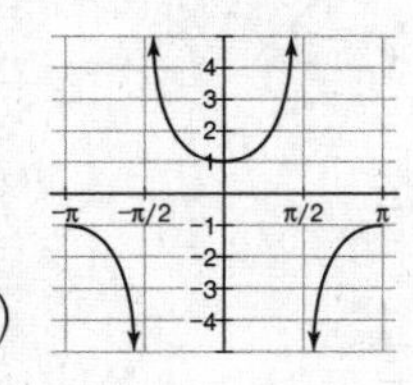

25 _____

26 If a function is defined by the equation $y = 3x + 2$, which equation defines the inverse of this function?

(1) $x = \frac{1}{3}y + \frac{1}{2}$ (3) $y = \frac{1}{3}x - \frac{2}{3}$

(2) $y = \frac{1}{3}x + \frac{1}{2}$ (4) $y = -3x - 2$

26 _____

27 The mean of a normally distributed set of data is 56, and the standard deviation is 5. In which interval do approximately 95.4% of all cases lie?

(1) 46–56 (3) 51–61
(2) 46–66 (4) 56–71

27 _____

PART II

Answer all 8 questions in this part. Each correct answer will receive 2 credits. Clearly indicate the necessary steps, including appropriate formula substitutions, diagrams, graphs, charts, etc. For all questions in this part, a correct numerical answer with no work shown will receive only 1 credit. [16 credits]

28 Solve for x to the *nearest thousandth*: $e^{0.5x} = 8$

29 Write the equation of the circle $x^2 + y^2 - 8x + 7 = 0$ in center-radius form.

30 Gregory wants to build a garden in the shape of an obtuse isosceles triangle with one of the congruent sides equal to 12 yards. If the area of his garden will be 55 square yards, find, to the *nearest tenth of a degree*, the measure of the vertex angle of the triangle.

31 Ellyssa has 6 mystery books and 5 science fiction books on her shelf. She takes down 4 books at random. What is the probability she has 3 mysteries and 1 science fiction book?

32 Write the first four terms of the sequence defined recursively by

$$a_1 = 4 \text{ and } a_{n+1} = 2a_n - n.$$

33 Mr. and Mrs. Doran have a genetic history such that the probability that a child being born to them with a certain trait is $\frac{1}{8}$. If they have four children, what is the probability that exactly three of their four children will *not* have that trait?

34 The distance traveled by the weight on the end of a pendulum follows a geometric sequence. The weight traveled 20 centimeters on its first swing, 18 centimeters on its second swing, and 16.2 centimeters on its third swing. To the nearest *tenth of a centimeter*, what **total** distance had the weight traveled after 25 swings?

35 Conant High School has 17 students on its championship bowling team. Each student bowled one game. The scores are listed in the accompanying table.

Score (x_i)	Frequency (f_i)
140	4
145	3
150	2
160	3
170	2
180	2
194	1

Find, to the *nearest tenth*, the population standard deviation of these scores.

Conant's rival school, Springfield Central, has a mean of 157 and a population standard deviation of 5. How do their scores compare to Conant's?

PART III

Answer all 3 questions in this part. Each correct answer will receive 4 credits. Clearly indicate the necessary steps, including appropriate formula substitutions, diagrams, graphs, charts, etc. For all questions in this part, a correct numerical answer with no work shown will receive only 1 credit. [12 credits]

36 Sketch the graph of the function $f(x) = e^x$. Sketch the graph of $f(x)$ reflected over the line $y = x$, obtaining a new function, $g(x)$. Write the equation of $g(x)$ and its domain.

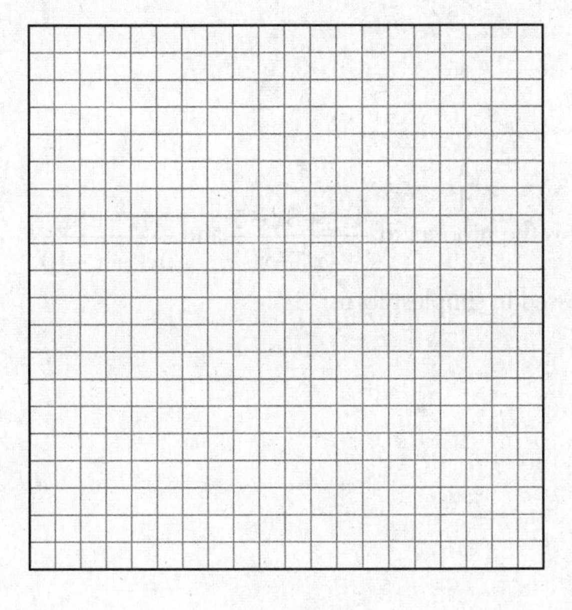

37 Express the solution in interval notation:

$$\frac{3x}{2x-6}-\frac{2x}{x-3}\leq 1$$

38 What is the product of $\frac{4x^2+8x-5}{4x^3-x}$ and $\frac{4x^2+8x}{2x^2+x-10}$ expressed in simplest form?

PART IV

Answer the question in this part. The correct answer will receive 6 credits. Clearly indicate the necessary steps, including appropriate formula substitutions, diagrams, graphs, charts, etc. For the question in this part, a correct numerical answer with no work shown will receive only 1 credit. [6 credits]

39 Find all values of θ in the interval $0° < \theta < 360°$ that satisfy the equation $3 \cos 2\theta + 2 \sin \theta + 1 = 0$. Round all answers to the nearest *hundredth of a degree.*

[Only an algebraic solution can receive full credit.]

Answers Sample Test 2

Algebra 2/Trigonometry

Answer Key

PART I

1. (4)	**6.** (1)	**11.** (1)	**16.** (4)	**21.** (2)	**26.** (3)
2. (2)	**7.** (2)	**12.** (4)	**17.** (4)	**22.** (3)	**27.** (2)
3. (4)	**8.** (3)	**13.** (2)	**18.** (2)	**23.** (1)	
4. (3)	**9.** (4)	**14.** (1)	**19.** (2)	**24.** (1)	
5. (3)	**10.** (4)	**15.** (1)	**20.** (3)	**25.** (4)	

PART II

28. 4.159

29. $(x-4)^2+y^2=9$

30. 130.2°

31. $\frac{100}{330}$ or $\frac{10}{33}$

32. {4, 7, 12, 21}

33. $\frac{343}{1024}$

34. 185.6 centimeters

35. $\sigma = 16.2$ *See the accompanying detailed solution.*

PART III

36. $g(x) = \ln x$; domain: $x > 0$. *See the accompanying detailed solution.*

37. $(-\infty, 2] \cup (3, \infty)$

38. $\frac{4(x+2)}{(2x+1)(x-2)}$

PART IV

39. {90°, 221.81°, 318.19°}

Answers Explained

PART I

1. To convert degrees to radians, multiply by $\frac{\pi}{180^\circ}$:

$$235^\circ \cdot \frac{\pi}{180^\circ} = \frac{235\pi}{180} = \frac{47\pi}{36}$$

CALC Check

Set the calculator to the mode you want for the answer. Put the symbol for the original unit (found in the Angle menu) after the angle and select Enter. Compare to the decimal form of the choices.

```
235°
            4.101523742
47π/36
            4.101523742
```

The correct choice is **(4)**.

2. Inverse variation means p and v always have the same product:

$$p_1v_1 = p_2v_2$$

$$(20)(500) = (p)(400)$$

$$1000 = 400p$$

$$p = 25$$

The correct choice is **(2)**.

3. A correlation of –1 indicates a perfect linear negative correlation; the points form a perfect straight line with a negative slope. Of the choices shown, (1) and (4) have negative slopes, but choice (4) is much closer to being a perfect straight line.

The correct choice is **(4)**.

4. The district wants a representative sample of the taxpayers for the survey.

- Choice (1): Not all the taxpayers may read the local paper. In addition, the district will receive responses only from people who take the time to fill out and return the survey. Often, these will only be the people with very strong opinions on the topic. This is probably *not* a good way to get a representative sample.
- Choice (2): This will sample only taxpayers who have children in the school. Their opinions may not be the same as people who do not have children in the school but still pay school taxes. This is *not* a good way to get a representative sample.
- Choice (3): Assuming that the last digit of the taxpayer's ID numbers is random, this should be a good way to get a representative sample.
- Choice (4): Selecting every fifth person should provide a random sample of people who attend the school concerts. However, those people may not be representative of all the taxpayers in the district. This is *not* a good way to get a representative sample.

The correct choice is **(3)**.

5. For a function to be one-to-one, each output must be associated with only one input. In choice (2), output 2 comes from both inputs 1 and 4. In choice (4), output 1 comes from both inputs 2 or 3. This eliminates choices (2) and (4).

To be onto, the function must take on all the values in its set of possible outputs. In choice (1), no input is associated with output 4, so this function is not onto. In choice (3), every possible output has an input associated with it. This function is both one-to-one and onto.

The correct choice is **(3)**.

6. To rationalize a binomial radical denominator, multiply by the conjugate:

$$\frac{4}{5-\sqrt{13}}=\frac{4}{\left(5-\sqrt{13}\right)}\cdot\frac{\left(5+\sqrt{13}\right)}{\left(5+\sqrt{13}\right)}=\frac{4\left(5+\sqrt{13}\right)}{25-13}=\frac{4\left(5+\sqrt{13}\right)}{12}=\frac{5+\sqrt{13}}{3}$$

CALC Check

```
4/(5-√(13))
         2.868517092
(5+√(13))/3
         2.868517092
```

The correct choice is **(1)**.

7. The first four powers of i are $i^0 = 1$, $i^1 = i$, $i^2 = -1$, and $i^3 = -i$. Successive powers just repeat the pattern: 1, i, -1, $-i$. For any whole number power of i, you can find the result by raising i to the remainder obtained when the exponent is divided by 4.

- Choice (1): $\frac{46}{4} = 11$ remainder 2, so $i^{46} = i^2 = -1$.
- Choice (2): $\frac{47}{4} = 11$ remainder 3, so $i^{47} = i^3 = -i$, which is the desired answer.
- Choice (3): $\frac{48}{4} = 12$ remainder 0, so $i^{48} = i^0 = 1$.
- Choice (4): $\frac{49}{4} = 12$ remainder 1, so $i^{49} = i^1 = i$.

CALC Caution

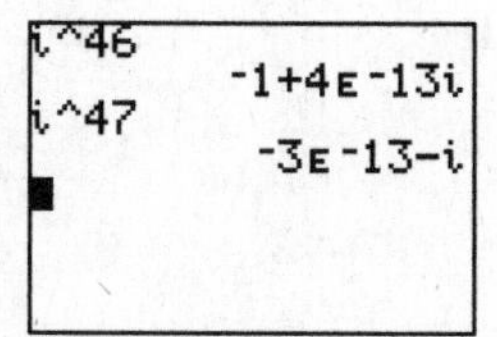

Remember, ⁻3E⁻13 is calculator notation for -3×10^{-13}, which is very close to 0. Interpret ⁻3E⁻13 as 0 to get the correct value of $0 - i$ or $-i$.

The correct choice is **(2)**.

8. The notation $(g \circ f)(3)$ means the same as $g(f(3))$. First evaluate $f(3)$, and then evaluate g at that answer:

$$f(3) = 5 \text{ and } g(5) = 8$$

Note: If you do the composition in the wrong order, you get $g(3) = 4$ and $f(4) = 6$, which is not correct.

The correct choice is **(3)**.

9. You can look up the formula for expanding the power of a binomial on the reference sheet: $(a+b)^n = \sum_{r=0}^{n} {}_nC_r a^{n-r} b^r$. Note that the summation starts at $r = 0$. This means the first term will have $r = 0$ (not 1), the second term will have $r = 1$, and the third term will have $r = 2$. In the given problem, $n = 5$, $a = \cos x$, and $b = 3$. Thus the third term of the expansion is

$${}_5C_2(\cos x)^{5-2}(3)^2 = 10\cos^3 x(9) = 90\cos^3 x$$

The correct choice is **(4)**.

10. The common ratio of a geometric series is the ratio of any two consecutive terms: $\frac{a_{n+1}}{a_n}$. In other words, divide the second term by the first term. Using the first two terms in this problem, the ratio for the given series is $\frac{-6}{18} = -\frac{1}{3}$. You can check this by dividing the next pair of terms: $\frac{2}{-6} = -\frac{1}{3}$, which is the same ratio. Note that if you invert the ratio (write it upside down) as $\frac{a_n}{a_{n+1}}$, you will get an incorrect answer (in this case -3 instead of $-\frac{1}{3}$). Be careful to divide the numbers in the correct order.

The correct choice is **(4)**.

11. Solve absolute value inequalities in two steps. First, solve for equality.

$$|3x+2|=1$$

Eliminate the absolute value by writing two separate equations.

$$3x+2=1 \quad \text{or} \quad -(3x+2)=1$$
$$-3x-2=1$$

Solve both equations.

$$3x=-1 \quad \text{or} \quad -3x=3$$

Solving absolute value equations can lead to extraneous roots so you *must* check your solutions in the original equation.

$$x=-\frac{1}{3} \quad \text{or} \quad x=-1$$

$$\text{Check } \left|3\left(-\frac{1}{3}\right)+2\right| \stackrel{?}{=} 1$$

$|1|=1$ ✔

or

$|3(-1)+2| \stackrel{?}{=} 1$

$|-1|=1$ ✔

Now solve the inequality. Plot the solutions on a number line. In this case, use open circles because of the strict inequality. Check a number from each *interval* in the original inequality. Shade the number line if a value in the interval checks.

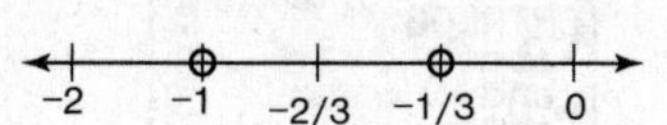

Check $x=-2$: $|3(-2)+2| \stackrel{?}{<} 1$

$|-4|<1$ No

Check $x=-\frac{2}{3}$: $\left|3\left(-\frac{2}{3}\right)+2\right| \stackrel{?}{<} 1$

$|0|<1$ Yes

Check $x=0$: $|3(0)+2| \stackrel{?}{<} 1$

$|2|<1$ No

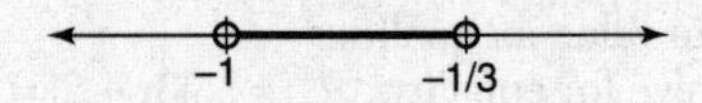

Read the solution from the number line.

The solution is $-1 < x < -\frac{1}{3}$. These values are all negative.

CALC Check

This problem is most easily solved graphically. Use ZoomDecimal. You want those x-values for which the absolute value graph, the V shape, is less than (below) the line $y = 1$. This only happens for negative values of x.

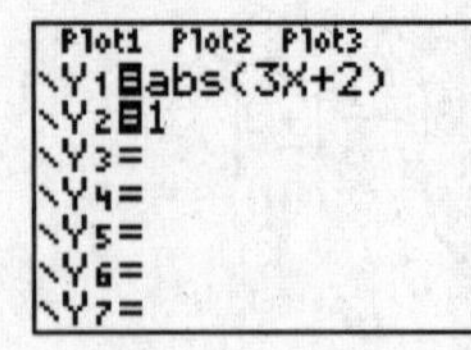

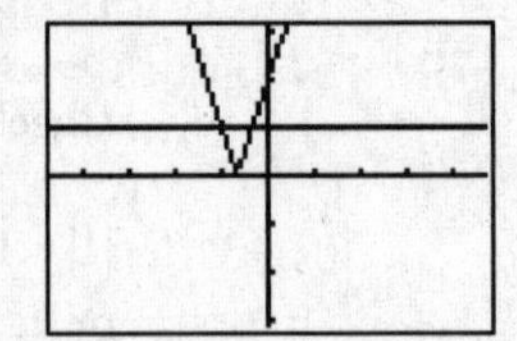

CALC Tip

In your calculator, Catalog, which you get by pressing 2nd 0, lists all calculator functions in alphabetical order. This is the easiest place to find abs(and is very handy if you have forgotten where to find a function.

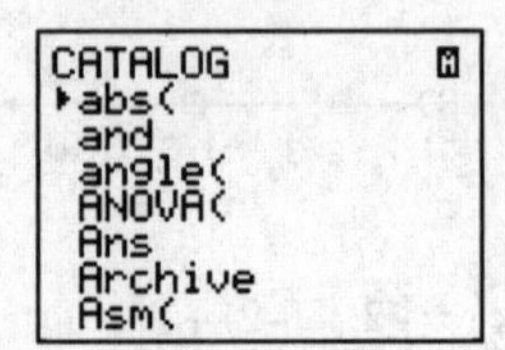

The correct choice is (**1**).

12. Work inside the parentheses first, and convert the radical to exponent notation.

$$\left(\sqrt{a^2b^{\frac{1}{2}}}\right)^{-1} = \left(\left(a^2b^{\frac{1}{2}}\right)^{\frac{1}{2}}\right)^{-1}$$

To simplify a power raised to a power, multiply the exponents.

$$\left(a^{2\left(\frac{1}{2}\right)}b^{\frac{1}{2}\left(\frac{1}{2}\right)}\right)^{-1} = \left(a^1b^{\frac{1}{4}}\right)^{-1}$$

Finally, a negative exponent means take the reciprocal of the base.

$$= \frac{1}{ab^{\frac{1}{4}}}$$

The correct choice is **(4)**.

13. Draw a rough diagram; do not worry about scale or accuracy for this.

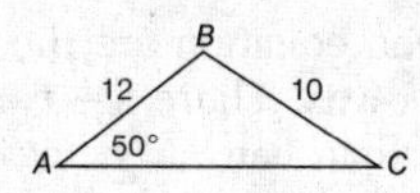

This is the Ambiguous Case Triangle problem. Use the Law of Sines found on the reference sheet to find possible values for m$\angle C$:

$$\frac{10}{\sin 50°} = \frac{12}{\sin C}$$

$$10 \sin C = 12 \sin 50°$$

$$\sin C = \frac{12 \sin 50°}{10} = 0.9192533317$$

Sine is positive in Quadrants I and II, so there are two possible angles. Use $\sin^{-1}$ on your calculator to find one angle; the second is 180° minus that answer:

$$\text{Acute angle } C\text{: m}\angle C = \sin^{-1}(0.9192533317) \approx 66.82°$$

$$\text{Obtuse angle } C\text{: m}\angle C = 180° - 66.82° = 113.18°$$

Finally, find m$\angle B$ for each of the values found for m$\angle C$. If m$\angle C$ = 66.82°, then m$\angle B$ = 180° – 50° – 66.82° = 63.18°. One triangle can be constructed. If m$\angle C$ = 113.18°, then m$\angle B$ = 180° – 50° – 113.18° = 16.82°. A second triangle can be constructed.

The correct choice is **(2)**.

14. The logarithmic regression is the best choice. It has the equation $y = a + b \ln x$ and the graph shown.

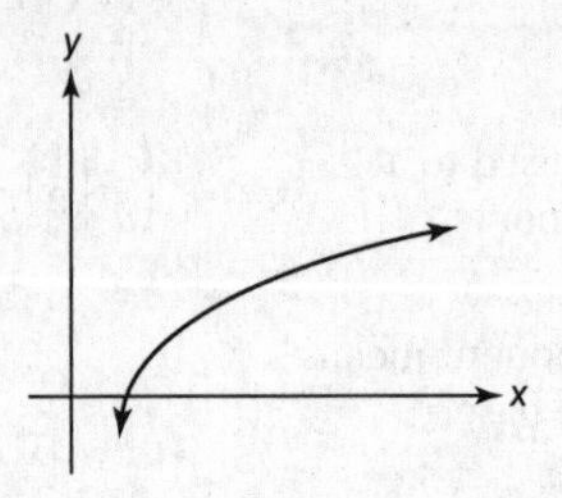

Review Section 15 of the Facts and Skills chapter to see graphs of the other types of regression models.

The correct choice is **(1)**.

15. The roots of the given equation are the x-values where the graph of the function intersects the x-axis. There are two roots. The one on the left appears to be slightly more than halfway between 0 and –1, say about –0.6. The one on the right appears to be very close to halfway between 2 and 3, say 2.5. Even if you do not agree exactly on the decimals, only one solution is even close. Do not include the y-intercept; it is *not* a root.

The correct choice is **(1)**.

16. Rewrite the equation in standard form, $5x^2 - 4x + 1 = 0$. Since all the choices given are imaginary, do not waste time trying to factor the equation. By using the quadratic formula, with $a = 5$, $b = -4$, and $c = 1$, you get:

$$x = \frac{-b \pm \sqrt{b^2 - 4ac}}{2a}$$

$$= \frac{-(-4) \pm \sqrt{(-4)^2 - 4(5)(1)}}{2(5)}$$

$$= \frac{4 \pm \sqrt{16 - 20}}{2(5)}$$

$$= \frac{4 \pm \sqrt{-4}}{10}$$

$$= \frac{4 \pm 2i}{10}$$

$$= \frac{4}{10} \pm \frac{2}{10}i$$

$$= \frac{2}{5} \pm \frac{1}{5}i$$

CALC Check

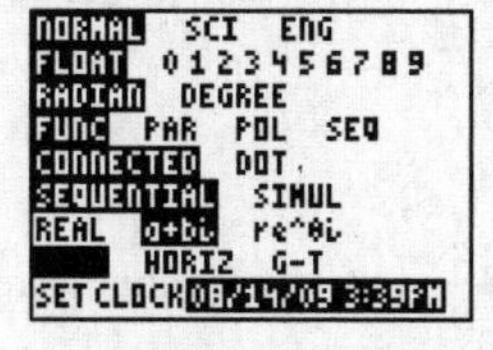

2/5+i/5→A
.4+.2i
5A²−4A+1
0

The correct choice is **(4)**.

17. Remember to convert radicals of negative numbers to $a + bi$ form before doing any arithmetic. Use the distributive property to multiply binomials:

$$(5+\sqrt{-36})(1-\sqrt{-49}) = (5+6i)(1-7i)$$
$$= 5 - 35i + 6i - 42i^2$$
$$= 5 - 29i - 42(-1)$$
$$= 47 - 29i$$

CALC Check

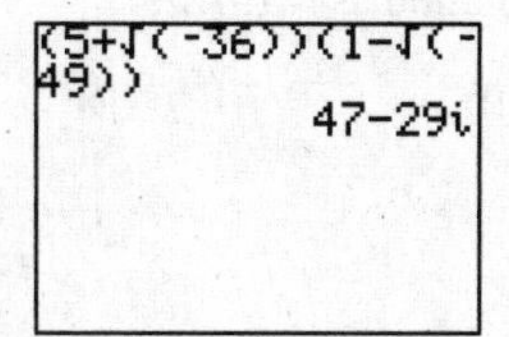

The correct choice is **(4)**.

18. In any base, the sum of two logs is the log of the product.

$$\log_x 4 + \log_x 9 = 2$$

$$\log_x(4 \cdot 9) = 2$$

$$\log_x 36 = 2$$

Now rewrite the log in exponential form and solve.

$$x^2 = 36$$

Reject the negative value because bases of logarithms are always positive.

$$x = 6 \text{ or } \cancel{x = -6}$$

CALC Check

With the change of base formula, $\log_b a = \dfrac{\log a}{\log b}$, you can evaluate and graph any logarithmic function.

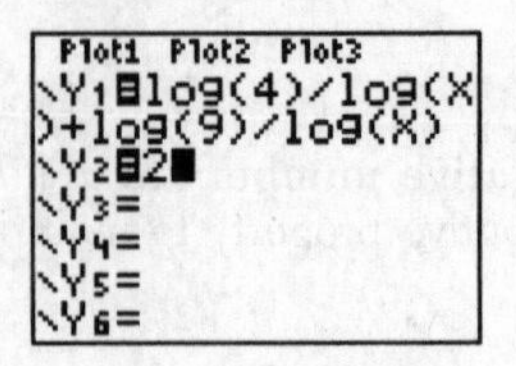

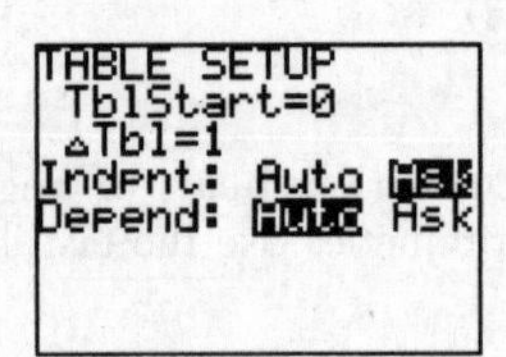

X	Y1	Y2
3.6056	2.7942	2
6	2	2
6.5	1.9145	2
10	1.2398	2

X=

The correct choice is **(2)**.

19. The domain of a function is the set of all possible inputs, x-values in this case, that the function can have. When looking at the graph, you see the smallest x-value is the leftmost point on the graph, where $x = -1$. You also see that the largest x-value is the rightmost point, where $x = 12$. The graph hits all x-values between those two numbers, so the domain is $-1 \le x \le 12$. This is represented in interval notation as $[-1, 12]$.

The correct choice is **(2)**.

20. One way to do this problem is with the half-angle formula for sine shown on the reference sheet, $\sin\frac{1}{2}A = \pm\sqrt{\frac{1-\cos A}{2}}$. This formula requires that you know cos *A*, but the problem gives tan *A*. Draw a diagram with *A* in Quadrant II. Label the opposite and adjacent sides so that $\tan A = -\frac{24}{7}$. The hypotenuse can be found using the Pythagorean Theorem.

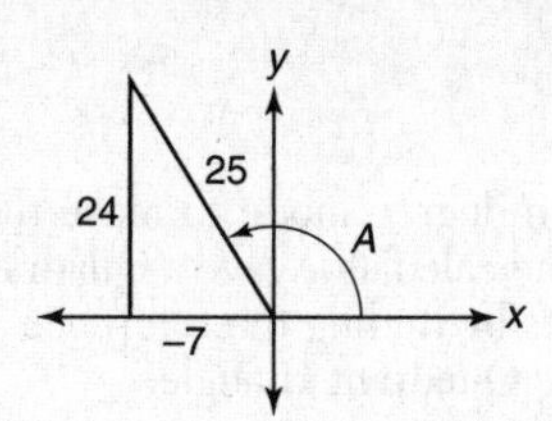

Now cos *A* can be read off of the diagram, $\cos A = -\frac{7}{25}$. Use the half-angle formula:

$$\sin\left(\frac{1}{2}A\right) = \pm\sqrt{\frac{1-\left(-\frac{7}{25}\right)}{2}}$$

$$= \pm\sqrt{\frac{\frac{32}{35}}{2}}$$

$$= \pm\sqrt{\frac{16}{25}}$$

$$= \pm\frac{4}{5}$$

Since *A* is in Quadrant II, $90° < A < 180°$, $\frac{1}{2}A$ will be in the interval $45° < \frac{1}{2}A < 90°$, which is in Quadrant I. In Quadrant I, sine is positive, so $\sin\left(\frac{1}{2}A\right) = \frac{4}{5}$.

CALC Caution

```
tan⁻¹(-24/7)
     -73.73979529
sin(.5Ans)
               -.6
```

CALC Check

The calculator is set in degree mode to make the quadrants of the angles easier to recognize. The calculator gave a Quadrant IV angle for A, but A is in Quadrant II. Fix this by finding the reference angle first and subtracting it from 180° to find the Quadrant II angle.

```
tan⁻¹(24/7)
      73.73979529
180-Ans
      106.2602047
sin(.5Ans)
                .8
```

The correct choice is **(3)**.

21. You should know the formula for arc length on a circle, $s = \theta r$, where s is the arc length, r is the radius of the circle, and θ is the central angle in radians. In the problem, you know $r = 20$ feet and $s = 55.8$ feet:

$$55.8 = 20\theta$$

$$\theta = \frac{55.8}{20}$$

$$= 2.79$$

Tip: If you cannot remember this formula but you know the formula for circumference of a circle, $C = 2\pi r$, set up a proportion. Let s be a fraction of the circumference and θ be a fraction of the rotation (a full circle has a rotation of 2π), $\frac{C}{2\pi} = \frac{s}{\theta}$.

The correct choice is **(2)**.

22. Using a diagram is helpful:

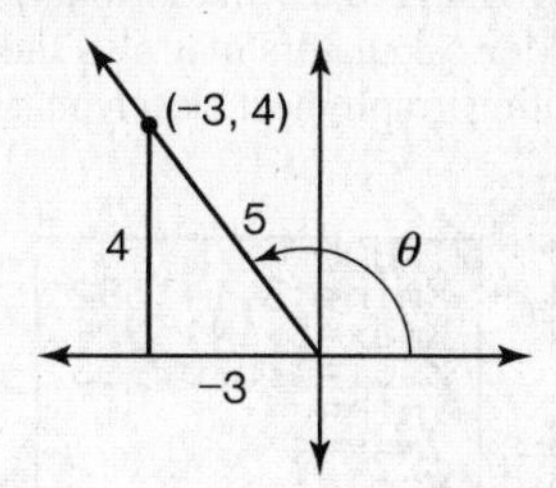

When a point $P(x, y)$ is on the terminal side of an angle in standard position, use the definition of the sine of the angle, $\sin\theta = \frac{y}{r}$ where $r = \sqrt{x^2 + y^2}$ by the Pythagorean Theorem. In this case, $r = \sqrt{(-3)^2 + 4^2} = 5$, so $\sin\theta = \frac{4}{5}$.

The correct choice is **(3)**.

23. Cofunctions, such as sine and cosine, of two angles will be equal when the angles are complementary (their sum equals 90°).

$$(6A + B) + (9A - B) = 90$$
$$15A = 90$$
$$A = 6$$

The correct choice is **(1)**.

24. The graph of $y = f(-x)$ is the reflection of the graph of $y = f(x)$ over the y-axis. This is choice (1).

WRONG CHOICES EXPLAINED:

(2) This is a vertical translation, $y = f(x) - 1$.

(3) This is a reflection over the x-axis, $y = -f(x)$.

(4) This is a reflection in the origin, $y = -f(-x)$.

The correct choice is **(1)**.

25. Use your graphing calculator; remember that secant is the reciprocal of cosine. Be sure you are in radian mode. Match your window to the scales shown in the answer choices. (Older calculators may also show vertical lines where the asymptotes are.) The resulting graph most closely matches choice (4).

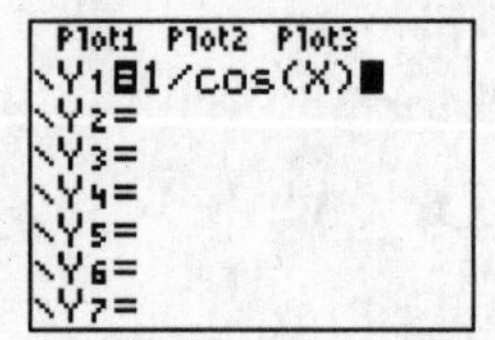

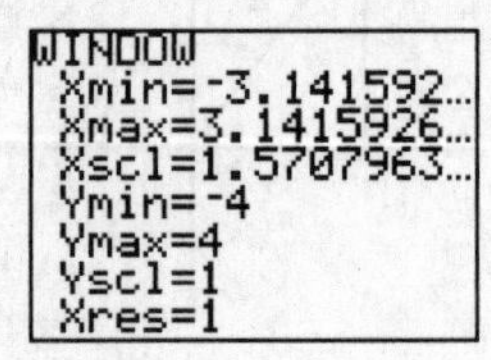

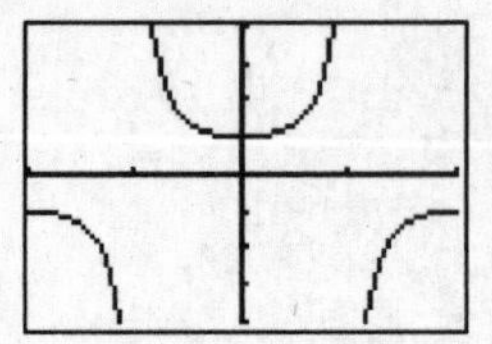

The correct choice is **(4)**.

26. To find the inverse of a function defined by an equation, switch the x and the y in the equation and then solve for the (new) y:

$$\text{Function: } y = 3x + 2 \quad \rightarrow \quad \text{Inverse: } x = 3y + 2$$

$$3y = x - 2$$

$$y = \frac{x-2}{3}$$

$$y = \frac{1}{3}x - \frac{2}{3}$$

The correct choice is **(3)**.

27. Use the normal curve provided on the reference sheet. Mark up the scale on the bottom with the values for mean and standard deviation for this problem. Then add up percents for the different intervals in the choices.

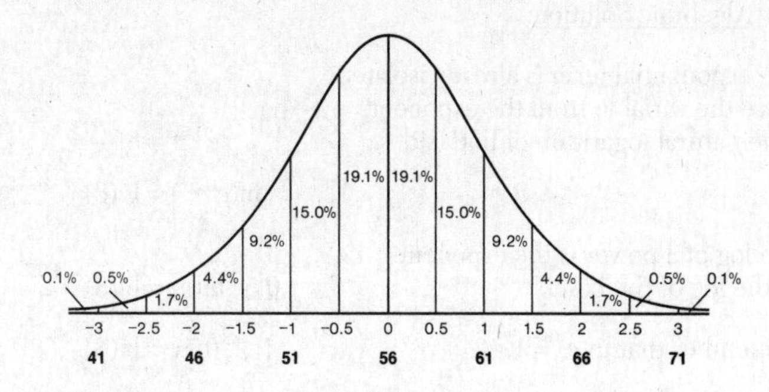

- Choice (1): 46–56: 4.4% + 9.2% +15.0% +19.1% = 47.7% No
- Choice (2): 46–66: 4.4% + 9.2% +15.0% +19.1% + 19.1% + 15.0% + 9.2% + 4.4% = 95.4% Yes
- Choice (3): 51–61: 15.0% + 19.1% + 19.1% + 15.0% = 68.2% No
- Choice (4) 56–71: 19.1% + 15.0% + 9.2% + 4.4% +1.7% + 0.5% = 49.9% No

It is worth remembering that in any normal distribution, about 68.2% of the data lie within one standard deviation of the mean and about 95.4% of the data lie within two standard deviations of the mean. The problem asks for an interval containing 95.4% of the data, which will be in the interval $\overline{x} - 2\sigma \le x \le \overline{x} + 2\sigma$.

$$56 - 2(5) \le x \le 56 + 2(5)$$

$$46 \le x \le 66$$

The correct choice is **(2)**.

PART II

28. Algebraic Solution:

The exponential term is already isolated.
To move the variable from the exponent, take the natural logarithm of both sides.

$$e^{0.5x} = 8$$

$$\ln(e^{0.5x}) = \ln(8)$$

The log of a power is the exponent times the log of the base.

$$0.5x \ln(e) = \ln(8)$$

Remember that $\ln(e) = 1$.

$$0.5x = \ln(8)$$

Solve for x.

$$x = \frac{\ln(8)}{0.5} = 4.158883$$

Graphical Solution:
This problem can also be solved graphically on your calculator. Graph $y = e^{0.5x}$ and $y = 8$ simultaneously. Then press Calc Intersect. If you choose to do the problem this way, you must include a scaled sketch of the graph with the point of intersection and equations labeled.

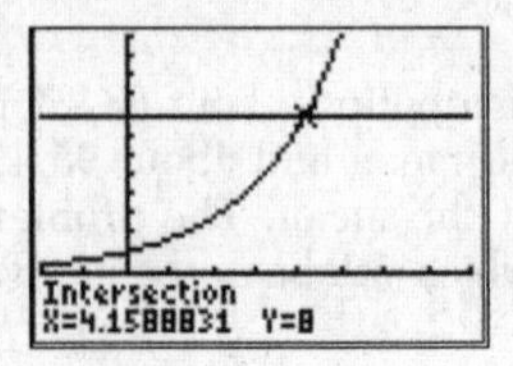

CALC Check

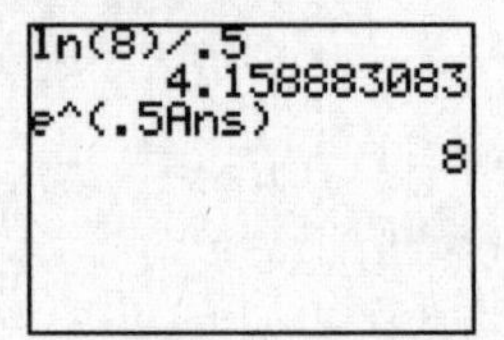

Tip: If you check the rounded answer, your result may not be exact. So check your answer before rounding.

The solution, to the *nearest thousandth*, is $x = 4.159$.

29. To write the circle in center-radius form, you need to complete the square for $x^2 - 8x$. Since there is no linear y term, you do not need to complete the square for y.

$$x^2 + y^2 - 8x + 7 = 0$$

Rearrange so the x terms are together.

$$x^2 - 8x + y^2 + 7 = 0$$

Complete the square in x. The constant in the perfect square is $\frac{1}{2}(-8) = -4$. The square of this, $(-4)^2 = 16$, is subtracted to balance the equation.

$$(x - 4)^2 - 16 + y^2 + 7 = 0$$

Add the constants, $-16 + 7 = -9$.

$$(x - 4)^2 + y^2 - 9 = 0$$

Rewrite in center-radius form.

$$(x - 4)^2 + y^2 = 9$$

In center-radius form, the equation of the given circle is $(x - 4)^2 + y^2 = 9$.

30. A diagram may be helpful.

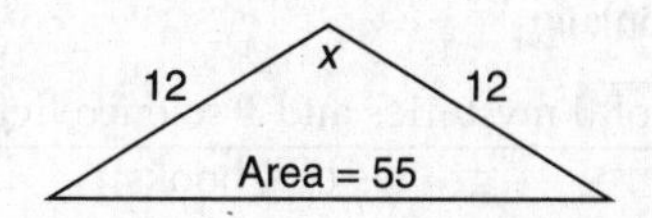

From the reference sheet, the area of a triangle is given by $K = \frac{1}{2}ab \sin C$, where C is the included angle between sides a and b. In this problem, $K = 55$, $a = b = 12$, and $m\angle C = x$.

$$55 = \frac{1}{2}(12)(12)\sin x$$

$$55 = 72 \sin x$$

$$\sin x = \frac{55}{72} = 0.7638888889$$

Sine is positive in Quadrants I and II, so there are two possible answers.

$$x = \sin^{-1}(0.7638888889) = 49.808°$$

or

$$x = 180° - 49.808° = 130.192°$$

The problem states that the isosceles triangle is obtuse, so choose the obtuse angle, 130.2°.

The vertex angle of Gregory's garden, to the *nearest tenth of a degree*, must be 130.2°.

31. The order in which Ellyssa picks her books off the shelf does not matter. So use combinations to count groups. The total number of groups of 4 books that could be selected from 11 total books is given by ${}_{11}C_4 = 330$; this is the size of the sample space. Of those 330 groups of 4 books, we need to know how many contain exactly 3 mysteries and 1 science fiction book. The number of groups of 3 mysteries that could be selected from a total of 6 mysteries is ${}_6C_3 = 20$. The number of groups of 1 science fiction book that can be selected from 5 total science fiction books is ${}_5C_1 = 5$. By the multiplication principle, the number of groups of four books containing 3 mysteries and 1 science fiction book is $20 \cdot 5 = 100$. So the probability of selecting exactly three mysteries and 1 science fiction book is

$$p = \frac{\text{groups of 3 mysteries and 1 science fiction book}}{\text{groups of 4 books}}$$

$$= \frac{{}_6C_3 \cdot {}_5C_1}{{}_{11}C_4}$$

$$= \frac{20 \cdot 5}{330}$$

$$= \frac{100}{330}$$

Note: Reducing the answer is optional. Do not take the chance of making a mistake.

The probability is $\frac{100}{330}$ or $\frac{10}{33}$.

32. The first term is $a_1 = 4$, and a_{n+1} is the notation for the next term after a_n. The recursive formula $a_{n+1} = 2a_n - n$ means that each new term is found by doubling the current term and then subtracting the number of the current term. It is helpful to set up a table as shown.

n	a_n	$a_{n+1} = 2a_n - n$
1	4	$a_2 = 2a_1 - 1 = 2(4) - 1 = 7$
2	7	$a_3 = 2a_2 - 2 = 2(7) - 2 = 12$
3	12	$a_4 = 2a_3 - 3 = 2(12) - 3 = 21$
4	21	

The first four terms of the sequence are {4, 7, 12, 21}.

33. This is an example of the binomial distribution. Use the formula:

$${}_{\text{number of trials}}C_{\text{number of successes}}(\text{probability of success})^{\text{number of successes}}(\text{probability of failure})^{\text{number of failures}}$$

In the given problem, there are four trials (children), 3 successes (not having the trait) and 1 failure (having the trait), the probability of success is 7/8 and the probability of failure is 1/8.

$$P(\text{3 children do } \textit{not} \text{ have the trait}) = {}_4C_3\left(\frac{7}{8}\right)^3\left(\frac{1}{8}\right)^1 = \frac{1372}{4096} \text{ or } \frac{343}{1024} \text{ or } 0.335.$$

Tip: Keep the probabilities in fraction form. Why lose points for possible rounding errors?

CALC Check

```
4 nCr 3(7/8)^3(1
/8)^1
      .3349609375
Ans▸Frac
        343/1024
```

$P(\text{3 children do } \textit{not} \text{ have the trait}) = \frac{1372}{4096}$ or $\frac{343}{1024}$ or 0.335.

34. The formula for the sum of a finite geometric sequence is given on the reference sheet: $S_n = \frac{a_1\left(1-r^n\right)}{1-r}$, where a_1 is the first term, 20; n is the number of terms, 25; and r is the common ratio between consecutive terms. Find r by dividing the distance of the second swing by that of the first swing, $r = \frac{18}{20} = 0.9$.

Evaluate $S_n = \frac{a_1\left(1-r^n\right)}{1-r}$ for $a_1 = 20$, $r = 0.9$, and $n = 25$:

$$S_{25} = \frac{20\left(1-(0.9)^{25}\right)}{1-0.9} = 185.642$$

After 25 swings, to the *nearest tenth of a centimeter*, the weight had traveled 185.6 centimeters.

35. The population standard deviation σ is found on your graphing calculator. Enter the scores in list L1 and the frequencies in list L2. Within STAT, Calc, press 1-Var Stats Enter L1, L2 Enter. The population standard deviation, σ, and the mean, $\bar{x}$, are shown on the screen:

$$\sigma = 16.22996502 \approx 16.2.$$

$$\bar{x} = 157$$

Since Springfield Central has a mean of 157, their scores will have the same average, or center, as Conant's. Because Springfield has a much smaller standard deviation, their scores will group closer to the mean than Conant's. Springfield's scores will not be spread out as much as Conant's. Springfield Central's bowlers are more consistent.

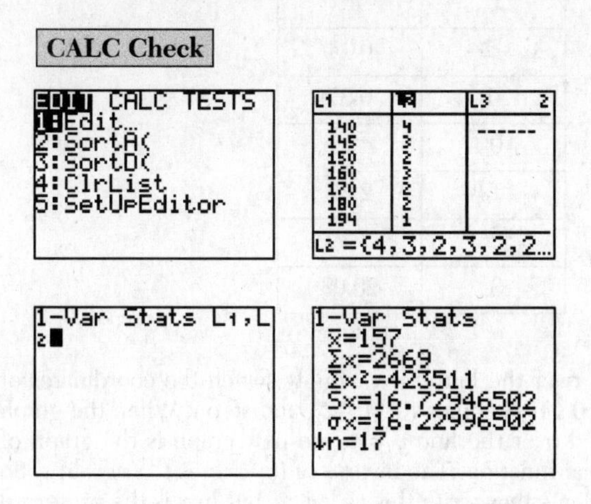

The population standard deviation for Conant High School is $\sigma = 16.2$. Springfield's scores are more consistent (less spread out) than Conant's, but the scores from both schools have the same mean.

PART III

36. Use the table feature on your graphing calculator to sketch the graph of $f(x) = e^x$. Label it. Draw arrows on each end to show that the curve continues.

x	e^x
–2	0.14
–1	0.37
0	1
1	2.72
2	7.39
3	20.09

To reflect the graph over the line $y = x$, simply switch the coordinates of each ordered pair: (–2, 0.14) becomes (0.14, –2), and so on. When the graph of a function is reflected over the line $y = x$, the new graph is the graph of the inverse of the original function. The inverse of $f(x) = e^x$ is $f^{-1}(x) = \ln x$. So $g(x) = \ln(x)$. (Some students may write this as $\log_e x$, but $\ln x$ is the preferred notation.)

The domain of $g(x)$ may be hard to determine from the graph or table. However, $x = 0$ is a vertical asymptote for the graph of $g(x)$. The domain of $g(x)$ is $x > 0$.

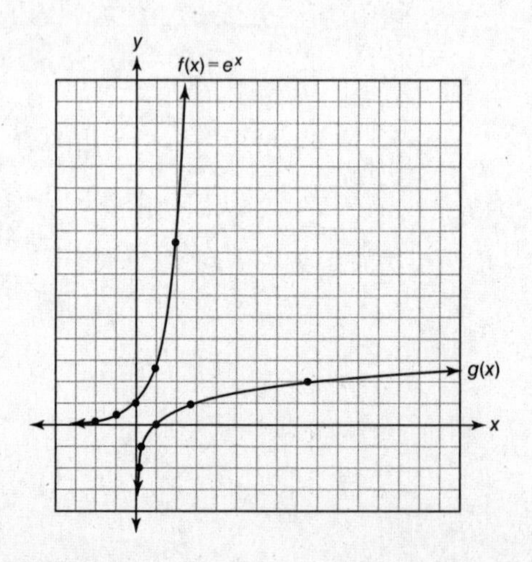

CALC Check

The graphs on the calculator may look like they just stop at the axes, but neither one does. They both continue infinitely along their asymptotes, the x-axis for the exponential function and the y-axis for the logarithmic function. You must know what the graphs of the functions look like so you can interpret the calculator graphs accurately.

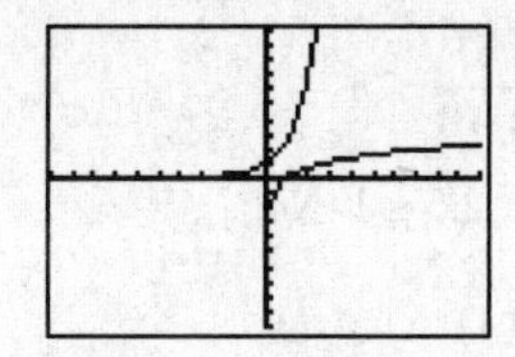

The graph is sketched above. The equation is $g(x) = \ln(x)$, and the domain of $g(x)$ is $x > 0$.

37. Algebraic Solution:

Find the roots by solving for equality.

$$\frac{3x}{2x-6} - \frac{2x}{x-3} = 1$$

Factor the denominators.

$$\frac{3x}{2(x-3)} - \frac{2x}{x-3} = 1$$

Rewrite with common denominators for all terms.

$$\frac{3x}{2(x-3)} - \frac{(2)2x}{2(x-3)} = (1)\frac{2(x-3)}{2(x-3)}$$

Equate and solve the numerators.

$$3x - (2)2x = 2(x - 3)$$

$$3x - 4x = 2x - 6$$

$$-3x = -6$$

$$x = 2$$

Find x-values that make any term undefined. This will occur at $x = 3$, only. Graph the roots and undefined points on a number line. Note that 2 is included (solid circle) because the original problem allows equality. However, 3 is not included (open circle) because 3 makes the problem undefined.

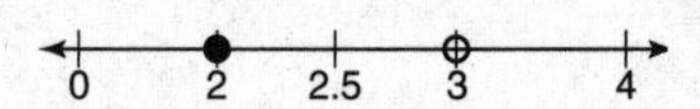

Check a value from each interval in the original inequality and shade the number line if the value checks.

Check $x = 0$: $$\frac{3(0)}{2(0)-6} - \frac{2(0)}{(0)-3} \overset{?}{\le} 1\,1$$

$$0 \le 1 \text{ Yes}$$

Check $x = 2.5$: $$\frac{3(2.5)}{2(2.5)-6} - \frac{2(2.5)}{(2.5)-3} \overset{?}{\le} 1$$

$$2.5 \le 1 \text{ No}$$

Check $x = 4$: $$\frac{3(4)}{2(4)-6} - \frac{2(4)}{(4)-3} \overset{?}{\le} 1$$

$$-2 \le 1 \text{ Yes}$$

2 3

Read the solution off the number line.

$$x \le 2 \text{ or } x > 3$$

Write the solution in interval notation.

$$(-\infty, 2] \cup (3, \infty)$$

Graphical Solution:

Use your calculator to graph $y = \frac{3x}{2x-6} - \frac{2x}{x-3}$ and $y = 1$ in separate equations.

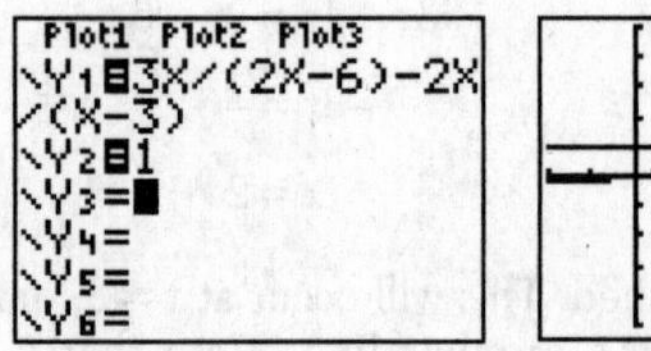

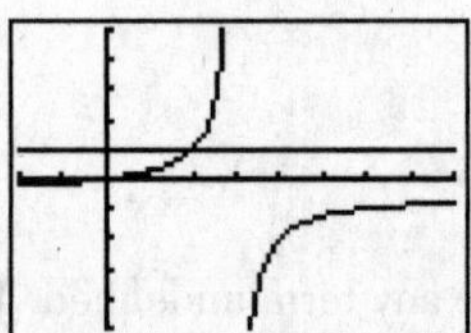

From either Table or Calc Intersect, find the intersection at $x = 2$. Check the original equation to see where the function is undefined by setting the denominators equal to 0:

$$2x - 6 = 0 \text{ when } x = 3$$

$$x - 3 = 0 \text{ when } x = 3$$

This tells you the discontinuity (vertical asymptote) in the graph occurs at $x = 3$. Finally, note that the curve is below (less than) the line on two intervals.

- $x \leq 2$ The number 2 is included because the original problem allows equality.
- $x > 3$ The number 3 is not included because the inequality is undefined there.

The solution is $x \leq 2$ or $x > 3$, which is written in interval notation as $(-\infty, 2] \cup (3, \infty)$.

Try a mix of these methods. Solve algebraically for the roots and undefined values. Construct a number line with the appropriate open and/or closed circles. Examine the graph to determine where the curve satisfies the inequality.

CALC Check

Use the table to check the intervals to see if they satisfy the inequality. ERROR is calculator notation for undefined, so $x = 3$ will be open.

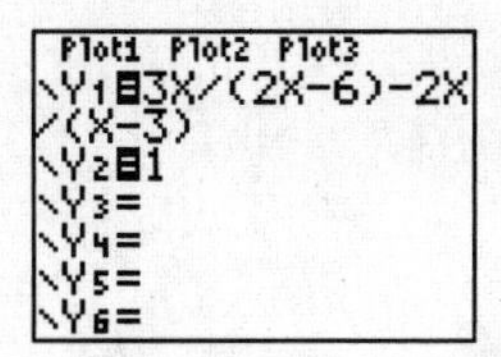

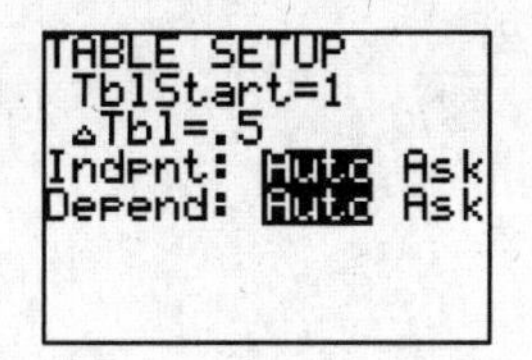

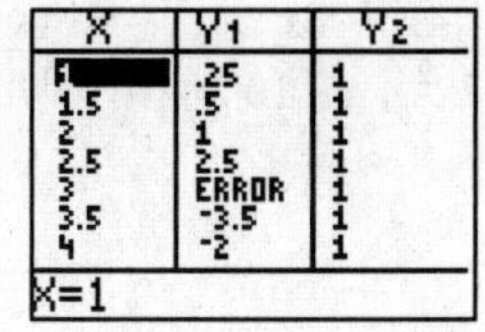

The solution is $(-\infty, 2] \cup (3, \infty)$.

38. Product means multiply.

$$\frac{4x^2+8x-5}{4x^3-x}\cdot\frac{4x^2+8x}{2x^2+x-10}$$

Factor completely all numerators and denominators. Start by factoring out common factors.

$$\frac{4x^2+8x-5}{x(4x^2-1)}\cdot\frac{4x(x+2)}{2x^2+x-10}$$

Then factor the binomials and trinomials.

$$\frac{(2x-1)(2x+5)}{x(2x-1)(2x+1)}\cdot\frac{4x(x+2)}{(2x+5)(x-2)}$$

Simplify by reducing like factors in the numerators and denominators.

$$\frac{\cancel{(2x-1)}\cancel{(2x+5)}}{\cancel{x}\cancel{(2x-1)}(2x+1)}\cdot\frac{4\cancel{x}(x+2)}{\cancel{(2x+5)}(x-2)}$$

Write the answer as a single fraction.

$$\frac{4(x+2)}{(2x+1)(x-2)}$$

Tip: Do not multiply out the final result. Why possibly lose points when you already have the answer?

When expressed in simplest form, the product of $\frac{4x^2+8x-5}{4x^3-x}$ and $\frac{4x^2+8x}{2x^2+x-10}$ is $\frac{4(x+2)}{(2x+1)(x-2)}$.

PART IV

39. The equation contains two different trigonometric functions, sine and cosine. Use the double angle formula shown on the reference sheet, $\cos 2A = 1 - 2\sin^2 A$, to write $\cos 2\theta$ in terms of $\sin\theta$.

$$3\cos 2\theta + 2\sin\theta + 1 = 0$$

$$3(1 - 2\sin^2\theta) + 2\sin + 1 = 0$$

The problem is now quadratic. Rewrite it in standard form by setting it equal to 0.

$$3 - 6\sin^2\theta + 2\sin\theta + 1 = 0$$

$$-6\sin^2\theta + 2\sin\theta + 4 = 0$$

Simplify by dividing by –2.

$$3\sin^2\theta - \sin\theta - 2 = 0$$

Factor this equation. Set each factor equal to 0 and solve for $\sin\theta$. You could also use the quadratic formula.

$$(3\sin\theta + 2)(\sin\theta - 1) = 0$$

$$3\sin\theta + 2 = 0$$

or

$$\sin\theta - 1 = 0$$

For $\sin\theta = -\frac{2}{3}$, θ can be in Quadrants III or IV. Find the reference angle and then find the appropriate angles.

$$\sin\theta = -\frac{2}{3} \text{ or } \sin\theta = 1$$

$$\theta_{ref} = \sin^{-1}\left(\frac{2}{3}\right) = 41.81°$$

$$\text{or } \theta = 90°$$

$$\theta = 180° + 41.81° = 221.81°$$

or

$$\theta = 360° - 41.81° = 318.19°$$

For $\sin\theta = 1$, there is only one solution, the quadrantal angle 90°.

Caution: $\sin^{-1}\left(-\frac{2}{3}\right) = -41.81°$ is not one of the solutions because the directions specify that all solutions be in the interval $0° < \theta < 360°$.

This problem requires the algebraic solution shown above. You can still use your calculator to check your work graphically. The advantage of a graphical solution over checking your answers in the original equation is that it will show all three solutions in the correct domain. Other values may check (such as –41.81°) but are not solutions. Or, you may have found only two of the three solutions in your algebraic work.

CALC Check

Solving trigonometric equations is the exception to the rule that trigonometric functions should be graphed in radian mode. Since the solution must be in degrees, the calculator must be in degree mode. To see the best graph, set the Window Xmin = 0 and Xmax = 360. Then select ZoomFit. Use Calc Zero to find the roots.

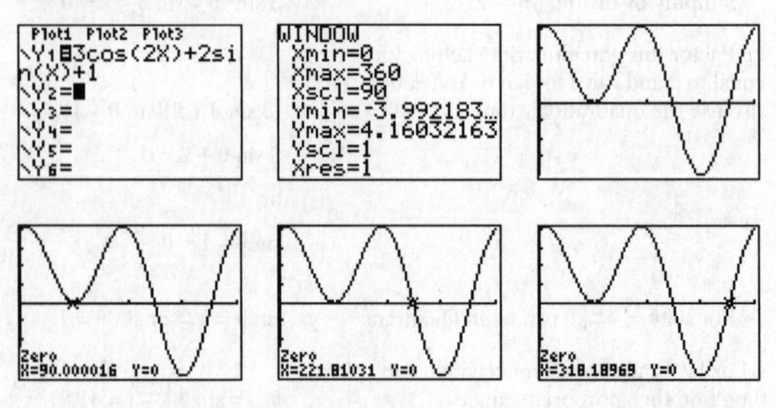

The first example has a rounding error in determining the root. The correct root is exactly 90°. The Calc Zero and Calc Intersect functions are accurate to about 4 decimal places. Since you need to round your answers to the hundredths place, this will not be an issue.

The solutions, to the *nearest hundredth of a degree*, are {90°, 221.81°, 318.19°}.

Topic	Question Numbers	Number of Points	Your Points	Your Percentage
1. Exponents and Radicals: operations, equivalent expressions, simplifying, rationalizing	6, 12	$2 + 2 = 4$		
2. Complex Numbers: operations, powers of *i*, rationalizing	7, 17	$2 + 2 = 4$		
3. Quadratics and Higher-Order Polynomials: operations, factoring, binomial expansion, formula, quadratic equations and inequalities, nature of the roots, sum and product of the roots, quadratic-linear systems, completing the square, higher-order polynomials	9, 15, 16	$2 + 2 + 2 = 6$		
4. Rationals: operations, equations	37, 38	$4 + 4 = 8$		
5. Absolute Value: equations and inequalities	11	2		
6. Direct and Inverse Variation	2	2		
7. Circles: center-radius equation, completing the square	29	2		
8. Functions: relations, functions, domain, range, one-to-one, onto, inverses, compositions, transformations of functions	5, 8, 19, 24, 26	$2 + 2 + 2 + 2 + 2 = 10$		
9. Exponential and Logarithmic Functions: equations, common bases, logarithm rules, e, ln	18, 28, 36	$2 + 2 + 4 = 8$		
10. Trigonometric Functions: radian measure, arc length, cofunctions, unit circle, inverses	1, 21, 22, 23	$2 + 2 + 2 + 2 = 8$		
11. Trigonometric Graphs: graphs of basic functions and inverse functions, domain, range, restricted domain, transformations of graphs, amplitude, frequency, period, phase shift	25	2		
12. Trigonometric Identities, Formulas, and Equations	20, 39	$2 + 6 = 8$		
13. Trigonometry Laws and Applications: area of a triangle, sine and cosine laws, ambiguous cases	13, 30	$2 + 2 = 4$		
14. Sequences and Series: sigma notation, arithmetic, geometric	10, 32, 34	$2 + 2 + 2 = 6$		

Topic	Question Numbers	Number of Points	Your Points	Your Percentage
15. Statistics: studies, central tendency, dispersion including standard deviation, normal distribution	4, 27, 35	$2 + 2 + 2 = 6$		
16. Regressions: linear, exponential, logarithmic, power, correlation coefficient	3, 14	$2 + 2 = 4$		
17. Probability: permutation, combination, geometric, binomial	31, 33	$2 + 2 = 4$		

HOW TO CONVERT YOUR RAW SCORE TO YOUR ALGEBRA 2/TRIGONOMETRY REGENTS EXAMINATION SCORE

Below is a sample conversion chart that can be used to determine your final score on Sample Test 2. To estimate your final exam score, locate in the column labeled "Raw Score" the total number of points you scored out of a possible 88 points. Since partial credit is allowed in Parts II, III, and IV of the test, you may need to approximate the credit you would receive for a solution that is not completely correct. Then locate in the adjacent column to the right the scaled score that corresponds to your raw score. The scaled score is your final sample Regents Examination score.

*__Important Note:__ These are inexact estimates, since no official conversion charts for Algebra 2/Trigonometry exist as of press time.

Raw Score	Scaled Score	Raw Score	Scaled Score	Raw Score	Scaled Score
88	**100**	58	**75**	28	**46**
87	**99**	57	**74**	27	**45**
86	**99**	56	**74**	26	**43**
85	**97**	55	**73**	25	**42**
84	**96**	54	**72**	24	**41**
83	**95**	53	**71**	23	**39**
82	**94**	52	**70**	22	**38**
81	**93**	51	**69**	21	**37**
80	**92**	50	**69**	20	**35**
79	**92**	49	**68**	19	**34**
78	**91**	48	**67**	18	**32**
77	**90**	47	**66**	17	**31**
76	**89**	46	**65**	16	**29**
75	**89**	45	**64**	15	**28**
74	**88**	44	**63**	14	**26**
73	**87**	43	**62**	13	**24**
72	**86**	42	**61**	12	**23**
71	**86**	41	**60**	11	**21**
70	**85**	40	**59**	10	**19**
69	**84**	39	**58**	9	**17**
68	**83**	38	**57**	8	**15**
67	**83**	37	**56**	7	**14**
66	**82**	36	**55**	6	**12**
65	**81**	35	**54**	5	**10**
64	**80**	34	**53**	4	**8**
63	**79**	33	**52**	3	**6**
62	**79**	32	**51**	2	**4**
61	**78**	31	**49**	1	**2**
60	**77**	30	**48**	0	**0**
59	**76**	29	**47**		

Official Test Sampler

Algebra 2/Trigonometry

REFERENCE SHEET

Area of a Triangle

$$K = \frac{1}{2}ab \sin C$$

Law of Cosines

$$a^2 = b^2 + c^2 - 2bc \cos A$$

Functions of the Sum of Two Angles

$$\sin(A + B) = \sin A \cos B + \cos A \sin B$$

$$\cos(A + B) = \cos A \cos B - \sin A \sin B$$

$$\tan(A + B) = \frac{\tan A + \tan B}{1 - \tan A \tan B}$$

Functions of the Double Angle

$$\sin 2A = 2 \sin A \cos A$$

$$\cos 2A = \cos^2 A - \sin^2 A$$

$$\cos 2A = 2 \cos^2 A - 1$$

$$\cos 2A = 1 - 2 \sin^2 A$$

$$\tan 2A = \frac{2 \tan A}{1 - \tan^2 A}$$

Functions of the Difference of Two Angles

$$\sin(A - B) = \sin A \cos B - \cos A \sin B$$

$$\cos(A - B) = \cos A \cos B + \sin A \sin B$$

$$\tan(A - B) = \frac{\tan A - \tan B}{1 + \tan A \tan B}$$

Functions of the Half Angle

$$\sin\frac{1}{2}A = \pm\sqrt{\frac{1-\cos A}{2}}$$

$$\cos\frac{1}{2}A = \pm\sqrt{\frac{1+\cos A}{2}}$$

$$\tan\frac{1}{2}A = \pm\sqrt{\frac{1-\cos A}{1+\cos A}}$$

Law of Sines

$$\frac{a}{\sin A} = \frac{b}{\sin B} = \frac{c}{\sin C}$$

Sum of a Finite Arithmetic Sequence

$$S_n = \frac{n\left(a_1 + a_n\right)}{2}$$

Sum of a Finite Geometric Sequence

$$S_n = \frac{a_1\left(1-r^n\right)}{1-r}$$

Binomial Theorem

$$(a+b)^n = {}_nC_0a^nb^0 + {}_nC_1a^{n-1}b^1 + {}_nC_2a^{n-2}b^2 + \cdots + {}_nC_na^0b^n$$

$$(a+b)^n = \sum_{r=0}^{n} {}_nC_ra^{n-r}b^r$$

Normal Curve Standard Deviation

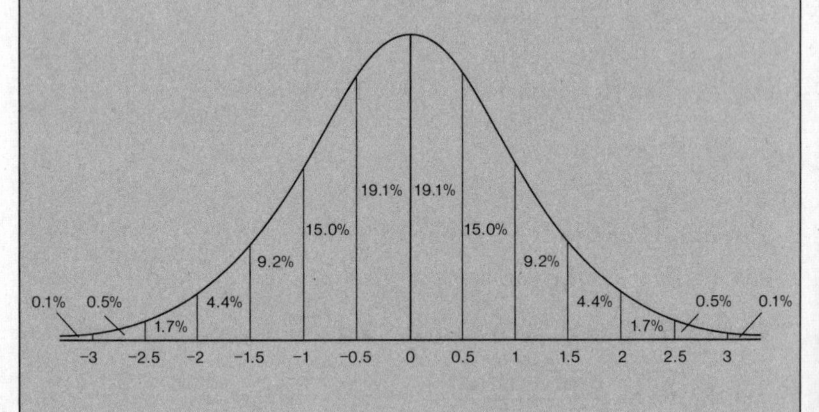

PART I

Answer all 27 questions in this part. Each correct answer will receive 2 credits. No partial credit will be allowed. For each question, write in the space provided the numeral preceding the word or expression that best completes the statement or answers the question. [54 credits]

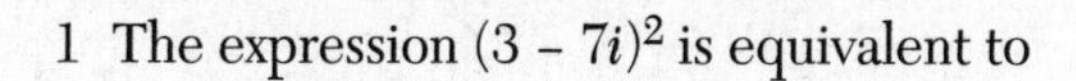

1 The expression $(3 - 7i)^2$ is equivalent to

(1) $-40 + 0i$ (3) $58 + 0i$
(2) $-40 - 42i$ (4) $58 - 42i$ 1 _____

2 If $f(x) = \frac{1}{2}x - 3$ and $g(x) = 2x + 5$, what is the value of $(g \circ f)(4)$?

(1) -13 (3) 3
(2) 3.5 (4) 6 2 _____

3 What are the values of θ in the interval $0° \leq \theta < 360°$ that satisfy the equation $\tan\theta - \sqrt{3} = 0$?

(1) 60°, 240°
(2) 72°, 252°
(3) 72°, 108°, 252°, 288°
(4) 60°, 120°, 240°, 300° 3 _____

4 A survey completed at a large university asked 2,000 students to estimate the average number of hours they spend studying each week. Every tenth student entering the library was surveyed. The data showed that the mean number of hours that students spend studying was 15.7 per week. Which characteristic of the survey could create a bias in the results?

(1) the size of the sample
(2) the size of the population
(3) the method of analyzing the data
(4) the method of choosing the students who were surveyed

4 ____

5 Which graph represents the solution set of $|6x - 7| \leq 5$?

(1) 

(2) 

(3) 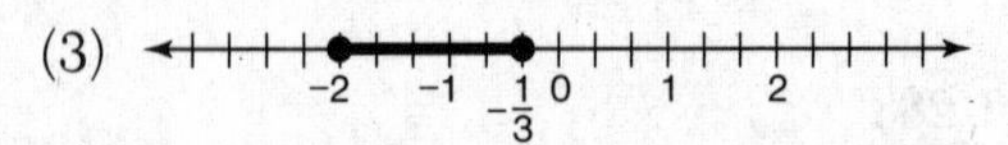

(4) 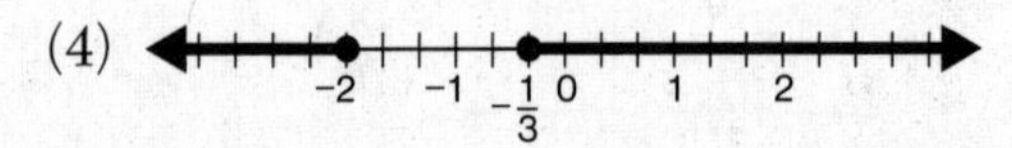

5 ____

6 Which function is *not* one-to-one?

(1) {(0,1), (1,2), (2,3), (3,4)}
(2) {(0,0), (1,1), (2,2), (3,3)}
(3) {(0,1), (1,0), (2,3), (3,2)}
(4) {(0,1), (1,0), (2,0), (3,2)} 6____

7 In $\triangle ABC$, $m\angle A = 120$, $b = 10$, and $c = 18$. What is the area of $\triangle ABC$ to the *nearest square inch*?

(1) 52 (3) 90
(2) 78 (4) 156 7____

8 Which graph does *not* represent a function?

(1)
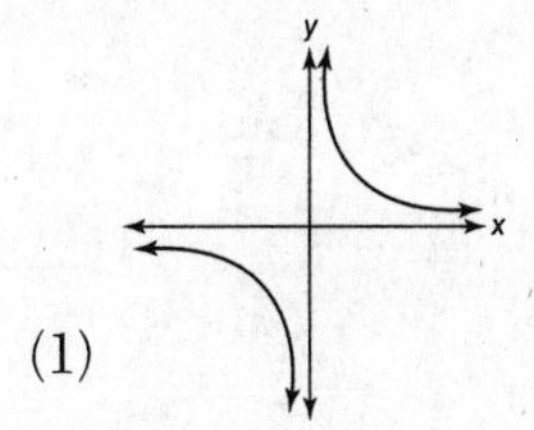

(3)
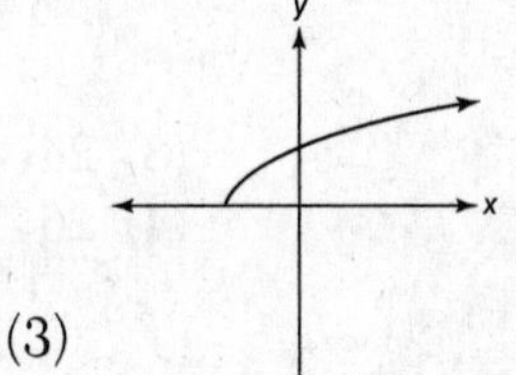

(2)
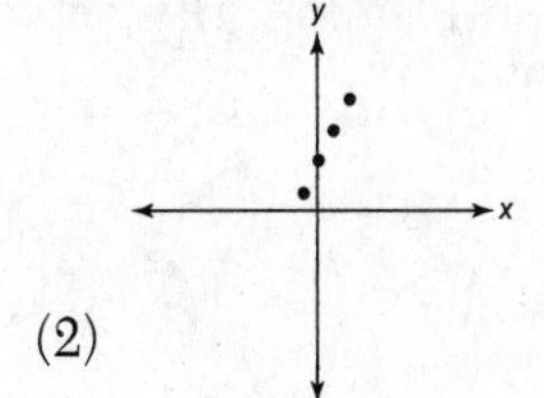

(4)
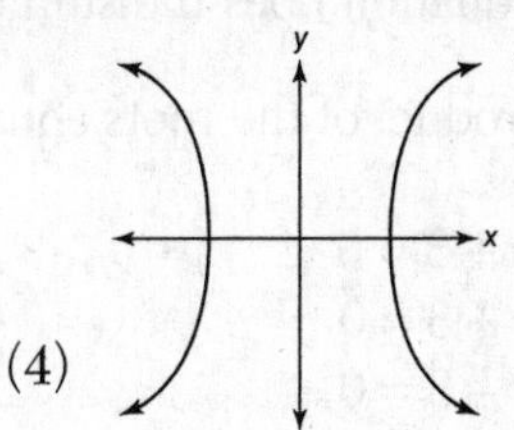

8____

9 The expression $\log_8 64$ is equivalent to

(1) 8
(2) 2
(3) $\frac{1}{2}$
(4) $\frac{1}{8}$

9 ______

10 The expression $\cos 4x \cos 3x + \sin 4x \sin 3x$ is equivalent to

(1) $\sin x$
(2) $\sin 7x$
(3) $\cos x$
(4) $\cos 7x$

10 ______

11 The value of the expression $2\sum_{n=0}^{2}\left(n^2 + 2^n\right)$ is

(1) 12
(2) 22
(3) 24
(4) 26

11 ______

12 For which equation does the sum of the roots equal $\frac{3}{4}$ and the product of the roots equal -2?

(1) $4x^2 - 8x + 3 = 0$
(2) $4x^2 + 8x + 3 = 0$
(3) $4x^2 - 3x - 8 = 0$
(4) $4x^2 + 3x - 2 = 0$

12 ______

13 Which graph represents the equation $y = \cos^{-1} x$?

(1) 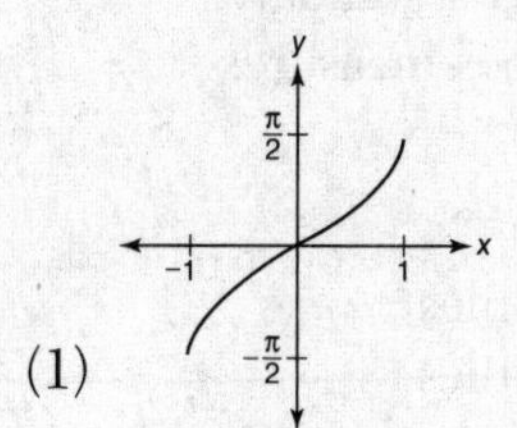

(3) 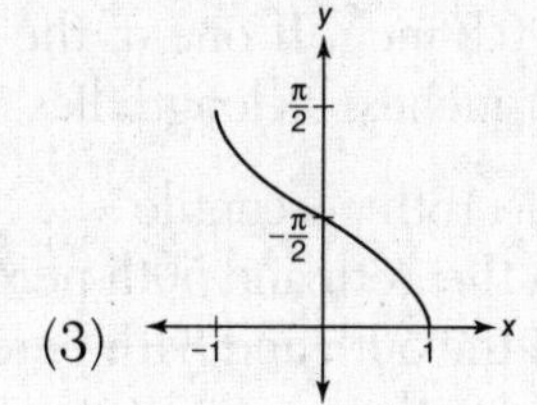

(2) 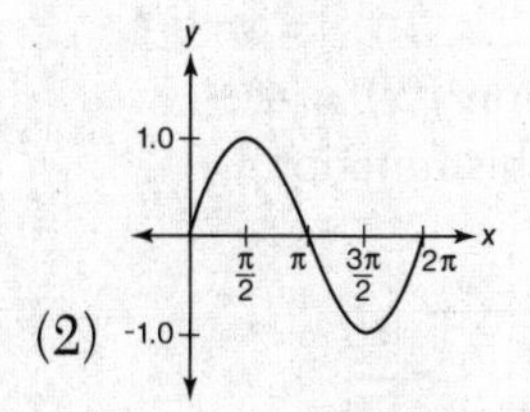

(4) 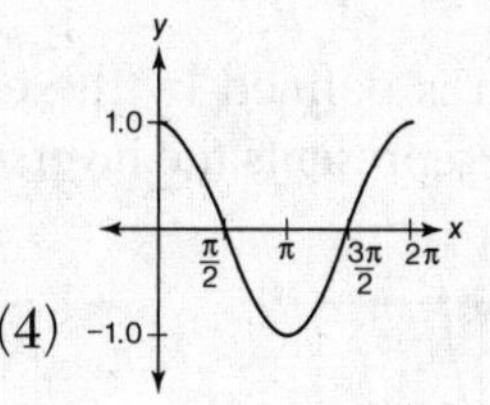

13 ____

14 The expression $\dfrac{a^2b^{-3}}{a^{-4}b^2}$ is equivalent to

(1) $\dfrac{a^6}{b^5}$ (3) $\dfrac{a^2}{b}$

(2) $\dfrac{b^5}{a^6}$ (4) $a^{-2}b^{-1}$

14 ____

15 The lengths of 100 pipes have a normal distribution with a mean of 102.4 inches and a standard deviation of 0.2 inch. If one of the pipes measures exactly 102.1 inches, its length lies

(1) below the 16th percentile
(2) between the 16th and 50th percentiles
(3) between the 50th and 84th percentiles
(4) above the 84th percentile

15 ______

16 If a function is defined by the equation $f(x) = 4^x$, which graph represents the inverse of this function?

(1) 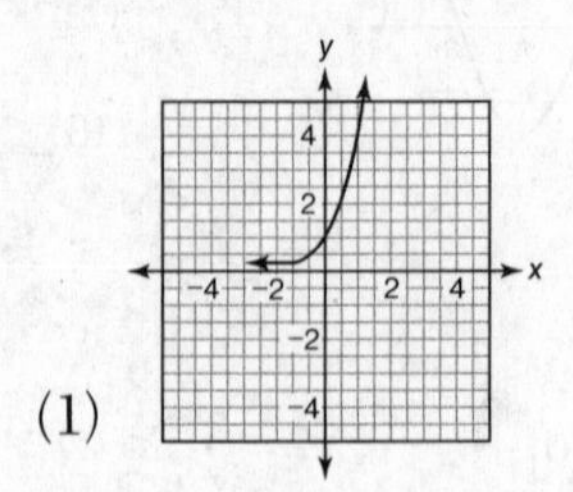

(3) 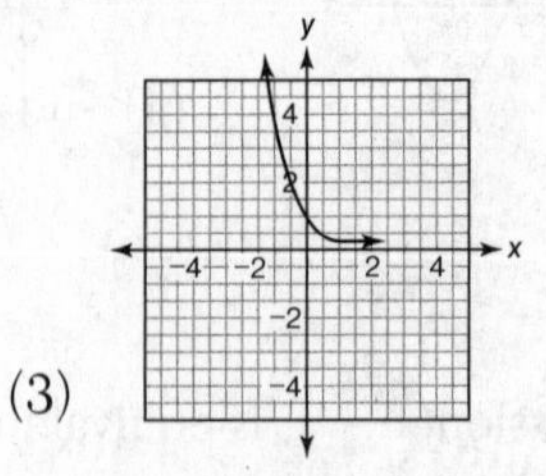

(2) 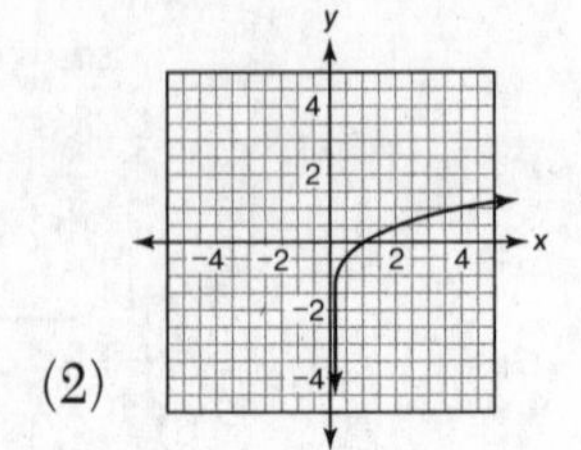

(4) 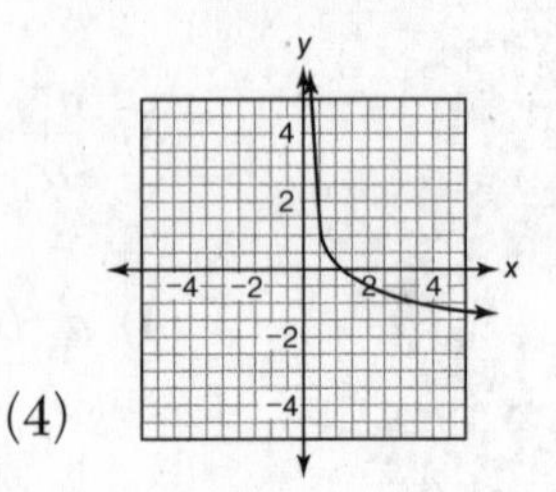

16 ______

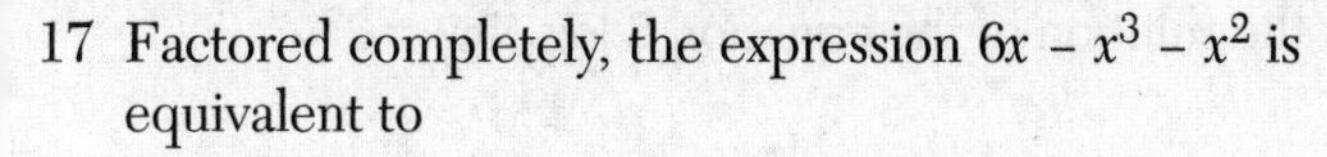

17 Factored completely, the expression $6x - x^3 - x^2$ is equivalent to

(1) $x(x + 3)(x - 2)$
(2) $x(x - 3)(x + 2)$
(3) $-x(x - 3)(x + 2)$
(4) $-x(x + 3)(x - 2)$

17 ______

18 The expression $4ab\sqrt{2b} - 3a\sqrt{18b^3} + 7ab\sqrt{6b}$ is equivalent to

(1) $2ab\sqrt{6b}$
(2) $16ab\sqrt{2b}$
(3) $-5ab + 7ab\sqrt{6b}$
(4) $-5ab\sqrt{2b} + 7ab\sqrt{6b}$

18 ______

19 What is the fourth term in the expansion of $(3x - 2)^5$?

(1) $-720x^2$
(2) $-240x$
(3) $720x^2$
(4) $1{,}080x^3$

19 ______

20 Written in simplest form, the expression $\dfrac{\frac{x}{4} - \frac{1}{x}}{\frac{1}{2x} + \frac{1}{4}}$ is equivalent to

(1) $x - 1$
(2) $x - 2$
(3) $\dfrac{x - 2}{2}$
(4) $\dfrac{x^2 - 4}{x + 2}$

20 ______

21 What is the solution of the equation $2 \log_4(5x) = 3$?

(1) 6.4 (3) $\frac{9}{5}$

(2) 2.56 (4) $\frac{8}{5}$ 21 ______

22 A circle has a radius of 4 inches. In inches, what is the length of the arc intercepted by a central angle of 2 radians?

(1) 2π (3) 8π
(2) 2 (4) 8 22 ______

23 What is the domain of the function $f(x) = \sqrt{x-2} + 3$?

(1) $(-\infty, \infty)$ (3) $[2, \infty)$
(2) $(2, \infty)$ (4) $[3, \infty)$ 23 ______

24 The table below shows the first-quarter averages for Mr. Harper's statistics class.

Statistics Class Averages

Quarter Averages	Frequency
99	1
97	5
95	4
92	4
90	7
87	2
84	6
81	2
75	1
70	2
65	1

What is the population variance for this set of data?

(1) 8.2
(2) 8.3
(3) 67.3
(4) 69.3

24 ____

25 Which formula can be used to determine the total number of different eight-letter arrangements that can be formed using the letters in the word *DEADLINE*?

(1) $8!$
(2) $\frac{8!}{4!}$
(3) $\frac{8!}{2!+2!}$
(4) $\frac{8!}{2! \cdot 2!}$

25 ____

26 The graph below shows the function $f(x)$.

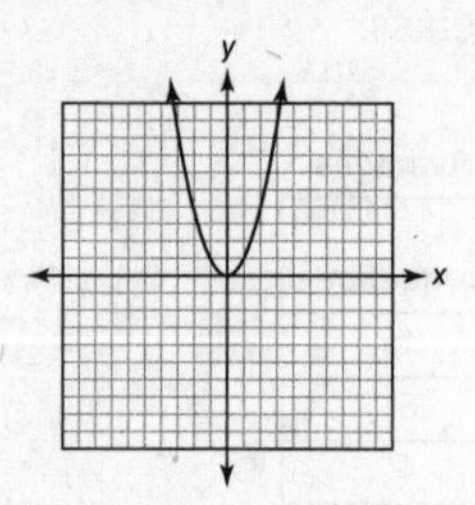

Which graph represents the function $f(x + 2)$?

(1)
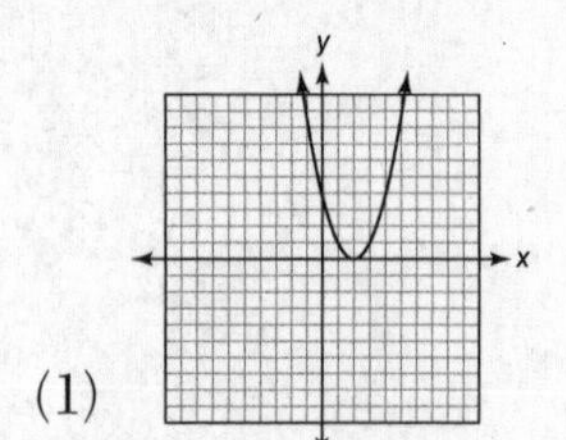

(3)
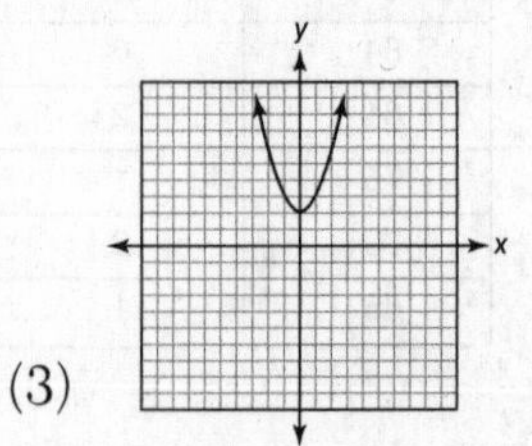

(2)
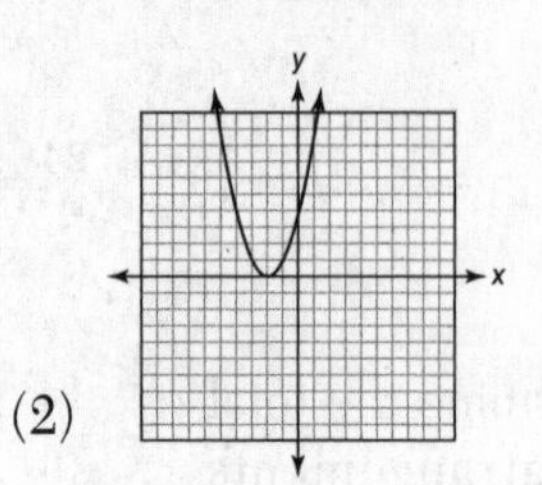

(4)
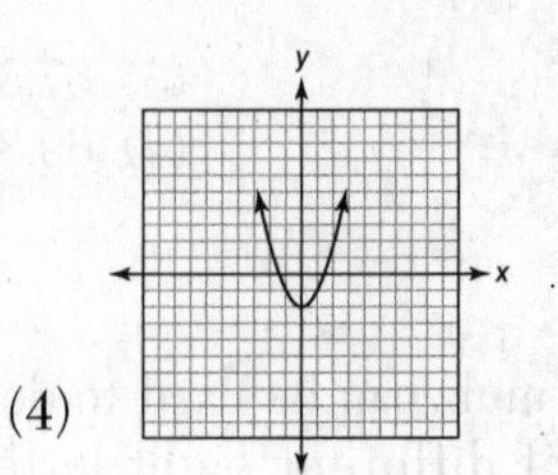

26 ______

27 The equation $y - 2 \sin \theta = 3$ may be rewritten as

(1) $f(y) = 2 \sin x + 3$
(2) $f(y) = 2 \sin \theta + 3$
(3) $f(x) = 2 \sin \theta + 3$
(4) $f(\theta) = 2 \sin \theta + 3$

27 ______

PART II

Answer all 8 questions in this part. Each correct answer will receive 2 credits. Clearly indicate the necessary steps, including appropriate formula substitutions, diagrams, graphs, charts, etc. For all questions in this part, a correct numerical answer with no work shown will receive only 1 credit. [16 credits]

28 Express $\frac{5}{3-\sqrt{2}}$ with a rational denominator, in simplest radical form.

29 Write an equation of the circle shown in the graph below.

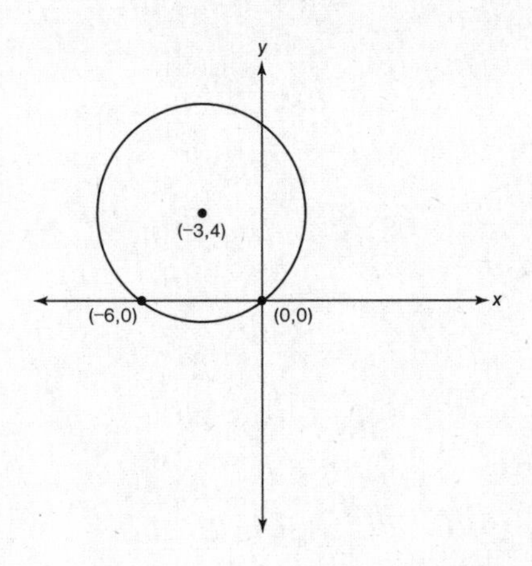

30 Solve for x: $\dfrac{4x}{x-3} = 2 + \dfrac{12}{x-3}$

31 Find, to the *nearest minute*, the angle whose measure is 3.45 radians.

32 Matt places $1,200 in an investment account earning an annual rate of 6.5%, compounded continuously. Using the formula $V = Pe^{rt}$, where V is the value of the account in t years, P is the principal initially invested, e is the base of a natural logarithm, and r is the rate of interest, determine the amount of money, to the *nearest cent*, that Matt will have in the account after 10 years.

33 If θ is an angle in standard position and its terminal side passes through the point $(-3, 2)$, find the exact value of $\csc \theta$.

34 Find the first four terms of the recursive sequence defined below.

$$a_1 = -3$$

$$a_n = a_{(n-1)} - n$$

35 A committee of 5 members is to be randomly selected from a group of 9 teachers and 20 students. Determine how many different committees can be formed if 2 members must be teachers and 3 members must be students.

PART III

Answer all 3 questions in this part. Each correct answer will receive 4 credits. Clearly indicate the necessary steps, including appropriate formula substitutions, diagrams, graphs, charts, etc. For all questions in this part, a correct numerical answer with no work shown will receive only 1 credit. [12 credits]

36 Solve $2x^2 - 12x + 4 = 0$ by completing the square, expressing the result in simplest radical form.

37 Solve the equation $8x^3 + 4x^2 - 18x - 9 = 0$ algebraically for all values of x.

38 The table below shows the results of an experiment involving the growth of bacteria.

Time (*x*) (in minutes)	1	3	5	7	9	11
Number of Bacteria (y)	2	25	81	175	310	497

Write a power regression equation for this set of data, rounding all values to *three decimal places*.

Using this equation, predict the bacteria's growth, to the *nearest integer*, after 15 minutes.

PART IV

Answer the question in this part. The correct answer will receive 6 credits. Clearly indicate the necessary steps, including appropriate formula substitutions, diagrams, graphs, charts, etc. A correct numerical answer with no work shown will receive only 1 credit. [6 credits]

39 Two forces of 25 newtons and 85 newtons acting on a body form an angle of 55°.

Find the magnitude of the resultant force, to the *nearest hundredth of a newton*.

Find the measure, to the *nearest degree*, of the angle formed between the resultant and the larger force.

Answers Official Test Sampler
Algebra 2/Trigonometry

Answer Key

PART I

1. (2)	**6.** (4)	**11.** (3)	**16.** (2)	**21.** (4)	**26.** (2)
2. (3)	**7.** (2)	**12.** (3)	**17.** (4)	**22.** (4)	**27.** (4)
3. (1)	**8.** (4)	**13.** (3)	**18.** (4)	**23.** (3)	
4. (4)	**9.** (2)	**14.** (1)	**19.** (1)	**24.** (3)	
5. (1)	**10.** (3)	**15.** (1)	**20.** (2)	**25.** (4)	

PART II

28. $\frac{5(3+\sqrt{2})}{7}$

29. $(x+3)^2 + (y-4)^2 = 25$

30. no solution

31. 197°40′

32. $2,298.65

33. $\frac{\sqrt{13}}{2}$

34. –3, –5, –8, –12

35. 41,040

PART III

36. $3 \pm \sqrt{7}$

37. $x = \pm\frac{3}{2}; x = -\frac{1}{2}$

38. $y = 2.001x^{2.298}$; 1009

PART IV

39. 101.43; 12°

Answers Explained

PART I

1. Square the binomial and simplify powers of i. Remember that $i^2 = -1$:

$$\begin{aligned}(3-7i)^2 &= (3-7i)(3-7i)\\ &= 9-21i-21i+49i^2\\ &= 9-42i+49(-1)\\ &= -40-42i\end{aligned}$$

This problem can also be done on the calculator set in $a + bi$ mode.

CALC Check

```
(3-7i)²
             -40-42i
```

The correct choice is **(2)**.

2. The notation $(g \circ f)(4)$ means the same as $g(f(4))$. First evaluate $f(4)$, and then evaluate g at that answer.

$$f(4) = \frac{1}{2}(4) - 3 = -1$$

$$g(-1) = 2(-1) + 5 = 3$$

Note that if you do the composition in the wrong order, you get $g(4) = 13$ and $f(13) = 3.5$, which is not correct.

The correct choice is **(3)**.

3. First isolate the trigonometric function.

$$\tan\theta - \sqrt{3} = 0$$

$$\tan\theta = \sqrt{3}$$

Use inverse tangent to find the reference angle that satisfies the equation.

$$\theta = \tan^{-1}(\sqrt{3}) = 60°$$

Since $\tan\theta > 0$, there is a second solution in Quadrant III with the same reference angle, 60°. Using a diagram is optional but may be helpful.

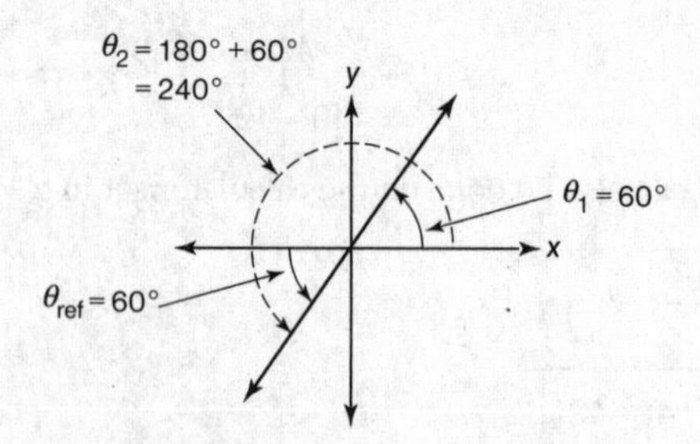

You can also solve this problem graphically. Graph $Y_1 = \tan(X) - \sqrt{3}$. Make sure your calculator is set to degree mode. Set the WINDOW to Xmin = 0 and Xmax = 360, and then select ZoomFit. (Ymin and Ymax may be adjusted if you like. Ymin = –5 and Ymax = 5 give a nice graph. However that scale is not necessary to get the solutions.) Use CALC zero to find the two roots.

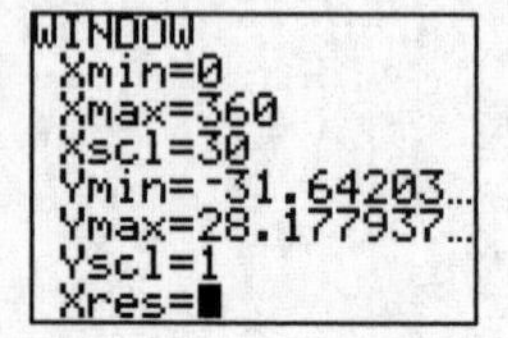

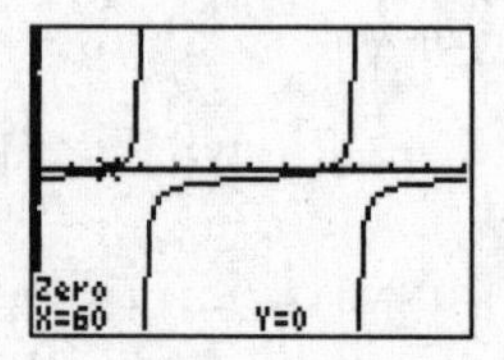

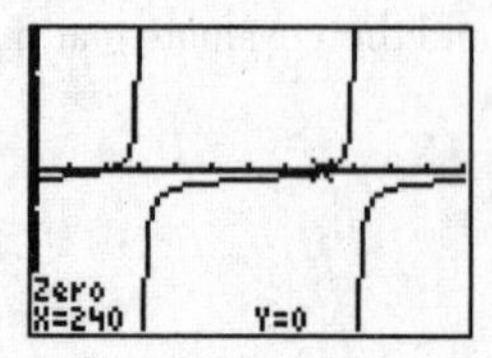

CALC Check

Check values in the multiple-choice solutions. If one value does not check, then the choice is not valid. The values in choice (1) are the only ones that check in the original equation.

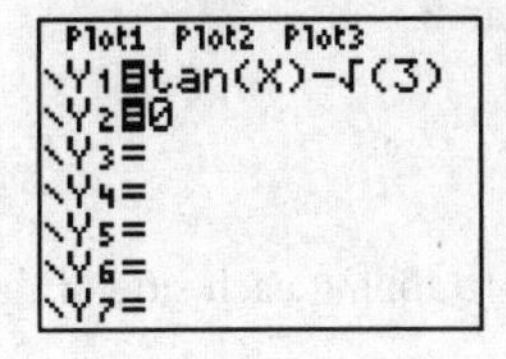

TABLE SETUP
TblStart=0
ΔTbl=1
Indpnt: Auto Ask
Depend: Auto Ask

X	Y1	Y2
60	0	0
240	0	0
120	-3.464	0
72	1.3456	0

X=

The correct choice is **(1)**.

4. The most likely source of bias is sampling bias, a bias introduced by the method of choosing the students who were surveyed. In this example, it is possible that students who spend time in the library study more than students who do not visit the library. If so, sampling only students entering the library would likely lead to an overestimate of the average number of hours studied by all students.

The correct choice is **(4)**.

5. To solve an absolute value inequality, first find the solutions for equality.

$$|6x - 7| = 5$$

$$6x - 7 = 5 \quad \text{or} \quad -(6x - 7) = 5$$

$$-6x + 7 = 5$$

$$x = 2 \quad \text{or} \quad x = \frac{1}{3}$$

Both check. So $x = \frac{1}{3}$ and $x = 2$ are the endpoints of the solution interval(s). This narrows the choices to either (1) or (2).

To see if choice (1) is correct, check a value of x between $\frac{1}{3}$ and 2 in the original inequality. Check $x = 1$:

$$|6(1) - 7| \overset{?}{\leq} 5$$

$$1 \leq 5 \text{ Yes}$$

Since $x = 1$ checks, all the numbers between $x = \frac{1}{3}$ and $x = 2$ will check and choice (1) is the correct answer. Had $x = 1$ not checked, choice (2) would have been the correct answer. This could have been confirmed by checking both a number less than $\frac{1}{3}$ and a number greater than 2.

CALC Check

This problem can be solved on the calculator by graphing each side of the inequality separately.

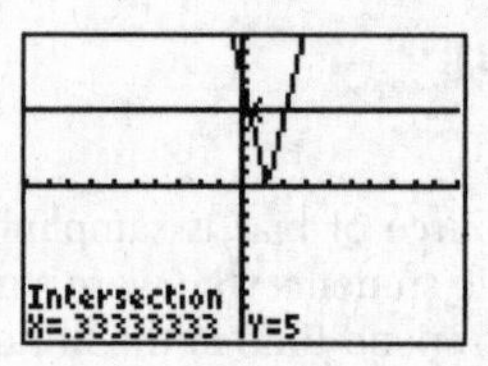

The roots can be found with CALC intersect (one is shown). The absolute value function is on or below (less than or equal to) the horizontal line $y = 5$ in the interval $\frac{1}{3} \leq x \leq 2$.

The correct choice is **(1)**.

6. A function is one-to-one if no two different x-values map to the same y-value. In choice (4), both $x = 1$ and $x = 2$ give $y = 0$. This function is not one-to-one. In the other choices, each y-value comes from a unique x-value.

The correct choice is **(4)**.

7. From the reference sheet, the area of a triangle is $K = \frac{1}{2}ab \sin C$. Remember that the labels of the sides and angle are not important as long as you know two sides and the included angle. In this problem, use $K = \frac{1}{2}bc \sin A = \frac{1}{2}(10)(18)\sin 120° = 77.9$, which is 78 to the nearest whole number.

The correct choice is **(2)**.

8. In a function, each x-value maps to only one y-value. Graphically, this means that a vertical line drawn anywhere on the graph of a function will never intersect the graph in more than one point. This is the case for the graphs shown in choices (1), (2), and (3). However, in choice (4), it is possible to draw vertical lines that intersect the graph in two places. This means that some x-values map to more than one y-value. Choice (4) does not represent a function.

The correct choice is **(4)**.

9. Let $\log_8 64 = n$. Rewriting this in exponential form gives $8^n = 64$, so $n = 2$.

If you know the change of base formula, $\log_b x = \frac{\log x}{\log b}$, you can also do this problem on the calculator:

$$\log_8 64 = \frac{\log 64}{\log 8} = 2$$

The correct choice is **(2)**.

10. The reference sheet shows that the formula with the given structure is $\cos(A - B) = \cos A \cos B + \sin A \sin B$. In this problem, let $A = 4x$ and $B = 3x$. So $\cos 4x \cos 3x + \sin 4x \sin 3x = \cos(4x - 3x) = \cos(x)$.

The correct choice is **(3)**.

11. Evaluate the terms for $n = 0$, 1, and 2. Then add the terms and multiply the result by 2:

$$2\sum_{n=0}^{2}\left(n^2 + 2^n\right) = 2\left[\left(0^2 + 2^0\right) + \left(1^2 + 2^1\right) + \left(2^2 + 2^2\right)\right]$$

$$= 2\left[(0+1) + (1+2) + (4+4)\right] = 2(12) = 24$$

The correct choice is **(3)**.

12. For any quadratic equation $ax^2 + bx + c = 0$, the sum of the roots is $-\frac{b}{a}$ and the product of the roots is $\frac{c}{a}$. Note that in all the choices, $a = 4$. Since the sum of the roots is $-\frac{b}{4} = \frac{3}{4}$, b must be -3. Choice (3) is the only possible choice. As confirmation, check that the product of the roots is $\frac{c}{a} = \frac{-8}{4} = -2$.

The correct choice is **(3)**.

13. Graph Y1 = cos^{-1}(X) on your calculator set to radian mode. The graph below uses ZoomDecimal, which is a better choice than ZoomTrig for inverse trigonometric graphs. The graph matches choice (3).

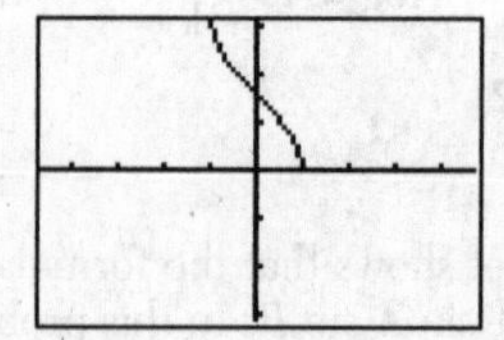

The correct choice is **(3)**.

14. When dividing powers of the same base, subtract the exponents. Pay careful attention to signs. In the last step, rewrite the negative exponent as the reciprocal of the base with a positive exponent:

$$\frac{a^2b^{-3}}{a^{-4}b^2} = a^{(2-(-4))}b^{(-3-2)}$$

$$= a^6b^{-5}$$

$$= \frac{a^6}{b^5}$$

Some students prefer to do the problem in a different order, first rewriting the negative exponents as the reciprocal with a positive exponent and then multiplying by adding exponents:

$$\frac{a^2b^{-3}}{a^{-4}b^2} = \frac{a^2a^4}{b^2b^3}$$

$$= \frac{a^{2+4}}{b^{2+3}}$$

$$= \frac{a^6}{b^5}$$

The correct choice is **(1)**.

15. Use the normal curve provided on the reference sheet. Mark the scale on the bottom with the values for mean and standard deviation for this problem. Then add up the percents for the intervals to the left of –1.5 standard deviations:

$$0.1\% + 0.5\% + 1.7\% + 4.4\% = 6.7\%$$

Since 6.7 is less than 16, the answer is below the 16th percentile.

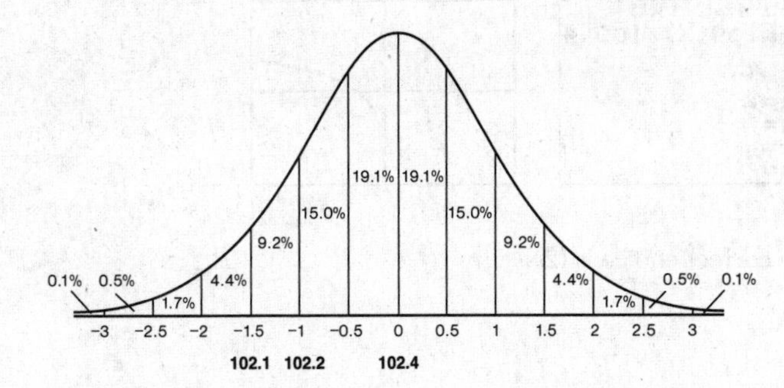

The correct choice is **(1)**.

16. $f(x) = 4^x$ is an exponential function. The graph is shown below. The graph of the inverse of the function is found by reflecting the graph of f over the line $y = x$ (also shown), which results in the graph in choice (3).

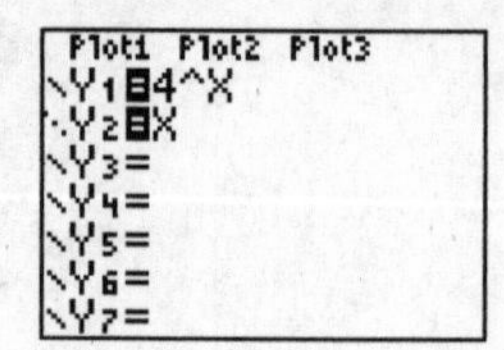

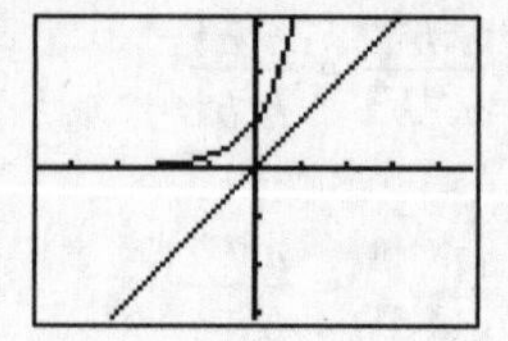

Many students have difficulty reflecting a graph over $y = x$. To help you do this accurately, you can rotate your calculator so the line $y = x$ is a vertical line. Vertical reflections are easier to visualize.

You can also identify the graph of the inverse function by switching the x- and y-values in the coordinate pairs. Find points on the original function graph, such as (0, 1) and (1, 4). Then look for the graph that goes through the points (1, 0) and (4, 1).

You should know that the inverse of an exponential function is a log function with the same base. So the inverse of $f(x) = 4^x$ is $f^{-1}(x) = \log_4 x$. If you also know the change of base formula, $\log_b x = \frac{\log x}{\log b}$, you can graph f^{-1} on your calculator.

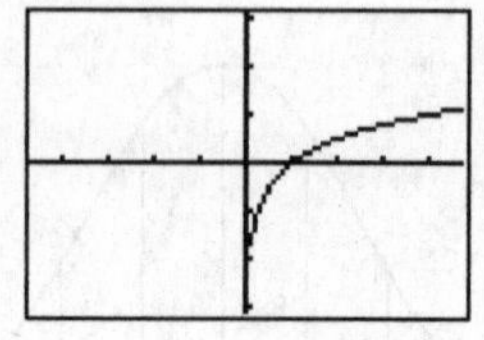

The correct choice is **(2)**.

17. Rearrange the terms in descending order, factor out the greatest common monomial, and then factor the remaining polynomial. Because the leading term, after rearranging, is negative, factor out –1 when factoring out the greatest common monomial.

$$\begin{aligned} 6x - x^3 - x^2 &= -x^3 - x^2 + 6x \\ &= -x(x^2 + x - 6) \\ &= -x(x+3)(x-2) \end{aligned}$$

CALC Check

```
Plot1 Plot2 Plot3
\Y1=6X-X³-X²
\Y2=-X(X+3)(X-2)
\Y3=
\Y4=
\Y5=
\Y6=
```

X	Y1	Y2
-2	-8	-8
-1	-6	-6
0	0	0
1	4	4
2	0	0
3	-18	-18
4	-56	-56

X=-2

The correct choice is **(4)**.

18. To add and subtract radical expressions, first simplify all the radicals and then combine like terms. In this problem, only $-3a\sqrt{18b^3}$ simplifies.

$$\begin{aligned} 4ab\sqrt{2b} - 3a\sqrt{18b^3} + 7ab\sqrt{6b} &= 4ab\sqrt{2b} - 3a\sqrt{9b^2}\sqrt{2b} + 7ab\sqrt{6b} \\ &= 4ab\sqrt{2b} - 3a(3b)\sqrt{2b} + 7ab\sqrt{6b} \\ &= 4ab\sqrt{2b} - 9ab\sqrt{2b} + 7ab\sqrt{6b} \\ &= -5ab\sqrt{2b} + 7ab\sqrt{6b} \end{aligned}$$

The correct choice is **(4)**.

19. You can look up the formula for expanding the power of a binomial on the reference sheet: $(a+b)^n = \sum_{r=0}^{n} {}_nC_r a^{n-r} b^r$. Note that the summation starts at $r = 0$. This means the first term will have $r = 0$ (not 1), the second term $r = 1$, the third term $r = 2$, and the fourth term $r = 3$. In the given problem, $n = 5$, $a = 3x$, and $b = -2$. The fourth term of the expansion is

$$ {}_5C_3(3x)^{5-3}(-2)^3 = 10(9x^2)(-8) = -720x^2 $$

The correct choice is **(1)**.

20. To simplify a complex rational expression, first rewrite the numerator and denominator each as a single fraction.

$$\frac{\frac{x}{4}-\frac{1}{x}}{\frac{1}{2x}+\frac{1}{4}} = \frac{\frac{x}{4}\left(\frac{x}{x}\right)-\left(\frac{4}{4}\right)\frac{1}{x}}{\left(\frac{2}{2}\right)\frac{1}{2x}+\frac{1}{4}\left(\frac{x}{x}\right)}$$

$$= \frac{\frac{x^2-4}{4x}}{\frac{2+x}{4x}}$$

Multiply the numerator by the reciprocal of the denominator, factor, and reduce.

$$= \frac{x^2-4}{4x} \cdot \frac{2+x}{4x}$$

$$= \frac{(x-2)\cancel{(x+2)}}{\cancel{4x}} \cdot \frac{\cancel{4x}}{\cancel{(2+x)}}$$

$$= x-2$$

This problem can also be done by multiplying the numerator and denominator of the original fraction by the least common denominator (LCD) of all the separate simple fractions. The LCD of of 4, x, and $2x$ is $4x$.

$$\frac{\left(\frac{x}{4}-\frac{1}{x}\right)4x}{\left(\frac{1}{2x}+\frac{1}{4}\right)4x} = \frac{x^2-4}{2+x}$$

Factor and reduce.

$$= \frac{(x-2)\cancel{(x+2)}}{\cancel{(2+x)}} = x-2$$

The correct choice is **(2)**.

21. To solve the problem algebraically, first isolate the log term by dividing both sides by 2.

$$2\log_4(5x) = 3$$

$$\log_4(5x) = \frac{3}{2}$$

Rewrite the log in exponential form and solve.

$$4^{\frac{3}{2}} = 5x$$

$$\left(\sqrt{4}\right)^3 = 5x$$

$$8 = 5x$$

$$\frac{8}{5} = x$$

If you know the change of base formula, $\log_b x = \dfrac{\log x}{\log b}$, you can solve this problem on your calculator.

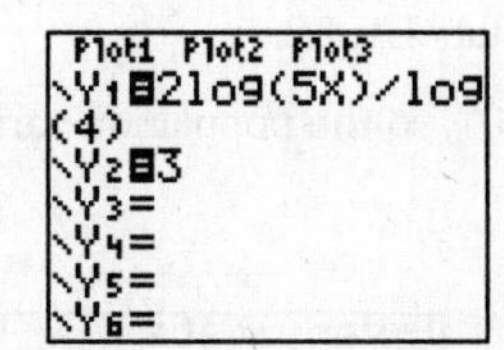

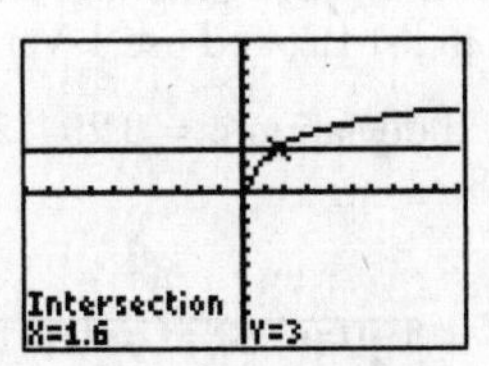

The correct choice is **(4)**.

22. Use the formula $s = \theta r$, where s is the arc intercepted by a central angle, θ, measured in radians, and r is the radius of the circle. For this problem, $s = (2)(4) = 8$.

The correct choice is **(4)**.

23. If the domain of a function is not given, it is assumed to be the largest set of real numbers for which the output of the function is defined and real. This means:

(A) Denominators must *not* equal 0 because division by 0 is undefined.

(B) Radicands (with an even index) must be greater than or equal to 0. Remember that even roots of negative numbers are imaginary.

(C) Arguments of logarithms must be greater than 0 because log(0) is undefined and the logarithm of a negative number is imaginary.

In this problem, only criterion (B) is relevant.

$$x - 2 \geq 0 \rightarrow x \geq 2$$

When written in interval notation, this is $[2, \infty)$.

The correct choice is **(3)**.

24. The population variance is the square of the population standard deviation σ, which can be found on your graphing calculator. Enter the averages in list L1, the frequencies in list L2, and use 1-Var Stats L1, L2.

The population standard deviation $\sigma = 8.20432937$, so the population variance is $8.20432937^2 \approx 67.3$.

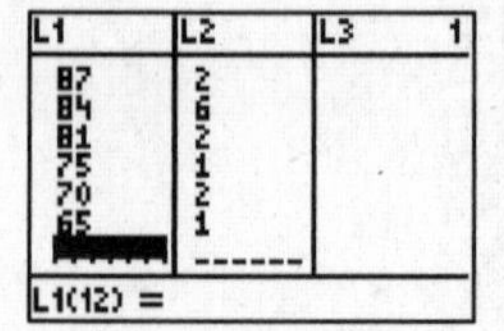

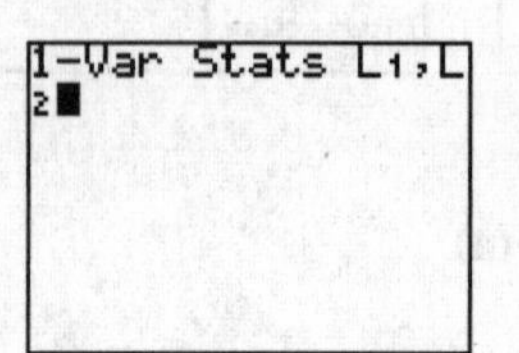

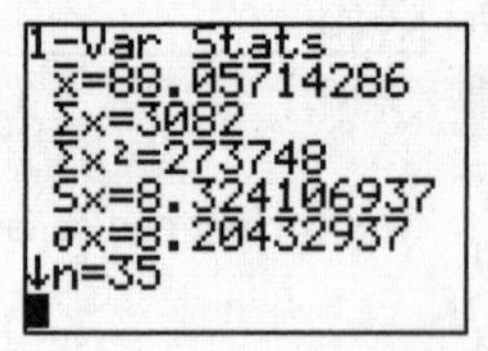

The correct choice is **(3)**.

25. An arrangement of letters is a permutation because the order of the letters matters. *DEADLINE* contains 8 letters, so there are ${}_8P_8$ orders. However, two letters are repeated; there are two *D*s and two *E*s. The total must be divided by the number of permutations for each of these letters, ${}_2P_2$, to eliminate the double counting of arrangements containing the repeated letters.

$$\frac{{}_8P_8}{{}_2P_2 \cdot {}_2P_2} = \frac{8!}{2! \cdot 2!}$$

The correct choice is **(4)**.

26. The transformation $f(x + a)$ is a horizontal translation of the graph of f to the left a units. Thus $f(x + 2)$ will translate the graph of f 2 units to the left.

The correct choice is **(2)**.

27. Solving for y gives $y = 2 \sin \theta + 3$. This implies that y is a function of the independent variable θ, $y = f(\theta)$. Using $f(\theta)$ in place of y gives $f(\theta) = 2 \sin \theta + 3$, which is choice (3). Notice how the other choices have a different variable for the independent variable in the function notation on the left than is represented in the equation on the right.

The correct choice is **(4)**.

PART II

28. Rationalize the denominator by multiplying by its conjugate. Simplify the radicals in the denominator.

$$\frac{5}{3-\sqrt{2}} = \frac{5}{(3-\sqrt{2})} \cdot \frac{(3+\sqrt{2})}{(3+\sqrt{2})}$$

$$= \frac{5(3+\sqrt{2})}{9+3\sqrt{3}-3\sqrt{3}-2}$$

$$= \frac{5(3+\sqrt{2})}{7}$$

The final answer may be written as either $\frac{5(3+\sqrt{2})}{7}$ or $\frac{15+5\sqrt{2}}{7}$.

Although you may use the calculator to confirm your answer, do not write the decimal from the calculator as an answer. The problem specifies the answer must be written in simplest radical form. If you write a decimal answer, you will receive no credit.

CALC Check

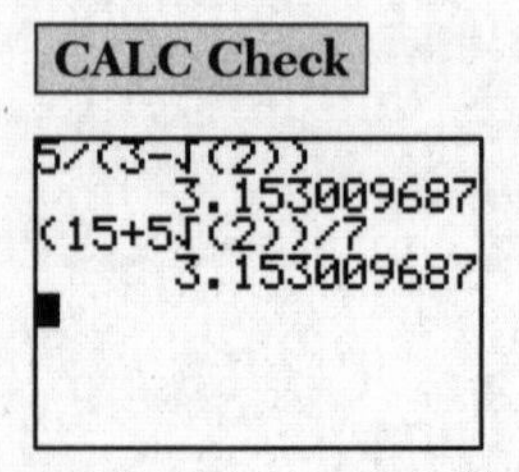

The answer is $\frac{5(3+\sqrt{2})}{7}$.

29. The equation of a circle with center at (h, k) and radius r is $(x-h)^2+(y-k)^2=r^2$. From the graph, the center is $(-3, 4)$. The radius is the distance from the center to either of the two known points on the circle. Finding the distance from the center to the origin $(0, 0)$, one of the known points, gives $r = \sqrt{(-3-0)^2+(4-0)^2} = 5$.

The equation of the circle is $(x-(-3))^2+(y-4)^2=5^2$ or $(x+3)^2+(y-4)^2 = 25$.

30. There are two ways you can begin this problem. With the first method, multiply both sides of the equation by $(x-3)$:

$$(x-3)\left(\frac{4x}{x-3}\right) = (x-3)\left(2+\frac{12}{x-3}\right)$$

$$4x = 2(x-3)+12$$

With the second method, rewrite each term with the common denominator. Then equate the numerators:

$$\frac{4x}{x-3} = \frac{2(x-3)}{x-3}+\frac{12}{x-3}$$

$$4x = 2(x-3)+12$$

Both methods yield the same equation. Solve for x:

$$4x = 2(x-3)+12$$
$$4x = 2x-6+12$$
$$4x = 2x+6$$
$$2x = 6$$
$$x = 3$$

Whenever you solve an equation containing a rational expression algebraically, you must check the candidate solutions in the original equation. In this problem, $x = 3$ makes two terms in the original equation undefined, so $x = 3$ must be rejected. Since it was the only candidate, the equation has no solution.

This problem can also be done graphically. If you graph each side of the equation separately and experiment with the window settings, you can get a graph like the one below. CALC intersect will not find any roots. However, it is difficult to justify from this graph that no roots exist.

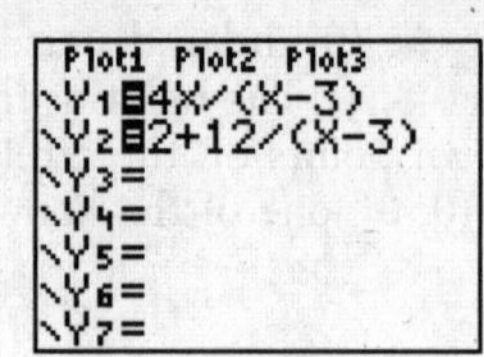

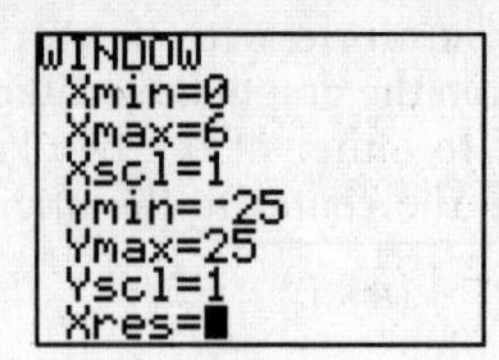

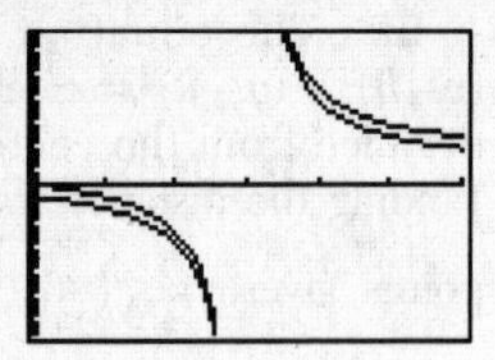

Instead, move all the terms to one side to set the equation equal to 0:

$$\frac{4x}{x-3}-2-\frac{12}{x-3}=0.$$

Graph this equation, shown below using the ZoomDecimal window. This graph appears to be a horizontal line (with one point at $x = 3$ missing). So it will never intersect the x-axis (equal 0.)

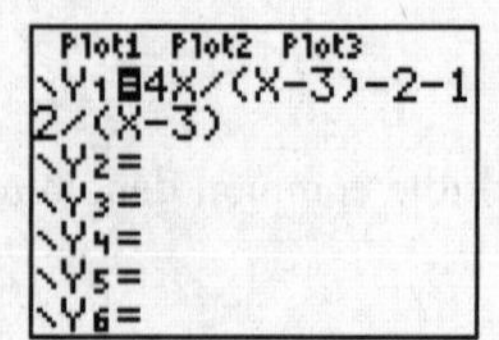

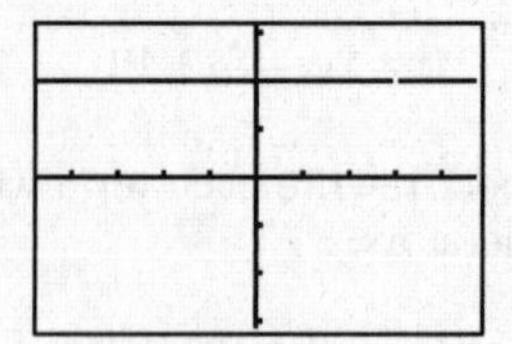

The equation has no solution. This can also be written as either { } or ∅. Note that using both symbols, {∅}, is incorrect.

31. To convert radians to degrees, multiply by $\frac{180°}{\pi}$:

$$3.45\left(\frac{180°}{\pi}\right)=197.6704393°$$

To convert decimal degrees to degrees and minutes, subtract the whole number of degrees and multiply the remaining decimal by 60.

```
3.45(180/π)
        197.6704393
Ans-197
       .6704393201
Ans*60
        40.22635921
```

To the nearest minute, 3.45 radians is 197°40′.

Alternatively, you can use the ▶DMS function in the ANGLE menu to convert to degrees-minutes-seconds.

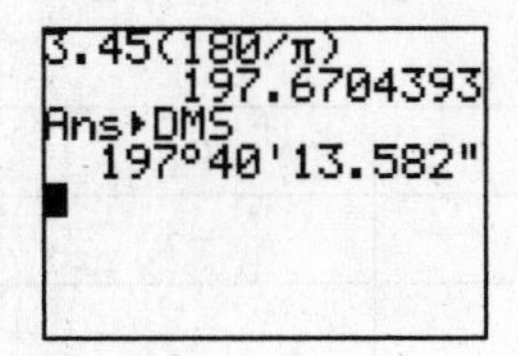

To the nearest minute, 3.45 radians is 197°40′.

32. Substitute the given values into the equation and evaluate. Remember that the interest rate must be converted to a decimal before it is used in the equation: 6.5% = 0.065. In this problem, $P = 1200$, $r = 0.065$, $t = 10$, and you are asked to find V.

$$V = Pe^{rt} = \$1{,}200e^{0.065(10)}$$

$$= \$2298.648995$$

To the nearest cent, Matt will have $2298.65.

33. You should know that csc θ is the reciprocal of sin θ. If θ is an angle in standard position passing through the point (x, y), $\sin\theta = \dfrac{\text{opposite}}{\text{hypotenuse}} = \dfrac{y}{r}$ where $r = \sqrt{x^2 + y^2}$. In this problem:

$$\sin\theta = \frac{2}{\sqrt{(-3)^2 + (2)^2}} = \frac{2}{\sqrt{13}}$$

$$\csc\theta = \frac{1}{\sin\theta} = \frac{\sqrt{13}}{2}$$

Note that the question asked for the "exact value." Do not use your calculator to find $\sin\theta = \dfrac{2}{\sqrt{(-3)^2 + (2)^2}} = 0.5547001962$ and $\csc\theta = \dfrac{1}{\sin\theta} = 1.802775638$.

If you give a decimal value for this answer, you will not receive full credit.

The exact value of csc θ is $\dfrac{\sqrt{13}}{2}$.

34. The first term is $a_1 = -3$. $a_{(n-1)}$ is the notation for the term immediately preceding a_n. The recursive formula $a_n = a_{(n-1)} - n$ says each new term is found by subtracting n, the index of the new term, from the value of the term before it. Setting up a table is helpful:

n	$a_n = a_{(n-1)} - n$	a_n
1	Given	–3
2	$a_2 = a_1 - 2 = -3 - 2 = -5$	–5
3	$a_3 = a_2 - 3 = -5 - 3 = -8$	–8
4	$a_4 = a_3 - 4 = -8 - 4 = -12$	–12

The first four terms of the sequence are –3, –5, –8, –12.

35. When selecting committees, the order of selection is not important so use combinations. There are ${}_9C_2$ different groups of 2 teachers that could be chosen from 9 and ${}_{20}C_3$ groups of 3 students that can be chosen from 20. To find the total number of committees that have 2 teachers and 3 students, multiply those two numbers:

$$ {}_9C_2 \cdot {}_{20}C_3 = 36 \cdot 1140 = 41{,}040 $$

The number of possible different committees is 41,040.

PART III

36. Divide both sides of the equation by 2.

$$2x^2 - 12x + 4 = 0$$

$$x^2 - 6x + 2 = 0$$

Subtract to move the constant to the right side.

$$x^2 - 6x = -2$$

Now complete the square on the left by adding $\left(\frac{b}{2}\right)^2 = \left(\frac{-6}{2}\right)^2 = 9$ to both sides.

$$x^2 - 6x + 9 = -2 + 9$$

Rewrite the left side as a perfect square.

$$(x - 3)^2 = 7$$

Solve for x.

$$x - 3 = \pm\sqrt{7}$$

$$x = 3 \pm \sqrt{7}$$

Although the solution for this equation can be found using the quadratic formula, the problem specifies that you must complete the square. If you solve this problem using the quadratic formula, you will not receive full credit.

The solution to $2x^2 - 12x + 4 = 0$ is $x = 3 \pm \sqrt{7}$.

37. The problem can be solved using factoring by grouping. Factor the common factor $4x^2$ out of the first two terms and –9 out of the remaining two terms. Be careful with the signs.

$$8x^3 + 4x^2 - 18x - 9 = 0$$

$$4x^2(2x + 1) - 9(2x + 1) = 0$$

Factor out the common binomial $2x + 1$.

$$(2x + 1)(4x^2 - 9) = 0$$

Continue factoring the difference of perfect squares.

$$(2x + 1)(2x + 3)(2x - 3) = 0$$

Set each factor equal to 0 and solve.

$$x = -\frac{1}{2};\ x = -\frac{3}{2};\ x = \frac{3}{2}$$

The question specifies that you must solve the problem algebraically. If you do not show the algebraic solution, you will not receive full credit. However, you can use a graphical solution on your calculator to check your answers. Use CALC zero to find all three roots. One root is shown below.

CALC Check

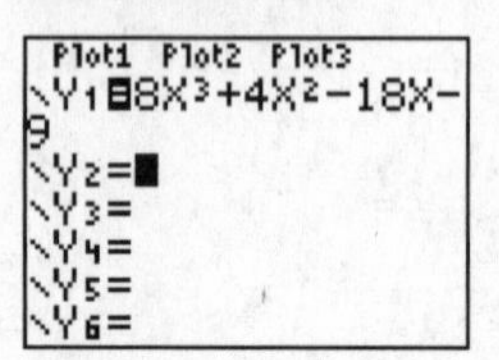

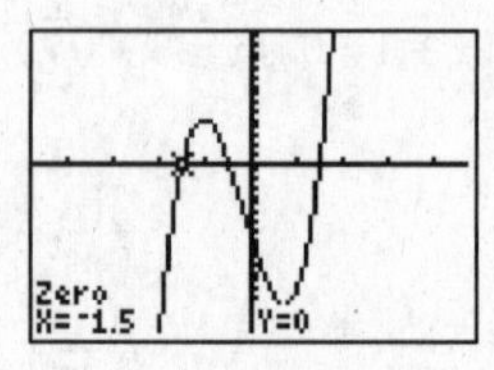

The solutions to $8x^3 + 4x^2 - 18x - 9 = 0$ are $x = \pm\frac{3}{2};\ x = -\frac{1}{2}$.

38. Enter the time, x, into list L1 on your graphing calculator and the number of bacteria, y, into list L2. Use the power regression function on your calculator.

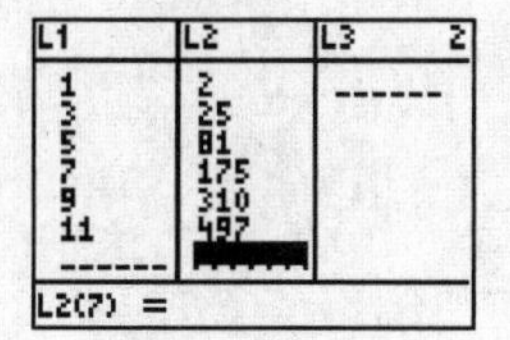

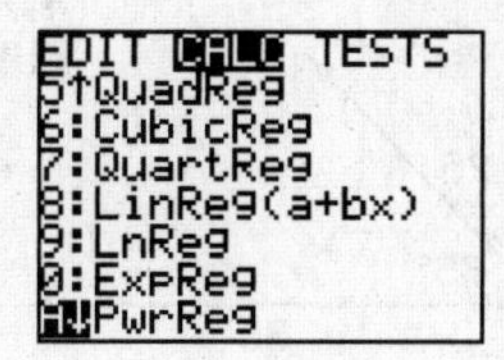

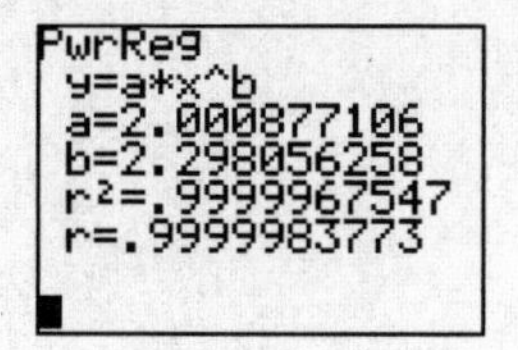

The power regression equation, rounded to three decimals, is $y = 2.001x^{2.298}$. Make sure you write $y =$ in your equation.

When $x = 15$, the equation gives $y = 2.001(15)^{2.298} = 1009.03$. You must write the substitution of 15 for x in your regression equation to receive full credit.

The power regression equation, rounded to three decimal places, is $y = 2.001x^{2.298}$. The estimated number of bacteria, to the nearest integer, after 15 minutes is 1009.

PART IV

39. Draw a diagram. Let one force act along the horizontal. Your initial diagram should look something like this.

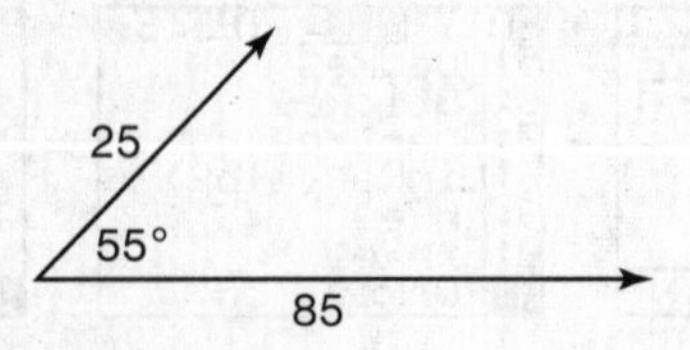

Use the parallelogram rule to add the vectors. Choose variables to represent the resultant force and the angle between the resultant and the larger force, r and θ. The 55° angle between the two forces is not part of either triangle. Because the two forces are not equal, the angle is not bisected by the resultant.

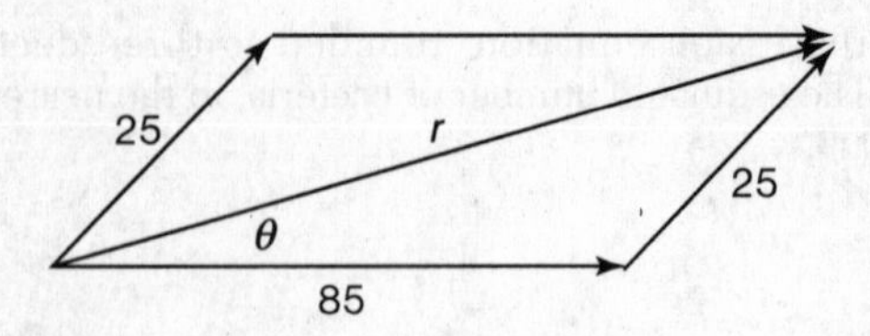

Since consecutive angles in a parallelogram are supplementary, find the measure of the obtuse angle, $180° - 55° = 125°$.

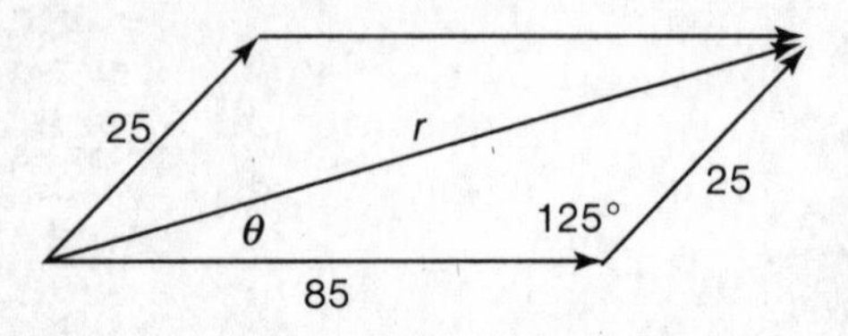

Since you know two sides and the included angle of a triangle, find the third side r using the Law of Cosines, shown on the reference sheet.

$$r^2 = 85^2 + 25^2 - 2(85)(25)\cos 125°$$

$$r^2 = 10287.69985$$

$$r = \sqrt{10287.69985} = 101.4282991$$

Angle θ can be found using the Law of Sines, shown on the reference sheet. Since you know θ is acute, you do not need to find a solution in Quadrant II.

$$\frac{101.4282991}{\sin 125°} = \frac{25}{\sin\theta}$$

$$101.4282991 \sin\theta = 25 \sin 125°$$

$$\sin\theta = \frac{25\sin 125}{101.4282991} = 0.201904215$$

$$\theta = \sin^{-1}(0.201904215) = 11.6483°$$

To the *nearest hundredth of a newton*, the resultant force is 101.43 N. To the *nearest degree*, the angle between the resultant and the larger force is 12°.

Topic	Question Numbers	Number of Points	Your Points	Your Percentage
1. Exponents and Radicals: operations, equivalent expressions, simplifying, rationalizing	14, 18, 28	$2+2+2=6$		
2. Complex Numbers: operations, powers of i, rationalizing	1	2		
3. Quadratics and Higher-Order Polynomials: operations, factoring, binomial expansion, formula, quadratic equations and inequalities, nature of the roots, sum and product of the roots, quadratic-linear systems, completing the square, higher-order polynomials	12, 17, 19, 36, 37	$2+2+2+4+4=14$		
4. Rationals: operations, equations	20, 30	$2+2=4$		
5. Absolute Value: equations and inequalities	5	2		
6. Direct and Inverse Variation	—	—		
7. Circles: center-radius equation, completing the square	29	2		
8. Functions: relations, functions, domain, range, one-to-one, onto, inverses, compositions, transformations of functions	2, 6, 8, 16 23, 26, 27	$2+2+2+2+2+2+2=14$		
9. Exponential and Logarithmic Functions: equations, common bases, logarithm rules, e, ln	9, 21, 32	$2+2+2=6$		
10. Trigonometric Functions: radian measure, arc length, cofunctions, unit circle, inverses	22, 31, 33	$2+2+2=6$		
11. Trigonometric Graphs: graphs of basic functions and inverse functions, domain, range, restricted domain, transformations of graphs, amplitude, frequency, period, phase shift	13	2		
12. Trigonometric Identities, Formulas, and Equations	3, 10	$2+2=4$		
13. Trigonometry Laws and Applications: area of a triangle, sine and cosine laws, ambiguous cases	7, 39	$2+6=8$		
14. Sequences and Series: sigma notation, arithmetic, geometric	11, 34	$2+2=4$		

Topic	Question Numbers	Number of Points	Your Points	Your Percentage
15. Statistics: studies, central tendency, dispersion including standard deviation, normal distribution	4, 15, 24	$2+2+2=6$		
16. Regressions: linear, exponential, logarithmic, power, correlation coefficient	38	4		
17. Probability: permutation, combination, geometric, binomial	25, 35	$2+2=4$		

HOW TO CONVERT YOUR RAW SCORE TO YOUR ALGEBRA 2/TRIGONOMETRY REGENTS EXAMINATION SCORE

Below is a sample conversion chart that can be used to determine your final score on the Official Test Sampler. To estimate your final exam score, locate in the column labeled "Raw Score" the total number of points you scored out of a possible 88 points. Since partial credit is allowed in Parts II, III, and IV of the test, you may need to approximate the credit you would receive for a solution that is not completely correct. Then locate in the adjacent column to the right the scaled score that corresponds to your raw score. The scaled score is your final sample Regents Examination score.

*__Important Note:__ These are inexact estimates, since no official conversion charts for Algebra 2/Trigonometry exist as of press time.

Raw Score	Scaled Score	Raw Score	Scaled Score	Raw Score	Scaled Score
88	**100**	58	**75**	28	**46**
87	**99**	57	**74**	27	**45**
86	**99**	56	**74**	26	**43**
85	**97**	55	**73**	25	**42**
84	**96**	54	**72**	24	**41**
83	**95**	53	**71**	23	**39**
82	**94**	52	**70**	22	**38**
81	**93**	51	**69**	21	**37**
80	**92**	50	**69**	20	**35**
79	**92**	49	**68**	19	**34**
78	**91**	48	**67**	18	**32**
77	**90**	47	**66**	17	**31**
76	**89**	46	**65**	16	**29**
75	**89**	45	**64**	15	**28**
74	**88**	44	**63**	14	**26**
73	**87**	43	**62**	13	**24**
72	**86**	42	**61**	12	**23**
71	**86**	41	**60**	11	**21**
70	**85**	40	**59**	10	**19**
69	**84**	39	**58**	9	**17**
68	**83**	38	**57**	8	**15**
67	**83**	37	**56**	7	**14**
66	**82**	36	**55**	6	**12**
65	**81**	35	**54**	5	**10**
64	**80**	34	**53**	4	**8**
63	**79**	33	**52**	3	**6**
62	**79**	32	**51**	2	**4**
61	**78**	31	**49**	1	**2**
60	**77**	30	**48**	0	**0**
59	**76**	29	**47**		

Examination June 2010

Algebra 2/Trigonometry

REFERENCE SHEET

Area of a Triangle

$K = \frac{1}{2}ab \sin C$

Law of Cosines

$a^2 = b^2 + c^2 - 2bc \cos A$

Functions of the Sum of Two Angles

$\sin(A + B) = \sin A \cos B + \cos A \sin B$

$\cos(A + B) = \cos A \cos B - \sin A \sin B$

$\tan(A + B) = \frac{\tan A + \tan B}{1 - \tan A \tan B}$

Functions of the Double Angle

$\sin 2A = 2 \sin A \cos A$

$\cos 2A = \cos^2 A - \sin^2 A$

$\cos 2A = 2 \cos^2 A - 1$

$\cos 2A = 1 - 2 \sin^2 A$

$\tan 2A = \frac{2 \tan A}{1 - \tan^2 A}$

Functions of the Difference of Two Angles

$\sin(A - B) = \sin A \cos B - \cos A \sin B$

$\cos(A - B) = \cos A \cos B + \sin A \sin B$

$\tan(A - B) = \frac{\tan A - \tan B}{1 + \tan A \tan B}$

Functions of the Half Angle

$$\sin\frac{1}{2}A = \pm\sqrt{\frac{1-\cos A}{2}}$$

$$\cos\frac{1}{2}A = \pm\sqrt{\frac{1+\cos A}{2}}$$

$$\tan\frac{1}{2}A = \pm\sqrt{\frac{1-\cos A}{1+\cos A}}$$

Law of Sines

$$\frac{a}{\sin A} = \frac{b}{\sin B} = \frac{c}{\sin C}$$

Sum of a Finite Arithmetic Sequence

$$S_n = \frac{n(a_1 + a_n)}{2}$$

Sum of a Finite Geometric Sequence

$$S_n = \frac{a_1(1-r^n)}{1-r}$$

Binomial Theorem

$$(a+b)^n = {}_nC_0a^nb^0 + {}_nC_1a^{n-1}b^1 + {}_nC_2a^{n-2}b^2 + \cdots + {}_nC_na^0b^n$$

$$(a+b)^n = \sum_{r=0}^{n} {}_nC_ra^{n-r}b^r$$

Normal Curve Standard Deviation

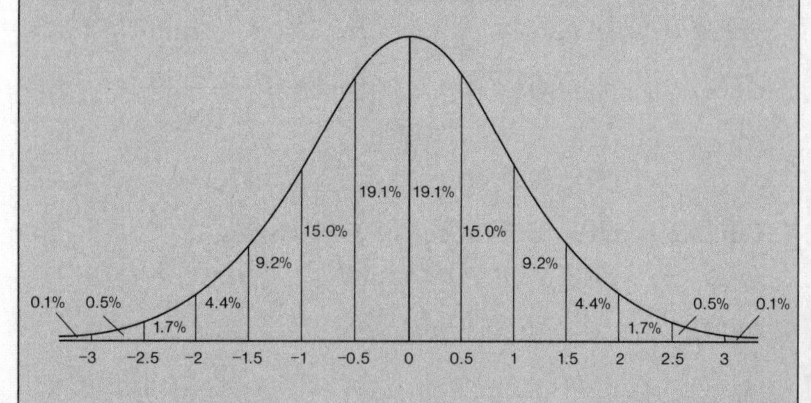

PART I

Answer all 27 questions in this part. Each correct answer will receive 2 credits. No partial credit will be allowed. For each question, write in the space provided the numeral preceding the word or expression that best completes the statement or answers the question. [54 credits]

1 What is the common difference of the arithmetic sequence 5, 8, 11, 14?

(1) $\frac{8}{5}$ (3) 3

(2) −3 (4) 9 1 _____

2 What is the number of degrees in an angle whose radian measure is $\frac{11\pi}{12}$?

(1) 150 (3) 330

(2) 165 (4) 518 2 _____

3 If $a = 3$ and $b = -2$, what is the value of the expression $\frac{a^{-2}}{b^{-3}}$?

(1) $-\frac{9}{8}$ (3) $-\frac{8}{9}$

(2) −1 (4) $\frac{8}{9}$ 3 _____

4 Four points on the graph of the function f(x) are shown below.

$$\{(0,1), (1,2), (2,4), (3,8)\}$$

Which equation represents f(x)?

(1) $f(x) = 2^x$
(2) $f(x) = 2x$
(3) $f(x) = x + 1$
(4) $f(x) = \log_2 x$

4 _____

5 The graph of $y = f(x)$ is shown below.

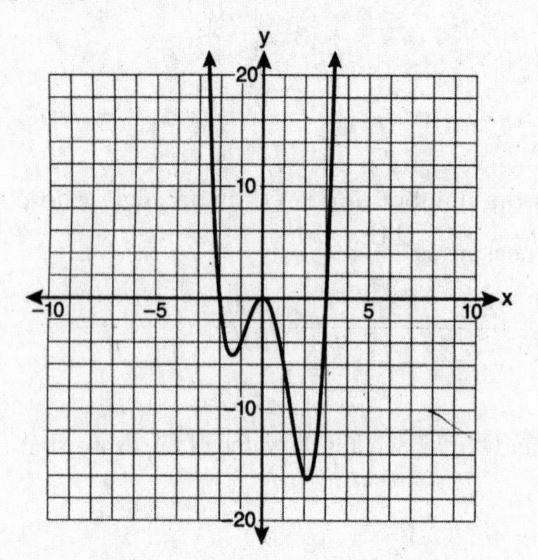

Which set lists all the real solutions of $f(x) = 0$?

(1) {–3, 2}
(2) {–2, 3}
(3) {–3, 0, 2}
(4) {–2, 0, 3}

5 _____

6 In simplest form, $\sqrt{-300}$ is equivalent to

(1) $3i\sqrt{10}$ (3) $10i\sqrt{3}$
(2) $5i\sqrt{12}$ (4) $12i\sqrt{5}$ 6 _____

7 Twenty different cameras will be assigned to several boxes. Three cameras will be randomly selected and assigned to box A. Which expression can be used to calculate the number of ways that three cameras can be assigned to box A?

(1) $20!$ (3) ${}_{20}C_3$
(2) $\frac{20!}{3!}$ (4) ${}_{20}P_3$ 7 _____

8 Factored completely, the expression $12x^4 + 10x^3 - 12x^2$ is equivalent to

(1) $x^2(4x + 6)(3x - 2)$
(2) $2(2x^2 + 3x)(3x^2 - 2x)$
(3) $2x^2(2x - 3)(3x + 2)$
(4) $2x^2(2x + 3)(3x - 2)$ 8 _____

9 The solutions of the equation $y^2 - 3y = 9$ are

(1) $\frac{3 \pm 3i\sqrt{3}}{2}$ (3) $\frac{-3 \pm 3\sqrt{5}}{2}$
(2) $\frac{3 \pm 3i\sqrt{5}}{2}$ (4) $\frac{3 \pm 3\sqrt{5}}{2}$ 9 _____

10 The expression $2 \log x - (3 \log y + \log z)$ is equivalent to

(1) $\log \frac{x^2}{y^3 z}$

(2) $\log \frac{x^2 z}{y^3}$

(3) $\log \frac{2x}{3yz}$

(4) $\log \frac{2xz}{3y}$

10 _____

11 The expression $(x^2 - 1)^{-\frac{2}{3}}$ is equivalent to

(1) $\sqrt[3]{(x^2 - 1)^2}$

(2) $\frac{1}{\sqrt[3]{(x^2 - 1)^2}}$

(3) $\sqrt{(x^2 - 1)^3}$

(4) $\frac{1}{\sqrt{(x^2 - 1)^3}}$

11 _____

12 Which expression is equivalent to $\frac{\sqrt{3}+5}{\sqrt{3}-5}$?

(1) $-\frac{14+5\sqrt{3}}{11}$

(2) $-\frac{17+5\sqrt{3}}{11}$

(3) $\frac{14+5\sqrt{3}}{14}$

(4) $\frac{17+5\sqrt{3}}{14}$

12 _____

13 Which relation is *not* a function?

(1) $(x-2)^2+y^2=4$
(2) $x^2+4x+y=4$
(3) $x+y=4$
(4) $xy=4$

13 ______

14 If $\angle A$ is acute and $\tan A=\frac{2}{3}$, then

(1) $\cot A=\frac{2}{3}$
(2) $\cot A=\frac{1}{3}$
(3) $\cot(90°-A)=\frac{2}{3}$
(4) $\cot(90°-A)=\frac{1}{3}$

14 ______

15 The solution set of $4^{x^2+4x}=2^{-6}$ is

(1) $\{1, 3\}$
(2) $\{-1, 3\}$
(3) $\{-1, -3\}$
(4) $\{1, -3\}$

15 ______

16 The equation $x^2+y^2-2x+6y+3=0$ is equivalent to

(1) $(x-1)^2+(y+3)^2=-3$
(2) $(x-1)^2+(y+3)^2=7$
(3) $(x+1)^2+(y+3)^2=7$
(4) $(x+1)^2+(y+3)^2=10$

16 ______

17 Which graph best represents the inequality $y + 6 \geq x^2 - x$?

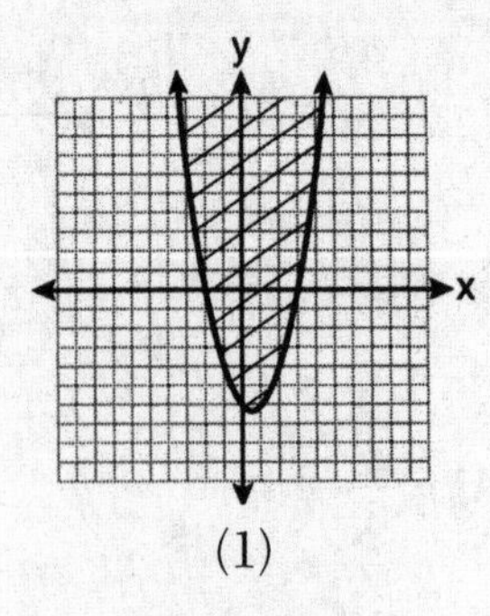

(1)

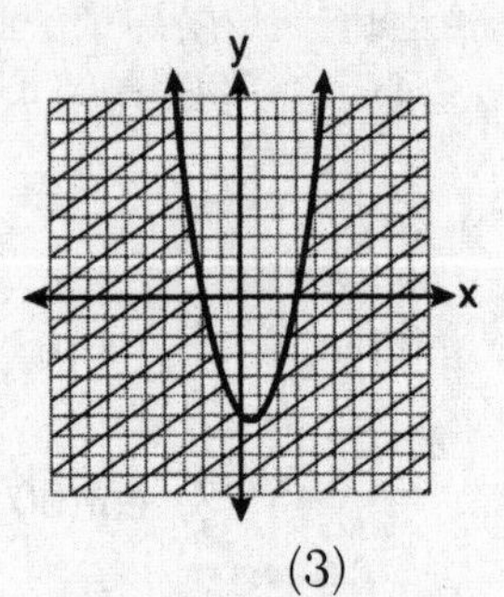

(3)

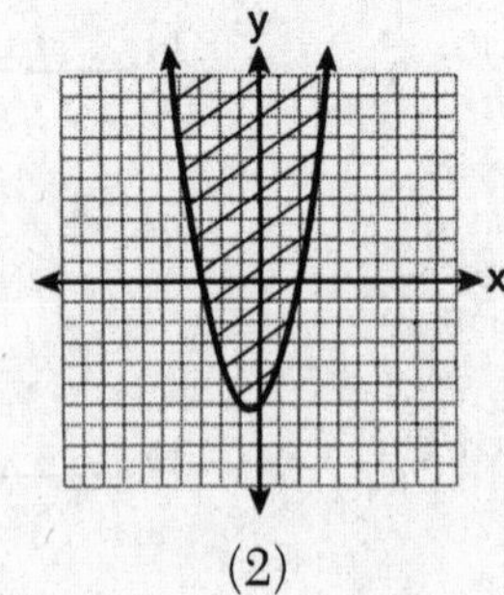

(2)

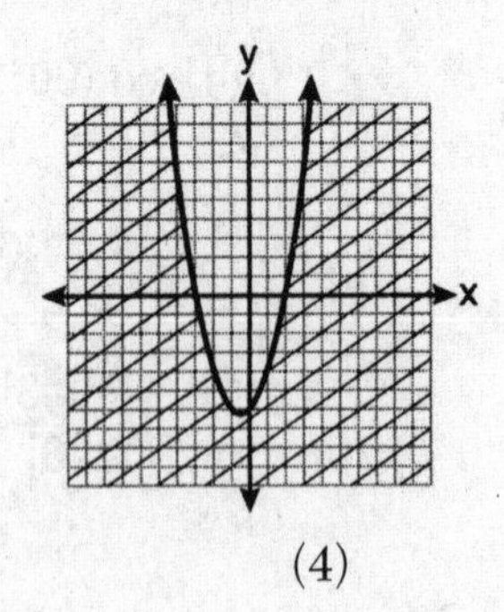

(4)

17 _____

18 The solution set of the equation $\sqrt{x+3} = 3 - x$ is

(1) $\{1\}$ (3) $\{1, 6\}$
(2) $\{0\}$ (4) $\{2, 3\}$

18 _____

19 The product of i^7 and i^5 is equivalent to

(1) 1 (3) i
(2) -1 (4) $-i$ 19 _____

20 Which equation is represented by the graph below?

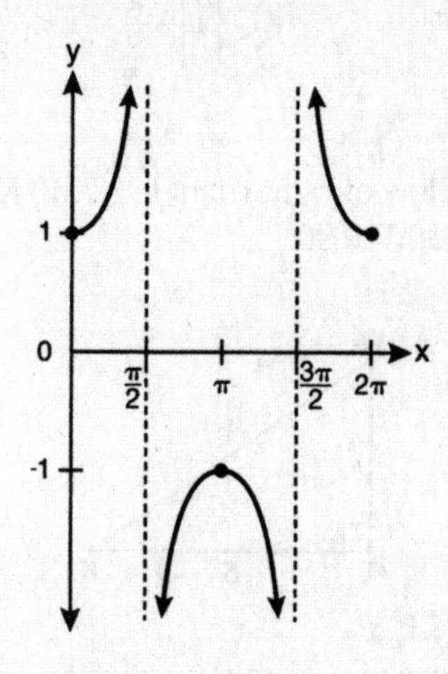

(1) $y = \cot x$ (3) $y = \sec x$
(2) $y = \csc x$ (4) $y = \tan x$ 20 _____

21 Which value of r represents data with a strong negative linear correlation between two variables?

(1) -1.07 (3) -0.14
(2) -0.89 (4) 0.92 21 _____

22 The function $f(x) = \tan x$ is defined in such a way that $f^{-1}(x)$ is a function. What can be the domain of $f(x)$?

(1) $\{x \mid 0 \le x \le \pi\}$

(2) $\{x \mid 0 \le x \le 2\pi\}$

(3) $\left\{x \middle| -\frac{\pi}{2} < x < \frac{\pi}{2}\right\}$

(4) $\left\{x \middle| -\frac{\pi}{2} < x < \frac{3\pi}{2}\right\}$

22 ______

23 In the diagram below of right triangle KTW, $KW = 6$, $KT = 5$, and $m\angle KTW = 90$.

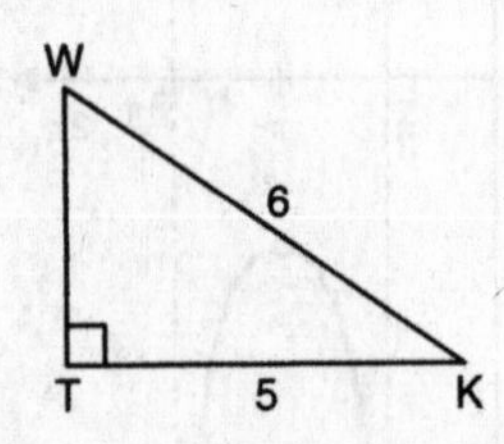

What is the measure of $\angle K$, to the *nearest minute*?

(1) 33°33'
(2) 33°34'
(3) 33°55'
(4) 33°56'

23 ______

24 The expression $\cos^2 \theta - \cos 2\theta$ is equivalent to

(1) $\sin^2 \theta$
(2) $-\sin^2 \theta$
(3) $\cos^2 \theta + 1$
(4) $-\cos^2 \theta - 1$

24 ______

25 Mrs. Hill asked her students to express the sum $1 + 3 + 5 + 7 + 9 + \ldots + 39$ using sigma notation. Four different student answers were given. Which student answer is correct?

(1) $\sum_{k=1}^{20} (2k-1)$

(2) $\sum_{k=2}^{40} (k-1)$

(3) $\sum_{k=-1}^{37} (k+2)$

(4) $\sum_{k=1}^{39} (2k-1)$

25 _____

26 What is the formula for the nth term of the sequence 54, 18, 6, . . . ?

(1) $a_n = 6\left(\frac{1}{3}\right)^n$

(2) $a_n = 6\left(\frac{1}{3}\right)^{n-1}$

(3) $a_n = 54\left(\frac{1}{3}\right)^n$

(4) $a_n = 54\left(\frac{1}{3}\right)^{n-1}$

26 _____

27 What is the period of the function $y = \frac{1}{2}\sin\left(\frac{x}{3} - \pi\right)$?

(1) $\frac{1}{2}$

(2) $\frac{1}{3}$

(3) $\frac{2}{3}\pi$

(4) 6π

27 _____

PART II

Answer all 8 questions in this part. Each correct answer will receive 2 credits. Clearly indicate the necessary steps, including appropriate formula substitutions, diagrams, graphs, charts, etc. For all questions in this part, a correct numerical answer with no work shown will receive only 1 credit. [16 credits]

28 Use the discriminant to determine all values of k that would result in the equation $x^2 - kx + 4 = 0$ having equal roots.

29 The scores of one class on the Unit 2 mathematics test are shown in the table below.

Unit 2 Mathematics Test

Test Score	Frequency
96	1
92	2
84	5
80	3
76	6
72	3
68	2

Find the population standard deviation of these scores, to the *nearest tenth.*

30 Find the sum and product of the roots of the equation $5x^2 + 11x - 3 = 0$.

31 The graph of the equation $y = \left(\frac{1}{2}\right)^x$ has an asymptote. On the grid below, sketch the graph of $y = \left(\frac{1}{2}\right)^x$ and write the equation of this asymptote.

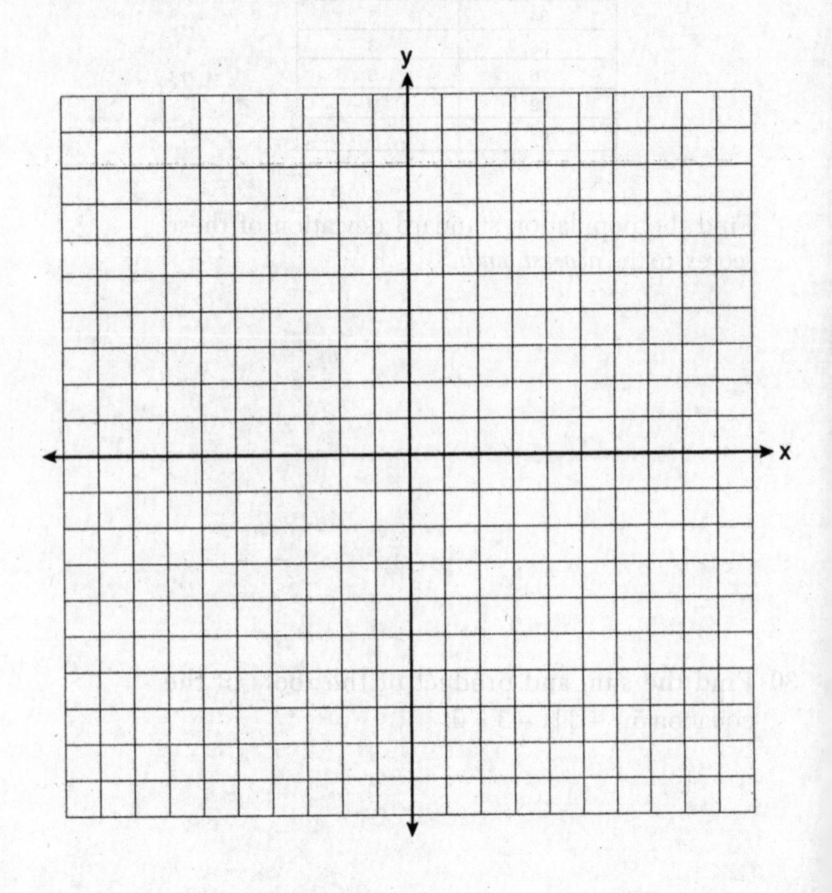

32 Express $5\sqrt{3x^3} - 2\sqrt{27x^3}$ in simplest radical form.

33 On the unit circle shown in the diagram below, sketch an angle, in standard position, whose degree measure is 240 and find the exact value of sin 240°.

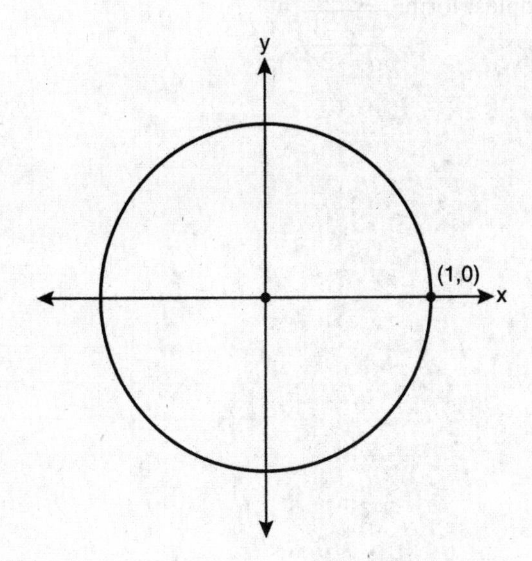

34 Two sides of a parallelogram are 24 feet and 30 feet. The measure of the angle between these sides is 57°. Find the area of the parallelogram, to the *nearest square foot.*

35 Express in simplest form: $\dfrac{\dfrac{1}{2}-\dfrac{4}{d}}{\dfrac{1}{d}+\dfrac{3}{2d}}$

PART III

Answer all 3 questions in this part. Each correct answer will receive 4 credits. Clearly indicate the necessary steps, including appropriate formula substitutions, diagrams, graphs, charts, etc. For all questions in this part, a correct numerical answer with no work shown will receive only 1 credit. [12 credits]

36 The members of a men's club have a choice of wearing black or red vests to their club meetings. A study done over a period of many years determined that the percentage of black vests worn is 60%. If there are 10 men at a club meeting on a given night, what is the probability, to the *nearest thousandth*, that *at least* 8 of the vests worn will be black?

37 Find all values of θ in the interval $0° \le \theta < 360°$ that satisfy the equation $\sin 2\theta = \sin \theta$.

38 The letters of any word can be rearranged. Carol believes that the number of different 9-letter arrangements of the word "TENNESSEE" is greater than the number of different 7-letter arrangements of the word "VERMONT." Is she correct? Justify your answer.

PART IV

Answer the question in this part. The correct answer will receive 6 credits. Clearly indicate the necessary steps, including appropriate formula substitutions, diagrams, graphs, charts, etc. A correct numerical answer with no work shown will receive only 1 credit. [6 credits]

39 In a triangle, two sides that measure 6 cm and 10 cm form an angle that measures 80°. Find, to the *nearest degree*, the measure of the smallest angle in the triangle.

Answers June 2010

Algebra 2/Trigonometry

Answer Key

PART I

1. (3)	**6.** (3)	**11.** (2)	**16.** (2)	**21.** (2)	**26.** (4)
2. (2)	**7.** (3)	**12.** (1)	**17.** (1)	**22.** (3)	**27.** (4)
3. (3)	**8.** (4)	**13.** (1)	**18.** (1)	**23.** (1)	
4. (1)	**9.** (4)	**14.** (3)	**19.** (1)	**24.** (1)	
5. (4)	**10.** (1)	**15.** (3)	**20.** (3)	**25.** (1)	

PART II

28. 4 and –4

29. 7.4

30. sum = $-\frac{11}{5}$ and product = $-\frac{3}{5}$

31. correct graph and $y = 0$

32. $-x\sqrt{3x}$

33. angle correctly drawn and $-\frac{\sqrt{3}}{2}$

34. 604

35. $\frac{d-8}{5}$

PART III

36. 0.167

37. 0°, 60°, 180°, and 300°

38. no, with justification

PART IV

39. 33°

Answers Explained

PART I

1. In an arithmetic sequence, the common difference is found by subtracting consecutive terms. In this problem, $8 - 5 = 3$. The common difference is the same for each set of consecutive terms: $11 - 8 = 3$ and $14 - 11 = 3$. Note that it is important to subtract in the correct order. For example, $5 - 8 = -3$ is incorrect.

The correct choice is **(3)**.

2. To convert radians to degrees, multiply by $\frac{180^\circ}{\pi}$:

$$\frac{11\pi}{12} \cdot \frac{180^\circ}{\pi} = \frac{1980^\circ}{12}$$

$$= 165^\circ$$

CALC Check

The conversion can also be done on your calculator. Set the calculator to the mode you want for the answer, degrees. Make sure the expression for the angle is in parentheses. Then put the symbol for the original unit (found in the ANGLE menu) after the expression and select ENTER.

The correct choice is **(2)**.

3. Substituting the given values into the expression gives $\frac{(3)^{-2}}{(-2)^{-3}}$. A negative exponent means to use the reciprocal of the base:

$$\frac{(3)^{-2}}{(-2)^{-3}} = \frac{(-2)^{3}}{(3)^{2}}$$

$$= \frac{-8}{9}$$

$$= -\frac{8}{9}$$

CALC Check

This problem is easily done on the calculator. Just remember to use parentheses around negative bases. To convert a decimal to a fraction, use MATH 1: ▶Frac.

```
3^-2/(-2)^-3
     -.8888888889
Ans▶Frac
             -8/9
```

The correct choice is **(3)**.

4. Check each point in each candidate function.

- Choice (1): $2^0 = 1$, $2^1 = 2$, $2^2 = 4$, and $2^3 = 8$ are all correct. This is the correct answer.
- Choice (2): $2(0) \neq 1$. You need not check the rest. This choice is wrong.
- Choice (3): $0 + 1 = 1$, $1 + 1 = 2$, and $2 + 1 \neq 3$. This choice is wrong.
- Choice (4): $\log_2 0$ is undefined. This choice is wrong.

Choice (3) illustrates an important rule. Check all the points in the function. Just because one or two answers check does not mean you have found the right answer.

CALC Check

You can also use a graphing calculator to find the answer. Use Y= and then TABLE.

Plot1 Plot2 Plot3
\Y1=2^X
\Y2=
\Y3=
\Y4=
\Y5=
\Y6=
\Y7=

X	Y1
0	1
1	2
2	4
3	8
4	16
5	32
6	64

X=6

The correct choice is **(1)**.

5. The roots of a function are the x-values where the graph of the function intersects the x-axis. This function has three roots: $x = -2$, $x = 0$, and $x = 3$. Remember that to intersect does not mean the graph must cross the x-axis. The graph only has to touch the x-axis. Hence $x = 0$ is a root. The solution set is $\{-2, 0, 3\}$.

The correct choice is **(4)**.

6. The square root of a negative number is imaginary. Rewrite it in terms of i, and then simplify the radical:

$$\sqrt{-300} = i\sqrt{300}$$

$$= i\sqrt{100}\sqrt{3}$$

$$= 10i\sqrt{3}$$

Note that $5i\sqrt{12}$ is equivalent to $\sqrt{-300}$ but is not in simplest form.

CALC Check

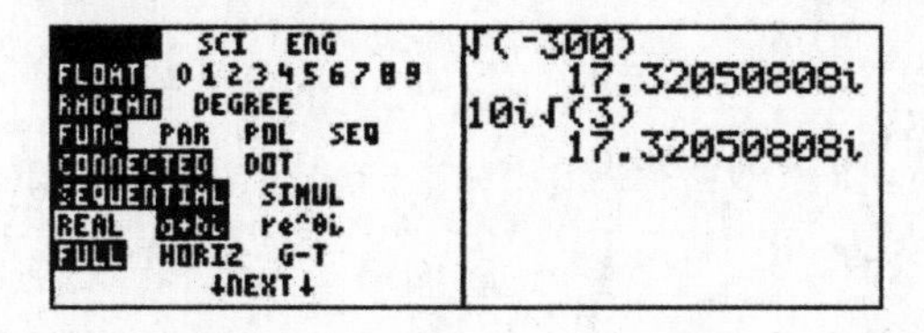

The correct choice is **(3)**.

7. The question is asking how many different groups of three cameras can be selected from a set of twenty. Since the problem does not refer to the order or arrangement of the cameras in the box, use combinations. The number of ways three cameras can be selected from a set of twenty is ${}_{20}C_3$.

- Choice (1) is the number of arrangements of all twenty cameras.
- Choice (2) is the number of arrangements of all twenty cameras where seventeen are different kinds and three are of the same type.
- Choice (4) is the number of arrangements of three cameras from the set of twenty.

The correct choice is **(3)**.

8. When factoring a polynomial, first factor out the greatest common factor. Then try to factor the remaining polynomial term. In this problem, the GCF is $2x^2$.

$$12x^4 + 10x^3 - 12x^2 = 2x^2(6x^2 + 5x - 6)$$

Then factor the trinomial:

$$2x^2(6x^2 + 5x - 6) = 2x^2(3x - 2)(2x + 3)$$

You should check your factoring by multiplying the trinomial. See the solution to problem 3.5 in the "Key Algebra 2/Trigonometry Facts and Skills" section for a method to factor a trinomial with your graphing calculator.

Choices (1) and (2) are equivalent expressions but are not factored completely.

The correct choice is **(4)**.

9. The equation is quadratic. Since all the answer choices are either imaginary or irrational, you will not be able to factor it. Rewrite it in standard form. Set one side equal to zero, and write the variables in descending order. Then use the quadratic formula.

$$y^2 - 3y = 9$$

$$y^2 - 3y - 9 = 0$$

so $a = 1$, $b = -3$, and $c = -9$

$$y = \frac{-b \pm \sqrt{b^2 - 4ac}}{2a}$$

$$= \frac{-(-3) \pm \sqrt{(-3)^2 - 4(1)(-9)}}{2(1)}$$

$$= \frac{3 \pm \sqrt{45}}{2}$$

$$= \frac{3 \pm \sqrt{9}\sqrt{5}}{2}$$

$$= \frac{3 \pm 3\sqrt{5}}{2}$$

The correct choice is **(4)**.

10. Use the properties of logarithms and exponents.

$$2 \log x - (3 \log y + \log z)$$

$n \log a = \log a^n$	$= \log x^2 - (\log y^3 + \log z)$
$\log a + \log b = \log(ab)$	$= \log x^2 - \log y^3 z$
$\log a - \log b = \log \dfrac{a}{b}$	$= \log \dfrac{x^2}{y^3 z}$

The correct choice is **(1)**.

11. For a negative exponent, take the reciprocal of the base. The fraction in the exponent means the expression will contain a radical. The numerator of the fraction will become the exponent; the denominator will become the index of the radical.

$$(x^2 - 1)^{-\frac{2}{3}} = \frac{1}{(x^2 - 1)^{\frac{2}{3}}} = \frac{1}{\sqrt[3]{(x^2 - 1)^2}}$$

CALC Check

You can use a graphing calculator to confirm your result. Put the original expression in Y_1 and your answer in Y_2. Then use TABLE to see if their values are identical.

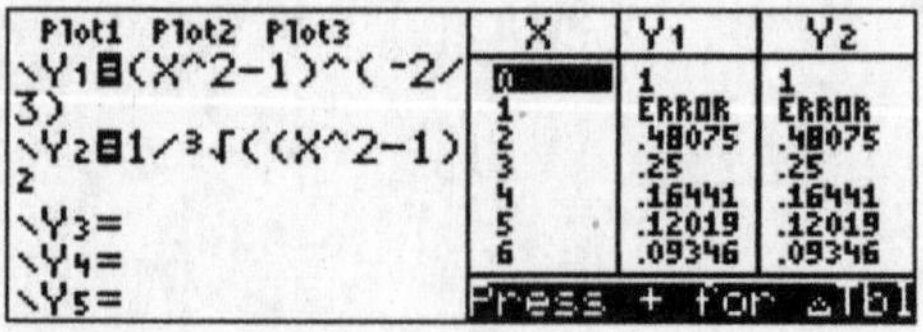

```
Plot1 Plot2 Plot3
\Y1=(X^2-1)^(-2/
3)
\Y2=1/³√((X^2-1)
2
\Y3=
\Y4=
\Y5=
```

X	Y1	Y2
0	1	1
1	ERROR	ERROR
2	.48075	.48075
3	.25	.25
4	.16441	.16441
5	.12019	.12019
6	.09346	.09346

Press + for ΔTbl

The correct choice is **(2)**.

12. Rationalize the denominator by multiplying by the conjugate.

$$\frac{\sqrt{3}+5}{\sqrt{3}-5} = \frac{\left(\sqrt{3}+5\right)}{\left(\sqrt{3}-5\right)} \cdot \frac{\left(\sqrt{3}+5\right)}{\left(\sqrt{3}+5\right)}$$

$$= \frac{3+5\sqrt{3}+5\sqrt{3}+25}{3+5\sqrt{3}-5\sqrt{3}-25}$$

$$= \frac{28+10\sqrt{3}}{-22}$$

$$= -\frac{14+5\sqrt{3}}{11}$$

CALC Check

Using the MathPrint mode on the TI-84 makes it easy to type complicated fractions accurately. First put in the original expression. Then put in your answer to see if they are identical.

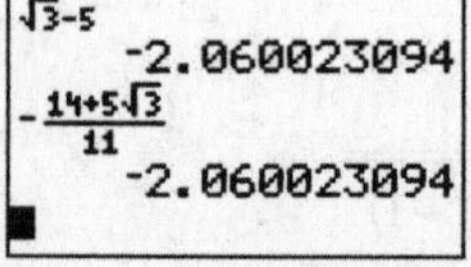

$$\frac{\sqrt{3}+5}{\sqrt{3}-5}$$
-2.060023094

$$-\frac{14+5\sqrt{3}}{11}$$
-2.060023094

The correct choice is **(1)**.

13. In a function, each input (x-value) has only one output (y-value). An equation that has either y^2 or no y will not represent a function. Choice (1) has y^2, which means it is quadratic in y. In other words, some x-values will lead to two different y-values. It is not a function.

Alternatively, you can use a graphical representation for each equation. Any vertical line will intersect the graph of a function in at most one point (vertical line test).

- Choice (1): You should recognize this as the equation of a circle. Some vertical lines will intersect the circle twice. This is not a function.
- Choice (2): This is the equation of a parabola opening down:

  $$y = -x^2 - 4x + 4.$$

  Each x-value will give only one value for y. Each vertical line will intersect the graph exactly once. This relation is a function.
- Choice (3): This is the equation of a slant line. All lines except vertical lines are functions.
- Choice (4): This is the equation of an inverse variation. The equation can be rewritten as $y = \frac{4}{x}$. Any nonzero x-value will give exactly one y-value. This is a function.

The correct choice is **(1)**.

14. Cotangent is the reciprocal of the tangent function. If $\tan A = \frac{2}{3}$, then $\cot A = \frac{3}{2}$. This eliminates choices (1) and (2). The "co" in cotangent (and in cosine and cosecant) comes from "complementary." The tangent of an angle equals the cotangent of the complement of the angle: $\tan A = \cot(90° - A)$. In this problem, $\tan A = \frac{2}{3}$. So $\cot(90° - A) = \frac{2}{3}$.

CALC Check

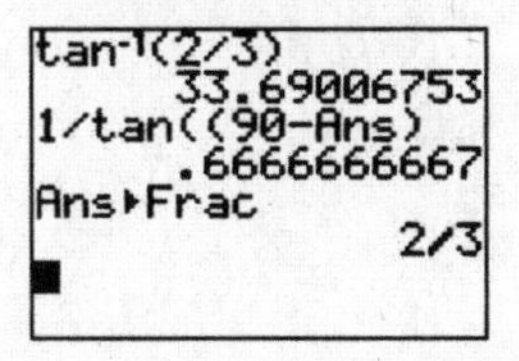

The correct choice is **(3)**.

15. Since 2 and 4 are both powers of 2, the equation can be rewritten with a common base by substituting 2^2 for 4.

$$4^{x^2+4x} = 2^{-6}$$
$$(2^2)^{x^2+4x} = 2^{-6}$$

When a power is raised to a power, exponents are multiplied.

$$2^{2(x^2+4x)} = 2^{-6}$$

If two powers of the same base are equal, the exponents must be equal.

$$2(x^2 + 4x) = -6$$
$$x^2 + 4x = -3$$
$$x^2 + 4x + 3 = 0$$
$$(x + 1)(x + 3) = 0$$
$$x = -1 \text{ or } x = -3$$

CALC Check

Set the Table to Ask. Then enter all the possible x-values from the choices given in the original problem. Only $x = -1$ and $x = -3$ give the same value for both expressions.

```
Plot1 Plot2 Plot3
\Y1=4^(X²+4X)
\Y2=2^-6
\Y3=
\Y4=
\Y5=
\Y6=
\Y7=
```

X	Y1	Y2
3	4.4E12	.01563
1	1024	.01563
-1	.01563	.01563
-3	.01563	.01563

X=

The correct choice is **(3)**.

16. The given equation of a circle must be rewritten in center-radius form by completing the square. Rearrange the terms to get the x-terms together and the y-terms together. Then complete the square for each variable.

$$x^2 - 2x \quad + y^2 + 6y \quad + 3 = 0$$
$$\underbrace{x^2 - 2x + 1}_{(x-1)^2} - 1 + \underbrace{y^2 + 6y + 9}_{(y+3)^2} - 9 + 3 = 0$$
$$(x - 1)^2 - 1 + (y + 3)^2 - 9 + 3 = 0$$
$$(x - 1)^2 + (y + 3)^2 - 7 = 0$$
$$(x - 1)^2 + (y + 3)^2 = 7$$

The correct choice is **(2)**.

17. Solve the inequality $y + 6 \geq x^2 - x$ for y, resulting in $y \geq x^2 - x - 6$. Compare each graph. The key differences are the location of the roots and the shading of the solution.

To find the roots for this problem, factor the equation, set the factors equal to zero, and solve.

$$y = x^2 - x - 6$$
$$y = (x + 2)(x - 3)$$

This is a parabola with roots at $x = -2$ and $x = 3$. This eliminates choices (2) and (4) since they each have roots at $x = -3$ and $x = 2$.

To determine where the solution should be shaded, consider the inequality symbol. The inequality $y \geq x^2 - x - 6$ means you want y-values that are on or above the graph of the parabola. These values have been shaded in choice (1). Choice (3) represents $y \leq x^2 - x - 6$, y-values on or below the parabola.

CALC Check

When using the graphing calculator, remember to change the symbol to the left of Y_1 to indicate greater than or equal to.

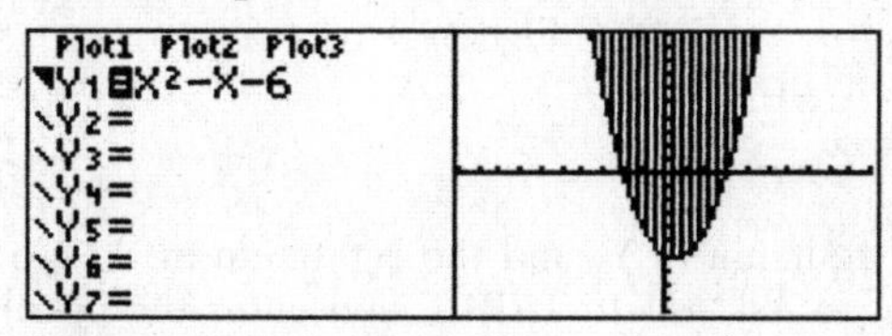

The correct choice is **(1)**.

18. Because this is a multiple-choice problem, check each of the given solutions in the equation. Try $x = 1$:

$$\sqrt{1+3} = 3 - 1$$
$$2 = 2$$

Since $2 = 2$, $x = 1$ is a solution. This eliminates choices (2) and (4). Now try $x = 6$:

$$\sqrt{6+3} = 3 - 6$$
$$3 \neq -3$$

Since $3 \neq -3$, $x = 6$ is not a solution. This eliminates choice (3).

Algebraic Solution:

Isolate the radical term (already done.) $\sqrt{x+3} = 3 - x$

Square both sides. $\left(\sqrt{x+3}\right)^2 = (3-x)^2$

Solve.

$$x + 3 = 9 - 6x + x^2$$
$$x^2 - 7x + 6 = 0$$
$$(x-1)(x-6) = 0$$
$$x = 1 \text{ or } x = 6$$

Always check answers in the original equation (see below for checking on the calculator).

Check $x = 1$: $\sqrt{1+3} = 3 - 1$

$2 = 2$ ✔

Check $x = 3$: $\sqrt{6+3} = 3 - 6 \rightarrow 3 = -3$

Reject.

CALC Check

Put the left side of the equation in Y_1 and the right side in Y_2. Go to TBLSET and set Indpnt to Ask. Go to TABLE and enter the possible x-values from the choices given. $x = 1$ is the only one that gives the same value for both expressions.

```
Plot1 Plot2 Plot3
\Y1=√(X+3)
\Y2=3-X
\Y3=
\Y4=
\Y5=
\Y6=
\Y7=
```

X	Y1	Y2
1	2	2
0	1.7321	3
6	3	-3
2	2.2361	1
3	2.4495	0

X=

The correct choice is **(1)**.

19. To multiply powers of the same base, keep the base and add the exponents: $(i^7)(i^5) = i^{12}$. The first four powers of i are $i^0 = 1$, $i^1 = i$, $i^2 = -1$, and $i^3 = -i$. Successive powers repeat the pattern: 1, i, −1, −i. For any whole number power of i, you can find the result by raising i to the remainder obtained when the exponent is divided by 4.

$$\frac{12}{4} = 3 \text{ remainder } 0$$

$$i^{12} = i^0 = 1$$

CALC Check

This problem can be done on the calculator. However, you need to be aware that the calculator has round-off errors on some calculations. −3E⁻13i is calculator notation for $(-3 \times 10^{-13})i$, which is very close to $0i$. This is calculator error, not calculator accuracy. The correct answer is $1 + 0i$ or just 1.

```
i^7*i^5
            1-3E-13i
■
```

The correct choice is **(1)**.

20. Graph each of the choices (one at a time) on your calculator. Make sure you are in radian mode. Set WINDOW to $0 \leq x \leq 2\pi$ and $-2 \leq y \leq 2$ to match the given graphs. The third choice, $y = \sec x$, best matches the given graph.

```
Plot1 Plot2 Plot3        WINDOW
\Y1=1/tan(X)              Xmin=0
\Y2=1/sin(X)              Xmax=6.2831853...
\Y3=1/cos(X)              Xscl=1
\Y4=tan(X)                Ymin=-2
\Y5=■                     Ymax=2
\Y6=                      Yscl=1
\Y7=                     ↓Xres=1■
```

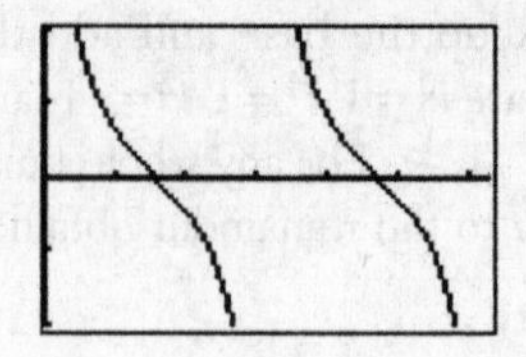

Choice (1): $y = \cot x = \dfrac{1}{\tan x}$

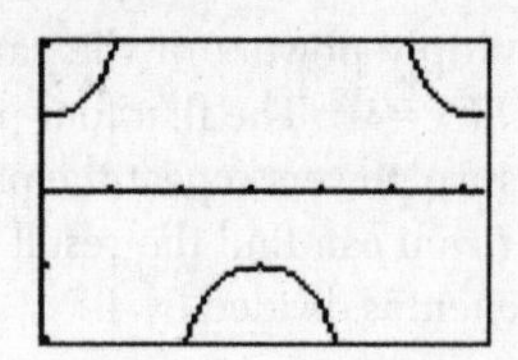

Choice (3): $y = \sec x = \dfrac{1}{\cos x}$

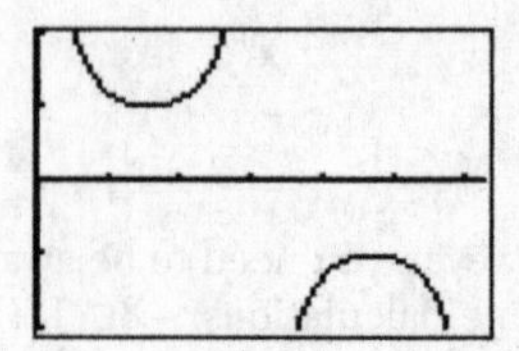

Choice (2): $y = \csc x = \dfrac{1}{\sin x}$

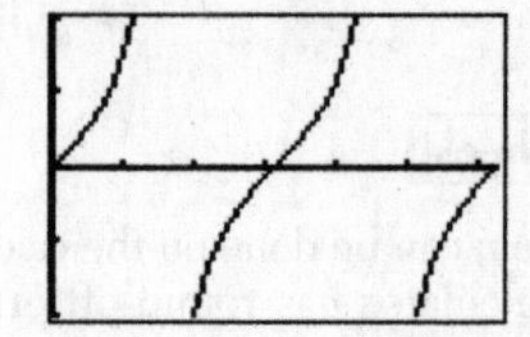

Choice (4): $y = \tan x$

The correct choice is **(3)**.

21. The linear correlation coefficient r will always be in the interval $-1 \le r \le 1$. Correlation is strongest when r is close to 1 or –1 and weakest when r is close to 0.

- Choice (1): $r = -1.07$ is outside the range of possible r-values.
- Choice (2): $r = -0.89$ shows a strong negative correlation.
- Choice (3): $r = -0.14$ shows a weak negative correlation.
- Choice (4): $r = 0.92$ shows a strong positive correlation.

The correct choice is **(2)**.

22. To have an inverse function, a function must be one-to-one. Graphically, this means horizontal lines must intersect the function only once (the horizontal line test, or HLT). Make sure the calculator is set to radian mode. The function $y = \tan x$ is graphed below with WINDOW set to the domains given in the choices. The x-axis has been marked at intervals of $\frac{\pi}{2}$.

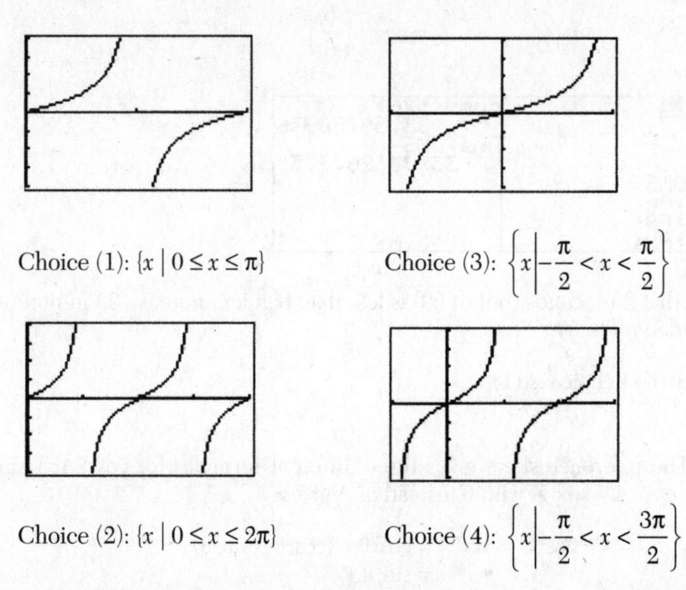

Choice (1): $\{x \mid 0 \le x \le \pi\}$

Choice (3): $\left\{x \middle| -\frac{\pi}{2} < x < \frac{\pi}{2}\right\}$

Choice (2): $\{x \mid 0 \le x \le 2\pi\}$

Choice (4): $\left\{x \middle| -\frac{\pi}{2} < x < \frac{3\pi}{2}\right\}$

From the graphs, you can see that in the intervals $0 \le x \le 2\pi$ and $-\frac{\pi}{2} < x < \frac{3\pi}{2}$, the tangent graph is not one-to-one (fails the HLT) and will not have an inverse function. That eliminates choices (2) and (4). The tangent graph passes the HLT on the intervals in choices (1) and (3), $0 \le x \le \pi$ and $-\frac{\pi}{2} < x < \frac{\pi}{2}$. However, tangent is undefined when $x = \frac{\pi}{2}$, so $\frac{\pi}{2}$ should be excluded from the domain. This eliminates choice (1). Note that in choice (3), neither $x = \frac{\pi}{2}$ nor $x = -\frac{\pi}{2}$ is included in the domain.

The correct choice is **(3)**.

23. The diagram provides the lengths of the hypotenuse and adjacent angle relative to ∠K, so use cosine. Make sure you are in degree mode and use ▶DMS in the ANGLE menu to convert to degrees, minutes, and seconds.

$$\cos K = \frac{5}{6}$$

$$K = \cos^{-1}\left(\frac{5}{6}\right)$$

```
ANGLE             cos⁻¹(5/6)
1:°                     33.55730976
2:'               Ans▶DMS
3:r                     33°33'26.315"
4:▶DMS
5:R▶Pr(
6:R▶Pθ(
7↓P▶Rx(
```

Note that 26 seconds (out of 60) is less than half a minute, so 33 minutes is not rounded up to 34.

The correct choice is **(1)**.

24. The reference sheet gives three different formulas for cos 2*A*. Using $\cos 2A = \cos^2 A - \sin^2 A$ with θ instead of *A* gives:

$$\cos^2\theta - \cos 2\theta = \cos^2\theta - (\cos^2\theta - \sin^2\theta)$$
$$= \sin^2\theta$$

Choice (1) is $\sin^2 \theta$. Note that choice (2) is the result if you forget to distribute the subtraction through the parentheses.

CALC Check

You can use a graphing calculator to confirm your result. Put the original expression in Y_1 and your answer in Y_2. Then use TABLE to see if their values are identical.

```
Plot1 Plot2 Plot3
\Y1■cos(X)²-cos(
2X)
\Y2■sin(X)²
\Y3=
\Y4=
\Y5=
\Y6=
```

X	Y1	Y2
-2	.00122	.00122
-1	3E-4	3E-4
0	0	0
1	3E-4	3E-4
2	.00122	.00122
3	.00274	.00274
4	.00487	.00487

X=0

The correct choice is **(1)**.

25. Try each of the choices.

- Choice (1): $\sum_{k=1}^{20}(2k-1)=1+3+5+7+9+\cdots+39$, which is correct.
- Choice (2): $\sum_{k=2}^{40}(k-1)=1+2+3+4+\cdots+39$, which includes even numbers and is incorrect.
- Choice (3): $\sum_{k=-1}^{37}(k+2)=1+2+3+4+\cdots+39$, which includes even numbers and is incorrect.
- Choice (4): $\sum_{k=1}^{39}(2k-1)=1+3+5+7+9+\cdots+77$, which is incorrect because the last term is not 39.

The correct choice is **(1)**.

26. Check each of the choices. In this particular problem, checking the first term of each is sufficient.

- Choice (1): $a_1=6\left(\frac{1}{3}\right)^1=2$
- Choice (2): $a_1=6\left(\frac{1}{3}\right)^{1-1}=6\left(\frac{1}{3}\right)^0=6(1)=6$
- Choice (3): $a_1=54\left(\frac{1}{3}\right)^1=18$
- Choice (4): $a_1=54\left(\frac{1}{3}\right)^{1-1}=54\left(\frac{1}{3}\right)^0=54(1)=54$

Only choice (4) has the correct first term.

CALC Check

You can easily check more terms with the TABLE feature of your calculator using X in place of n and Y instead of a_n.

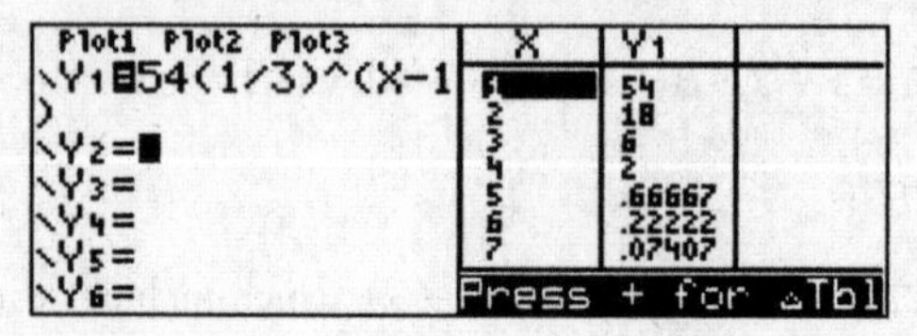

It may be worth memorizing the explicit rule for a geometric sequence: $a_n = a_1(r)^{n-1}$. In this problem, $a_1 = 54$ and $r = \frac{a_2}{a_1} = \frac{18}{54} = \frac{1}{3}$. So the sequence is given by $a_n = 54\left(\frac{1}{3}\right)^{n-1}$.

The correct choice is **(4)**.

27. The period of the function $y = a \sin(bx + c)$ is given by the formula period $= \frac{2\pi}{|b|}$. You should have this memorized. In this problem, $b = \frac{1}{3}$. So

$$\text{period} = \frac{2\pi}{\frac{1}{3}} = 6\pi.$$

CALC Check

Graphically, the period is the length of the interval to complete one cycle. With your calculator in radian mode, graph the equation, changing WINDOW to see the different choices for the period.

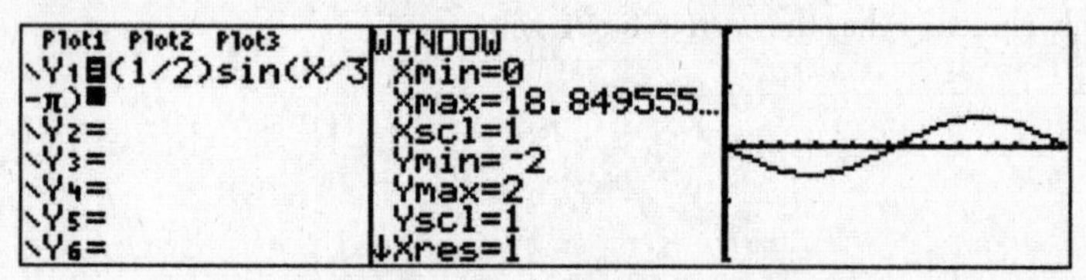

The correct choice is **(4)**.

PART II

28. The discriminant of the quadratic equation $ax^2 + bx + c = 0$ is $b^2 - 4ac$. This is the expression inside the square root of the quadratic formula. The two roots of the equation will be equal only when the discriminant is exactly zero. The equation $x^2 - kx + 4 = 0$ has $a = 1$, $b = -k$, and $c = 4$.

$$(-k)^2 - 4(1)(4) = 0$$
$$k^2 - 16 = 0$$
$$k^2 = 16$$
$$k = \pm 4$$

CALC Check

When a quadratic equation has equal roots, its graph is tangent to the x-axis.

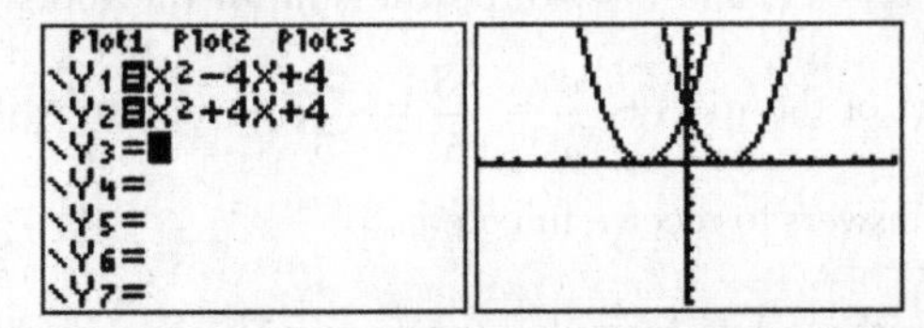

The equation $x^2 - kx + 4 = 0$ will have equal roots when $k = 4$ or $k = -4$.

29. Carefully enter the test scores into list L1 and the frequencies into list L2 in your calculator. Then use the 1-variable statistics key in STAT ▶ CALC 1:1 – Var Stats ENTER. Then press L1, L2 ENTER.

Note: You must key in L1, L2 after 1-Var Stats so the calculator will use the frequencies in L2. If you do not, you will get the wrong answer as shown in the screen below right. Check the number of scores, n. There are 22 total scores on the list.

Correct	**Wrong**
1-Var Stats	1-Var Stats
$\bar{x}$=79.45454545	$\bar{x}$=81.14285714
Σx=1748	Σx=568
Σx²=140080	Σx²=46720
Sx=7.538645026	Sx=10.25391911
σx=7.365319673	σx=9.493284415
↓n=22	↓n=7

The population standard deviation is σ_x. From the screen on the left, the population standard deviation is $\sigma = 7.365319673$. The New York State guidelines for using graphing calculators states that students should write the mean ($\overline{x}$), and the number of scores (n), as well as the standard deviation to show work for this type of problem. You may want to check with your teacher about your school's expectations. Some teachers instruct their students to copy the calculator screen.

When rounded to the *nearest tenth*, the population standard deviation is $\sigma = 7.4$.

30. Algebraic Solution:

For any quadratic equation in the form $ax^2 + bx + c = 0$, the sum of the roots is $r_1 + r_2 = -\frac{b}{a}$ and the product of the roots is $r_1 r_2 = \frac{c}{a}$. For the equation $5x^2 + 11x - 3 = 0$, $a = 5$, $b = 11$, and $c = -3$. So the sum of the roots is $-\frac{b}{a} = -\frac{11}{5}$, and the product of the roots is $\frac{c}{a} = \frac{-3}{5} = -\frac{3}{5}$. Be sure to write the formulas as well as the answers to receive full credit.

If you do not remember these two formulas, you can, with considerably more work, find the answers. First solve the equation. Then store the roots in A and B on your calculator, and then find A + B and AB.

Solve the equation using the quadratic formula. Store the roots in A and B without rounding.

$$x = \frac{-b \pm \sqrt{b^2 - 4ac}}{2a}$$

$$x = \frac{-11 \pm \sqrt{11^2 - 4(5)(-3)}}{2(5)}$$

The following shows the calculator in MathPrint mode.

```
(-11+√(11²-4*5*-3))/(2*5)→A
              .2453624047
(-11-√(11²-4*5*-3))/(2*5)→B
             -2.445362405
A+B
                     -2.2
AB
                      -.6
```

Graphical Solution:

If you solve it graphically, use CALC to find each root. Use QUIT to return to the home screen. Store the root by selecting X STO ▶ A and X STO ▶ B. Using only the digits displayed on the graph screen will lead to round-off error.

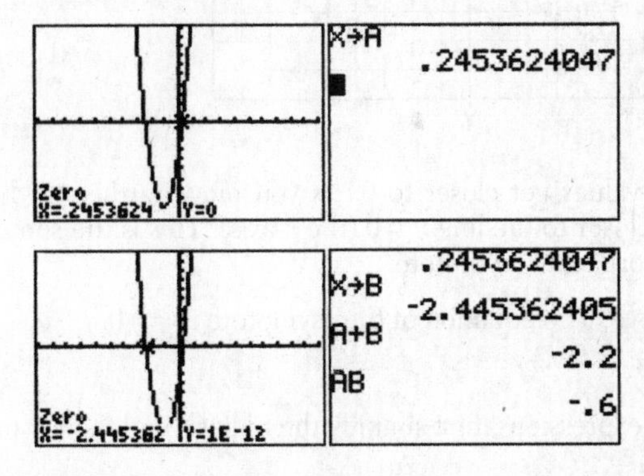

The sum of the roots is $-\frac{11}{5}$, and the product of the roots is $-\frac{3}{5}$.

31. Use your calculator to make a table of values for the function. Use both positive and negative x-values. Then neatly sketch the graph.

```
Plot1 Plot2 Plot3
\Y1=(1/2)^X
\Y2=
\Y3=
\Y4=
\Y5=
\Y6=
```

X	Y1
-3	8
-2	4
-1	2
0	1
1	.5
2	.25
3	.125

X=-3

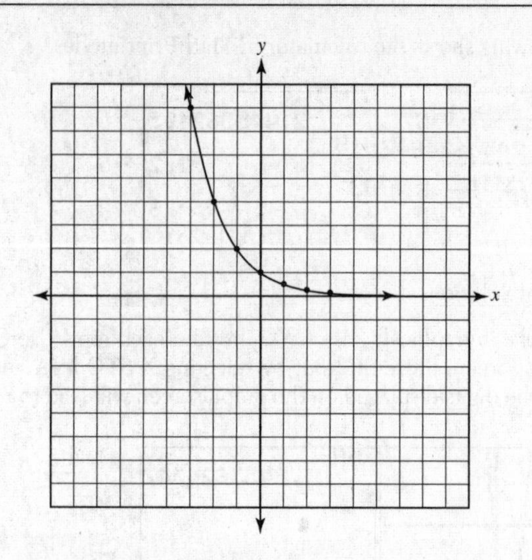

As x gets larger, the y-values get closer to 0. As you move farther to the right, the graph becomes closer to the line $y = 0$ (the x-axis). This is the somewhat simplified idea of a horizontal asymptote.

The graph is shown above. The equation of the asymptote is $y = 0$.

32. To subtract radical expressions, first simplify the radicals and then combine like terms.

$$\begin{aligned} 5\sqrt{3x^3} - 2\sqrt{27x^3} &= 5\sqrt{x^2}\sqrt{3x} - 2\sqrt{9x^2}\sqrt{3x} \\ &= 5x\sqrt{3x} - 2(3x)\sqrt{3x} \\ &= 5x\sqrt{3x} - 6x\sqrt{3x} \\ &= -x\sqrt{3x} \end{aligned}$$

Note: Strictly speaking, $\sqrt{x^2} = |x|$, not just x. This allows for the possibility that x could be negative in the expression. Some teachers prefer the answer $-|x|\sqrt{3x}$. In Algebra 2/Trigonometry, the usual, though not always stated, rule is that functions are assumed to have real answers unless the problem specifically indicates that imaginary numbers are to be considered. Under this assumption, the absolute value symbols may be omitted in this problem since negative values of x would make the original expression imaginary.

In simplest radical form, $5\sqrt{3x^3} - 2\sqrt{27x^3} = -x\sqrt{3x}$ (assuming $x \geq 0$).

33. Since $180° < 240° < 270°$, an angle measuring 240° is in Quadrant III. Specifically, it is $240° - 180° = 60°$ into the third quadrant. Since 60° is 2/3 of 90°, divide the third-quadrant portion of the unit circle into thirds and sketch the terminal side of the angle.

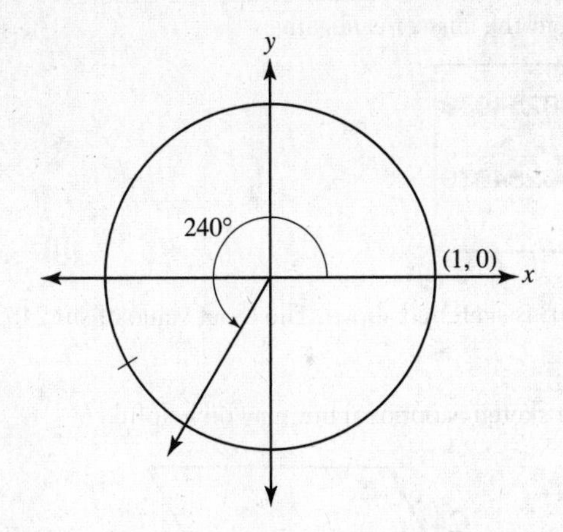

For angles in Quadrant III, sin θ is negative. The reference angle for 240° is 60°, so $\sin 240° = -\sin 60°$. You may have memorized the exact values of trigonometric functions of special angles such as 60°. You may instead prefer to use a special right triangle. In a 30°–60°–90° triangle, the shorter leg (across from the 30° angle) is half the hypotenuse. The longer leg can be either remembered or found using the Pythagorean theorem. Letting the hypotenuse equal 2 and the shorter leg 1, the longer leg is $\sqrt{3}$, and $\sin 60° = -\frac{\sqrt{3}}{2}$.

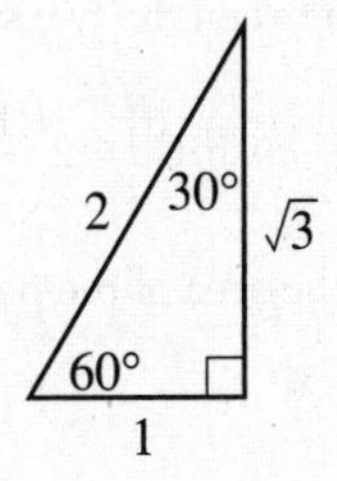

CALC Check

Because the question specifies the exact value, do not write the decimal approximation shown on your calculator. However, you can double-check your exact value with the calculator approximation. The calculator will remind you that the answer is negative.

```
sin(240)
      -.8660254038
-√3/2
      -.8660254038
```

The angle 240° is sketched above. The exact value of $\sin 240° = -\frac{\sqrt{3}}{2}$.

34. Making a sketch is optional but may be helpful.

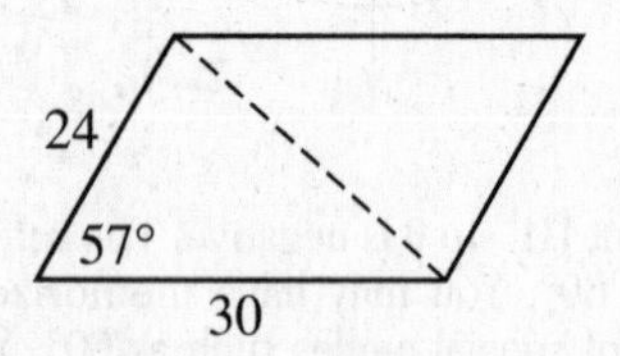

Find the area by recognizing that it is twice the area of one of the triangles created by drawing the diagonal as shown in the diagram. The formula for the area of a triangle is given on the reference sheet: $K = \frac{1}{2}ab\sin C$. The actual names of the sides and the angle are not important as long as the angle in the formula is the angle included between the two sides. In this problem, we have:

$$\text{Area of parallelogram} = 2\left(\frac{1}{2}ab\sin C\right) = 2\left(\frac{1}{2}\right)(24)(30)\sin 57° = 603.84$$

To the *nearest square foot*, the area of the parallelogram is 604 square feet.

35. The quickest way to solve this problem is to multiply both the numerator and denominator of the fraction by the least common denominator of all the individual fractions. The separate denominators are 2, d, and $2d$, so the LCD is $2d$.

$$\frac{\frac{1}{2}-\frac{4}{d}}{\frac{1}{d}+\frac{3}{2d}}=\frac{\left(\frac{1}{2}-\frac{4}{d}\right)(2d)}{\left(\frac{1}{d}+\frac{3}{2d}\right)(2d)}=\frac{d-8}{2+3}=\frac{d-8}{5}$$

Equivalently, you can rewrite the fractions with common denominators, divide, and reduce.

$$\frac{\frac{1}{2}-\frac{4}{d}}{\frac{1}{d}+\frac{3}{2d}}=\frac{\frac{d}{2d}-\frac{8}{2d}}{\frac{2}{2d}+\frac{3}{2d}}=\frac{\frac{d-8}{2d}}{\frac{5}{2d}}=\frac{d-8}{\cancel{2d}}\cdot\frac{\cancel{2d}}{5}=\frac{d-8}{5}$$

In simplest form, $\frac{\frac{1}{2}-\frac{4}{d}}{\frac{1}{d}+\frac{3}{2d}}=\frac{d-8}{5}$.

PART III

36. This is an example of the binomial distribution. To find the probability that at least 8 men wear black vests, you must add the individual probabilities that 8, 9, or 10 men wear black vests. Use the following formula:

$$_{\text{number of trials}}C_{\text{number of successes}}(\text{probability of success})^{\text{number of successes}}(\text{probability of failure})^{\text{number of failures}}$$

In the given problem, there are 10 trials (men wearing vests), with 8, 9, or 10 successes (black vests) and 2, 1, or 0 failures (red vests). The probability of success is 60% = 0.6, and the probability of failure is 1 – 0.6 = 0.4.

$$P(\text{at least 8 men wear black}) =$$
$$_{10}C_8(0.6)^8(0.4)^2 + {}_{10}C_9(0.6)^9(0.4)^1 + {}_{10}C_{10}(0.6)^{10}(0.4)^0 = 0.16729$$

Note that although the probability of wearing a black vest is given as a percent, the question did not ask for the answer to be given in percent form. Students who answered 16.729% lost a point for rounding incorrectly.

To the nearest thousandth, the probability that at least 8 men wear black vests is 0.167.

37. Algebraic Solution:

The equation contains both θ and 2θ. To solve the equation algebraically, use the formula sin 2*A* = 2 sin *A* cos *A* shown on the reference sheet to rewrite sin 2θ in terms of θ only.

Substitute the double angle formula. $\sin 2\theta = \sin\theta$

Set equal to 0 and factor out the common factor. Do not divide both sides by sinθ. You will lose some solutions.

$$2\sin\theta\cos\theta = \sin\theta$$
$$2\sin\theta\cos\theta - \sin\theta = 0$$
$$\sin\theta(2\cos\theta - 1) = 0$$

Set each factor equal to 0, and solve for sin θ and cos θ. $\sin\theta = 0$ or $\cos\theta = \frac{1}{2}$

$\sin\theta = 0$ when $\theta = 0°$ or $\theta = 180°$.

$\cos\theta = \frac{1}{2}$ has solutions in Quadrants I and IV.

$\theta_{ref} = \cos^{-1}\left(\frac{1}{2}\right) = 60°$, which is the Quadrant I solution. In Quadrant IV, $\theta = 360° - 60° = 300°$.

Graphical Solution:

Set your calculator degree mode. Graph each side of the equation. Set WINDOW to $0° \leq x \leq 360°$ and $-1.5 \leq y \leq 1.5$. Use CALC to find the intersections of the two graphs. In the graph below, the first intersection is at 0°, the second at 60°, the third at 180°, and the fourth at 300°. Note that there is a fifth intersection at 360°, but that is outside the interval specified in the problem.

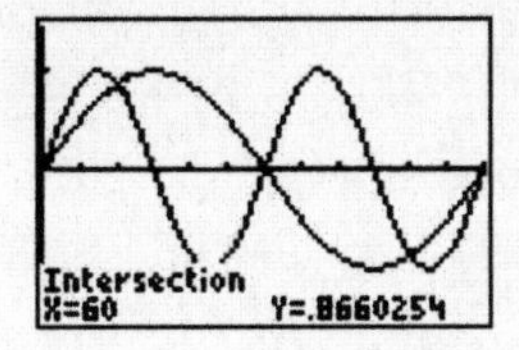

If you do the problem graphically, you need to make a neat sketch of the graphs, label the two graphs, indicate the scale, and show the intersections.

The solutions are 0°, 60°, 180°, and 300°.

38. An arrangement of the objects in a set is a permutation. Carol wants to compare the number of permutations of the letters in TENNESSEE to the number of permutations of the letters in VERMONT. Your justification should be mathematical. Compute and compare the number of arrangements for each set of letters.

VERMONT has 7 letters; none are repeated. The number of permutations is ${}_7P_7$ or $7! = 5040$.

TENNESSEE has 9 letters, but they include 4 Es, 2 Ns, and 2 Ss. When letters are repeated, the number of different permutations is found by dividing the total number of permutations by the number of possible permutations of each of the repeated letters:

$$\frac{9!}{(4!)(2!)(2!)} = 3780.$$

CALC Check

If you are using a TI-83 or an older TI-84 that has not been updated, be careful to use parentheses in the denominator or you will get the wrong answer. The new MathPrint mode on the TI-84 makes it easy to calculate the permutation for Tennessee accurately. The screen on the right is in MathPrint mode.

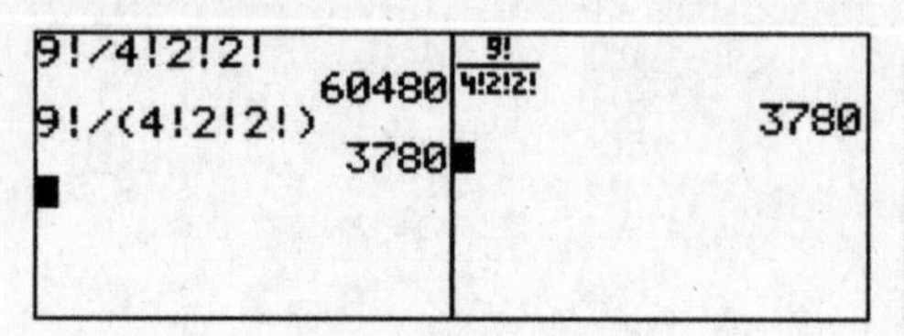

No, Carol is wrong because the letters in TENNESSEE can be arranged in only 3780 ways while the letters in VERMONT can be arranged in 5040 ways.

PART IV

39. Draw a diagram. Since the smallest angle of a triangle cannot be 80°, the smallest angle will be opposite the shorter of the two known sides. Label that angle x.

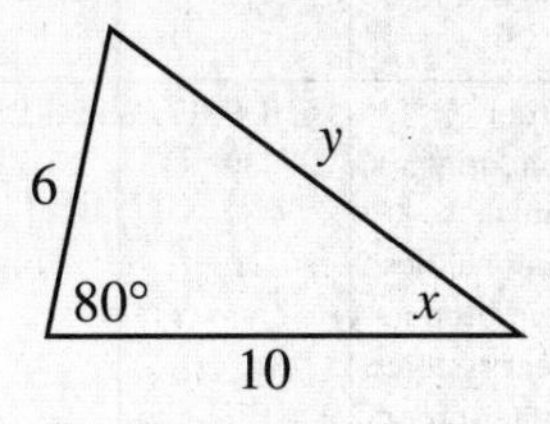

Solving for x is easier if you first find the length of the side opposite the 80° angle. Label that side y. Find y by using the Law of Cosines shown on the formula sheet.

$$y^2 = 6^2 + 10^2 - 2(6)(10)\cos 80°$$
$$y^2 = 115.1622187$$
$$y = 10.73136611$$

Consider storing this value in Y on your calculator. Now find x using the Law of Sines, also shown on the formula sheet.

$$\frac{6}{\sin x} = \frac{y}{\sin 80°}$$
$$y \sin x = 6 \sin 80°$$
$$\sin x = \frac{6 \sin 80°}{y}$$
$$\sin x = \frac{6 \sin 80°}{10.73136611}$$
$$\sin x = 0.5506145681$$
$$x = \sin^{-1}(0.5506145681)$$
$$x \approx 33.41°$$

To the *nearest degree*, the measure of the smallest angle of the triangle is 33°.

SELF-ANALYSIS CHART June 2010

Topic	Question Numbers	Number of Points	Your Points	Your Percentage
1. Exponents and Radicals: operations, equivalent expressions, simplifying, rationalizing	3, 11, 12, 18, 32	$2+2+2+2+2=10$		
2. Complex Numbers: operations, powers of i, rationalizing	6, 19	$2+2=4$		
3. Quadratics and Higher-Order Polynomials: operations, factoring, binomial expansion, formula, quadratic equations and inequalities, nature of the roots, sum and product of the roots, quadratic-linear systems, completing the square, higher-order polynomials	5, 8, 9, 17, 28, 30	$2+2+2+2+2+2=12$		
4. Rationals: operations, equations	35	2		
5. Absolute Value: equations and inequalities	—	—		
6. Direct and Inverse Variation	—	—		
7. Circles: center-radius equation, completing the square	16	2		
8. Functions: relations, functions, domain, range, one-to-one, onto, inverses, compositions, transformations of functions	13	2		
9. Exponential and Logarithmic Functions: equations, common bases, logarithm rules, e, ln	4, 10, 15, 31	$2+2+2+2=8$		
10. Trigonometric Functions: radian measure, arc length, cofunctions, unit circle, inverses	2, 14, 33	$2+2+2=6$		
11. Trigonometric Graphs: graphs of basic functions and inverse functions, domain, range, restricted domain, transformations of graphs, amplitude, frequency, period, phase shift	20, 22, 23, 27	$2+2+2+2=8$		
12. Trigonometric Identities, Formulas, and Equations	24, 37	$2+4=6$		
13. Trigonometry Laws and Applications: area of a triangle, sine and cosine laws, ambiguous cases	34, 39	$2+6=8$		
14. Sequences and Series: sigma notation, arithmetic, geometric	1, 25, 26	$2+2+2=6$		

Topic	Question Numbers	Number of Points	Your Points	Your Percentage
15. Statistics: studies, central tendency, dispersion including standard deviation, normal distribution	29	2		
16. Regressions: linear, exponential, logarithmic, power, correlation coefficient	21	2		
17. Probability: permutation, combination, geometric, binomial	7, 36, 38	2 + 4 + 4 = 10		

HOW TO CONVERT YOUR RAW SCORE TO YOUR ALGEBRA 2/TRIGONOMETRY REGENTS EXAMINATION SCORE

Below is the conversion chart that can be used to determine your final score on the June 2010 Regents Examination. To determine your final exam score, locate in the column labeled "Raw Score" the total number of points you scored out of a possible 88 points. Since partial credit is allowed in Parts II, III, and IV of the test, you may need to approximate the credit you would receive for a solution that is not completely correct. Then locate in the adjacent column to the right the scaled score that corresponds to your raw score. The scaled score is your final Regents Examination score.

Raw Score	Scaled Score	Raw Score	Scaled Score	Raw Score	Scaled Score
88	**100**	58	**78**	28	**42**
87	**99**	57	**77**	27	**40**
86	**99**	56	**76**	26	**39**
85	**99**	55	**75**	25	**37**
84	**98**	54	**74**	24	**36**
83	**98**	53	**73**	23	**34**
82	**97**	52	**72**	22	**33**
81	**97**	51	**71**	21	**32**
80	**96**	50	**70**	20	**30**
79	**96**	49	**69**	19	**29**
78	**95**	48	**68**	18	**27**
77	**94**	47	**66**	17	**26**
76	**94**	46	**65**	16	**24**
75	**93**	45	**64**	15	**23**
74	**92**	44	**63**	14	**21**
73	**92**	43	**61**	13	**20**
72	**91**	42	**60**	12	**18**
71	**90**	41	**59**	11	**17**
70	**89**	40	**58**	10	**15**
69	**88**	39	**56**	9	**14**
68	**88**	38	**55**	8	**12**
67	**87**	37	**54**	7	**11**
66	**86**	36	**52**	6	**9**
65	**85**	35	**51**	5	**8**
64	**84**	34	**50**	4	**6**
63	**83**	33	**48**	3	**5**
62	**82**	32	**47**	2	**3**
61	**84**	31	**46**	1	**2**
60	**80**	30	**44**	0	**0**
59	**79**	29	**43**		

Examination August 2010

Algebra 2/Trigonometry

REFERENCE SHEET

Area of a Triangle

$K = \frac{1}{2}ab \sin C$

Law of Cosines

$a^2 = b^2 + c^2 - 2bc \cos A$

Functions of the Sum of Two Angles

$\sin(A + B) = \sin A \cos B + \cos A \sin B$

$\cos(A + B) = \cos A \cos B - \sin A \sin B$

$\tan(A + B) = \frac{\tan A + \tan B}{1 - \tan A \tan B}$

Functions of the Double Angle

$\sin 2A = 2 \sin A \cos A$

$\cos 2A = \cos^2 A - \sin^2 A$

$\cos 2A = 2 \cos^2 A - 1$

$\cos 2A = 1 - 2 \sin^2 A$

$\tan 2A = \frac{2 \tan A}{1 - \tan^2 A}$

Functions of the Difference of Two Angles

$\sin(A - B) = \sin A \cos B - \cos A \sin B$

$\cos(A - B) = \cos A \cos B + \sin A \sin B$

$\tan(A - B) = \frac{\tan A - \tan B}{1 + \tan A \tan B}$

Functions of the Half Angle

$$\sin\frac{1}{2}A = \pm\sqrt{\frac{1-\cos A}{2}}$$

$$\cos\frac{1}{2}A = \pm\sqrt{\frac{1+\cos A}{2}}$$

$$\tan\frac{1}{2}A = \pm\sqrt{\frac{1-\cos A}{1+\cos A}}$$

Law of Sines

$$\frac{a}{\sin A} = \frac{b}{\sin B} = \frac{c}{\sin C}$$

Sum of a Finite Arithmetic Sequence

$$S_n = \frac{n(a_1 + a_n)}{2}$$

Sum of a Finite Geometric Sequence

$$S_n = \frac{a_1(1-r^n)}{1-r}$$

Binomial Theorem

$$(a+b)^n = {}_nC_0a^nb^0 + {}_nC_1a^{n-1}b^1 + {}_nC_2a^{n-2}b^2 + \cdots + {}_nC_na^0b^n$$

$$(a+b)^n = \sum_{r=0}^{n} {}_nC_ra^{n-r}b^r$$

Normal Curve Standard Deviation

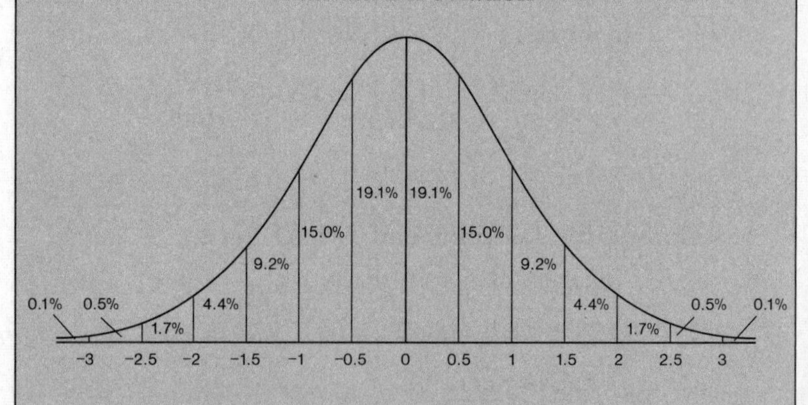

PART I

Answer all 27 questions in this part. Each correct answer will receive 2 credits. No partial credit will be allowed. For each question, write in the space provided the numeral preceding the word or expression that best completes the statement or answers the question. [54 credits]

1 The product of $(3 + \sqrt{5})$ and $(3 - \sqrt{5})$ is

(1) $4 - 6\sqrt{5}$ (3) 14

(2) $14 - 6\sqrt{5}$ (4) 4 1____

2 What is the radian measure of an angle whose measure is $-420°$?

(1) $-\frac{7\pi}{3}$ (3) $\frac{7\pi}{6}$

(2) $-\frac{7\pi}{6}$ (4) $\frac{7\pi}{3}$ 2____

3 What are the domain and the range of the function shown in the graph below?

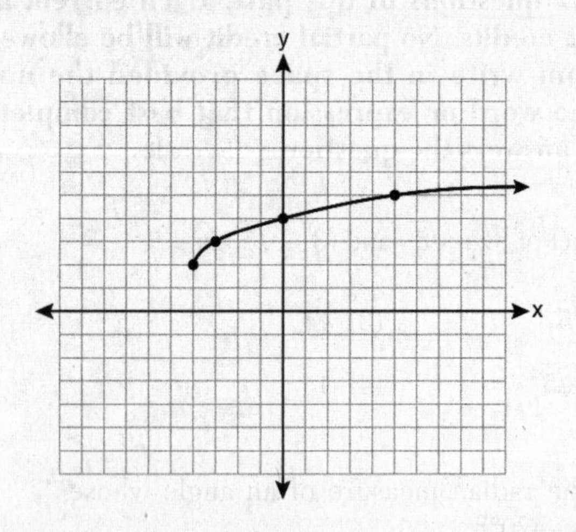

(1) $\{x \mid x > -4\}; \{y \mid y > 2\}$

(2) $\{x \mid x \geq -4\}; \{y \mid y \geq 2\}$

(3) $\{x \mid x > 2\}; \{y \mid y > -4\}$

(4) $\{x \mid x \geq 2\}; \{y \mid y \geq -4\}$ 3 _____

4 The expression $2i^2 + 3i^3$ is equivalent to

(1) $-2 - 3i$ (3) $-2 + 3i$

(2) $2 - 3i$ (4) $2 + 3i$ 4 _____

5 In which graph is θ coterminal with an angle of −70°?

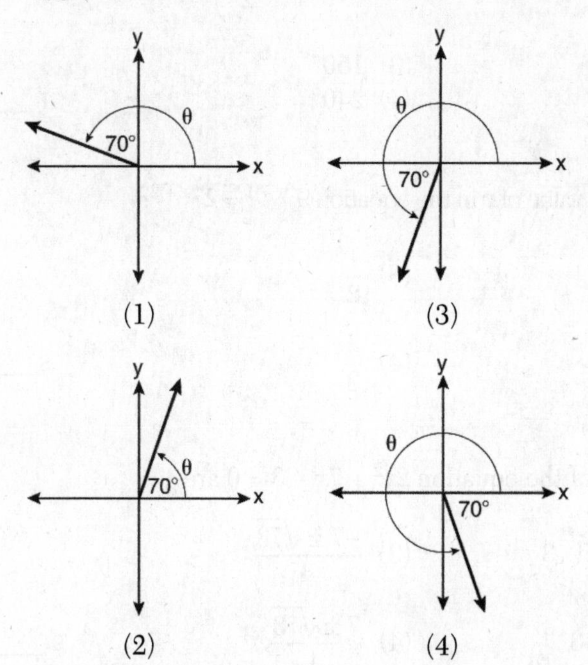

5 _____

6 In $\triangle ABC$, $m\angle A = 74$, $a = 59.2$, and $c = 60.3$. What are the two possible values for $m\angle C$, to the *nearest tenth*?

(1) 73.7 and 106.3
(2) 73.7 and 163.7
(3) 78.3 and 101.7
(4) 78.3 and 168.3

6 _____

7 What is the principal value of $\cos^{-1}\left(-\frac{\sqrt{3}}{2}\right)$?

(1) $-30°$
(2) $60°$
(3) $150°$
(4) $240°$

7 _____

8 What is the value of x in the equation $9^{3x+1} = 27^{x+2}$?

(1) 1
(2) $\frac{1}{3}$
(3) $\frac{1}{2}$
(4) $\frac{4}{3}$

8 _____

9 The roots of the equation $2x^2 + 7x - 3 = 0$ are

(1) $-\frac{1}{2}$ and -3
(2) $\frac{1}{2}$ and 3
(3) $\frac{-7 \pm \sqrt{73}}{4}$
(4) $\frac{7 \pm \sqrt{73}}{4}$

9 _____

10 Which ratio represents csc A in the diagram below?

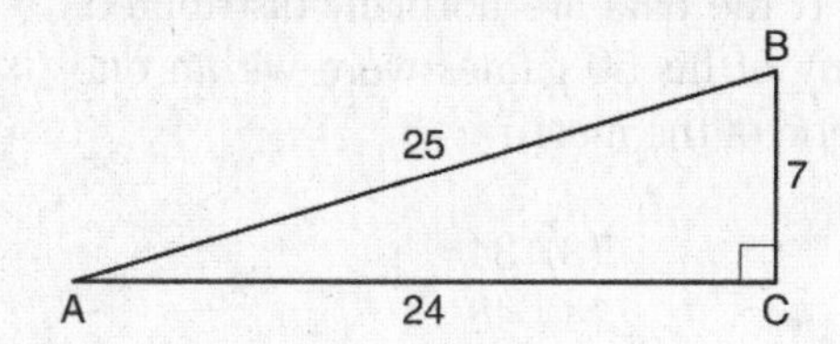

(1) $\frac{25}{24}$ (3) $\frac{24}{7}$

(2) $\frac{25}{7}$ (4) $\frac{7}{24}$ 10 ______

11 When simplified, the expression $\left(\frac{w^{-5}}{w^{-9}}\right)^{\frac{1}{2}}$ is equivalent to

(1) w^{-7} (3) w^7
(2) w^2 (4) w^{14} 11 ______

12 The principal would like to assemble a committee of 8 students from the 15-member student council. How many different committees can be chosen?

(1) 120 (3) 32,432,400
(2) 6,435 (4) 259,459,200 12 ______

13 An amateur bowler calculated his bowling average for the season. If the data are normally distributed, about how many of his 50 games were within one standard deviation of the mean?

(1) 14 (3) 34
(2) 17 (4) 48

13 ____

14 What is a formula for the nth term of sequence B shown below?

$$B = 10, 12, 14, 16, \ldots$$

(1) $b_n = 8 + 2n$ (3) $b_n = 10(2)^n$
(2) $b_n = 10 + 2n$ (4) $b_n = 10(2)^{n-1}$

14 ____

15 Which values of x are in the solution set of the following system of equations?

$$y = 3x - 6$$
$$y = x^2 - x - 6$$

(1) 0, −4 (3) 6, −2
(2) 0, 4 (4) −6, 2

15 ____

16 The roots of the equation $9x^2 + 3x - 4 = 0$ are

(1) imaginary
(2) real, rational, and equal
(3) real, rational, and unequal
(4) real, irrational, and unequal

16 ____

17 In $\triangle ABC$, $a = 3$, $b = 5$, and $c = 7$. What is m$\angle C$?

(1) 22
(2) 38
(3) 60
(4) 120

17 ____

18 When $x^{-1} - 1$ is divided by $x - 1$, the quotient is

(1) -1
(2) $-\frac{1}{x}$
(3) $\frac{1}{x^2}$
(4) $\frac{1}{(x-1)^2}$

18 ____

19 The fraction $\frac{3}{\sqrt{3a^2b}}$ is equivalent to

(1) $\frac{1}{a\sqrt{b}}$
(2) $\frac{\sqrt{b}}{ab}$
(3) $\frac{\sqrt{3b}}{ab}$
(4) $\frac{\sqrt{3}}{a}$

19 ____

20 Which graph represents a one-to-one function?

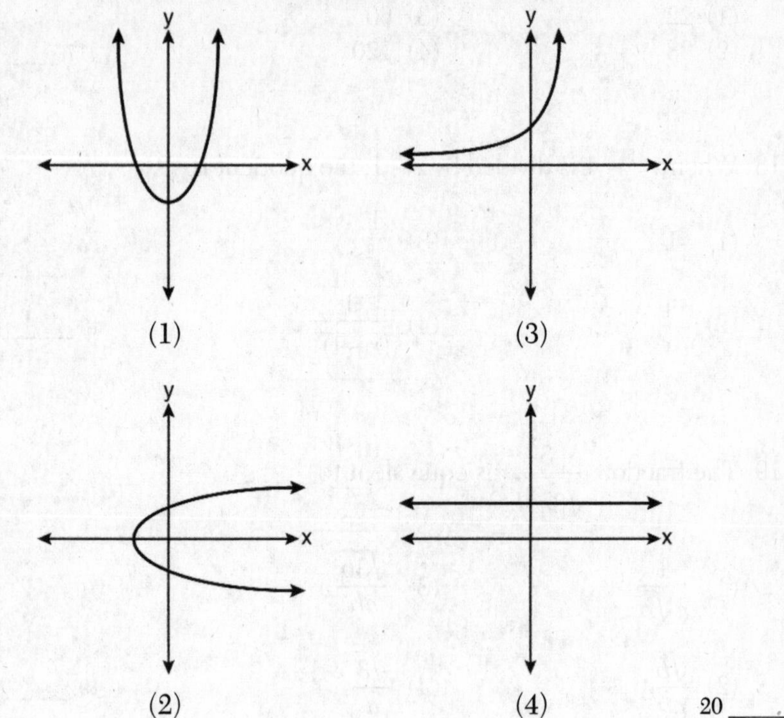

20 _____

21 The sides of a parallelogram measure 10 cm and 18 cm. One angle of the parallelogram measures 46 degrees. What is the area of the parallelogram, to the *nearest square centimeter*?

(1) 54 (3) 129
(2) 125 (4) 162

21 _____

22 The minimum point on the graph of the equation $y = f(x)$ is $(-1,-3)$. What is the minimum point on the graph of the equation $y = f(x) + 5$?

(1) $(-1,2)$ (3) $(4,-3)$
(2) $(-1,-8)$ (4) $(-6,-3)$

22 _____

23 The graph of $y = x^3 - 4x^2 + x + 6$ is shown below.

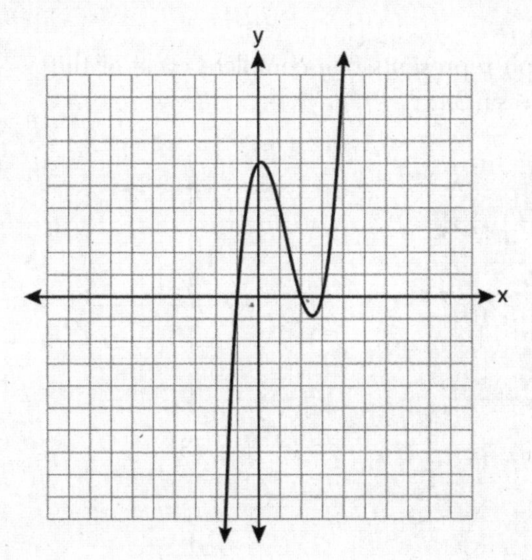

What is the product of the roots of the equation $x^3 - 4x^2 + x + 6 = 0$?

(1) -36 (3) 6
(2) -6 (4) 4

23 _____

24 What is the conjugate of $-2 + 3i$?

(1) $-3 + 2i$ (3) $2 - 3i$
(2) $-2 - 3i$ (4) $3 + 2i$

24 _____

25 What is the common ratio of the geometric sequence whose first term is 27 and fourth term is 64?

(1) $\frac{3}{4}$ (3) $\frac{4}{3}$

(2) $\frac{64}{81}$ (4) $\frac{37}{3}$

25 ______

26 Which graph represents one complete cycle of the equation $y = \sin 3\pi x$?

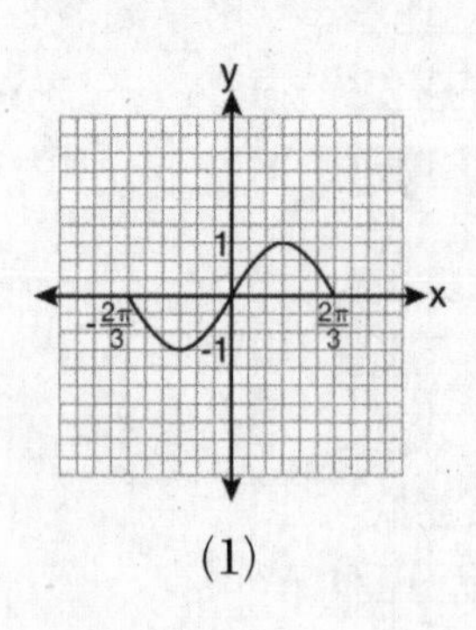

(1)

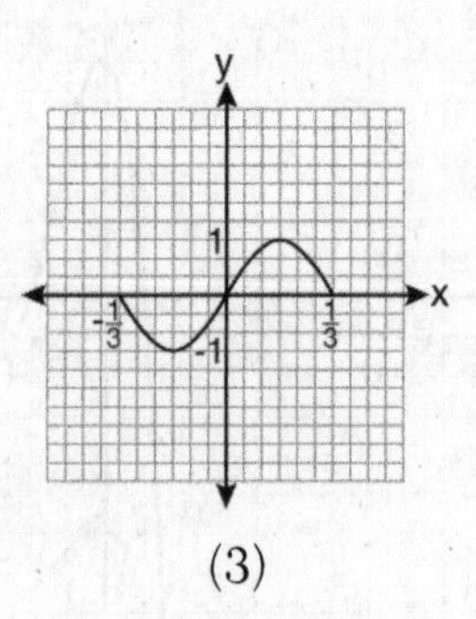

(3)

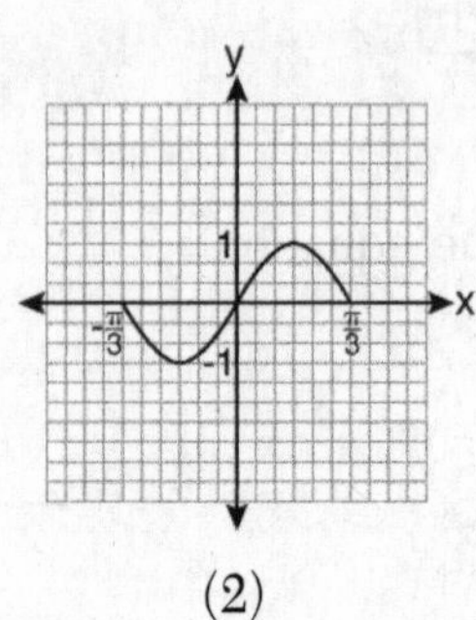

(2)

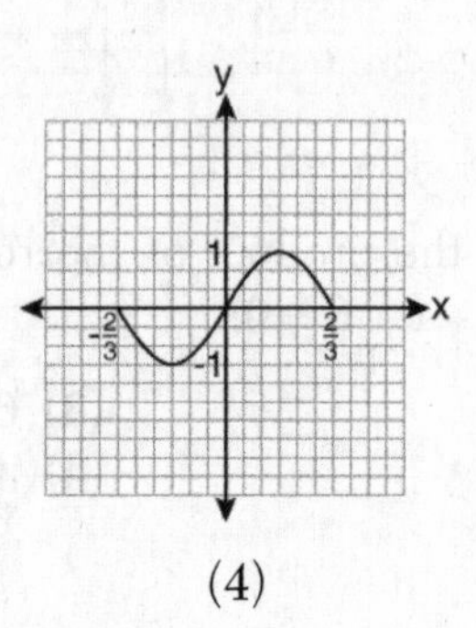

(4)

26 ______

27 Which two functions are inverse functions of each other?

(1) $f(x) = \sin x$ and $g(x) = \cos x$
(2) $f(x) = 3 + 8x$ and $g(x) = 3 - 8x$
(3) $f(x) = e^x$ and $g(x) = \ln x$
(4) $f(x) = 2x - 4$ and $g(x) = -\frac{1}{2}x + 4$

27 ______

PART II

Answer all 8 questions in this part. Each correct answer will receive 2 credits. Clearly indicate the necessary steps, including appropriate formula substitutions, diagrams, graphs, charts, etc. For all questions in this part, a correct numerical answer with no work shown will receive only 1 credit. [16 credits]

28 Factor completely: $10ax^2 - 23ax - 5a$

29 Express the sum $7 + 14 + 21 + 28 + \ldots + 105$ using sigma notation.

30 Howard collected fish eggs from a pond behind his house so he could determine whether sunlight had an effect on how many of the eggs hatched. After he collected the eggs, he divided them into two tanks. He put both tanks outside near the pond, and he covered one of the tanks with a box to block out all sunlight.

State whether Howard's investigation was an example of a controlled experiment, an observation, or a survey. Justify your response.

31 The table below shows the number of new stores in a coffee shop chain that opened during the years 1986 through 1994.

Year	Number of New Stores
1986	14
1987	27
1988	48
1989	80
1990	110
1991	153
1992	261
1993	403
1994	681

Using $x = 1$ to represent the year 1986 and y to represent the number of new stores, write the exponential regression equation for these data. Round all values to the *nearest thousandth*.

32 Solve the equation $2 \tan C - 3 = 3 \tan C - 4$ algebraically for all values of C in the interval $0° \leq C < 360°$.

$2\tan C - 3 = 3\tan C - 4$

$-3 = \tan C - 4$

$1 = \tan C$

$\tan^{-1}(1) = 45°$

$45 + 180$

225

S A

T C

33 A circle shown in the diagram below has a center of (–5,3) and passes through point (–1,7).

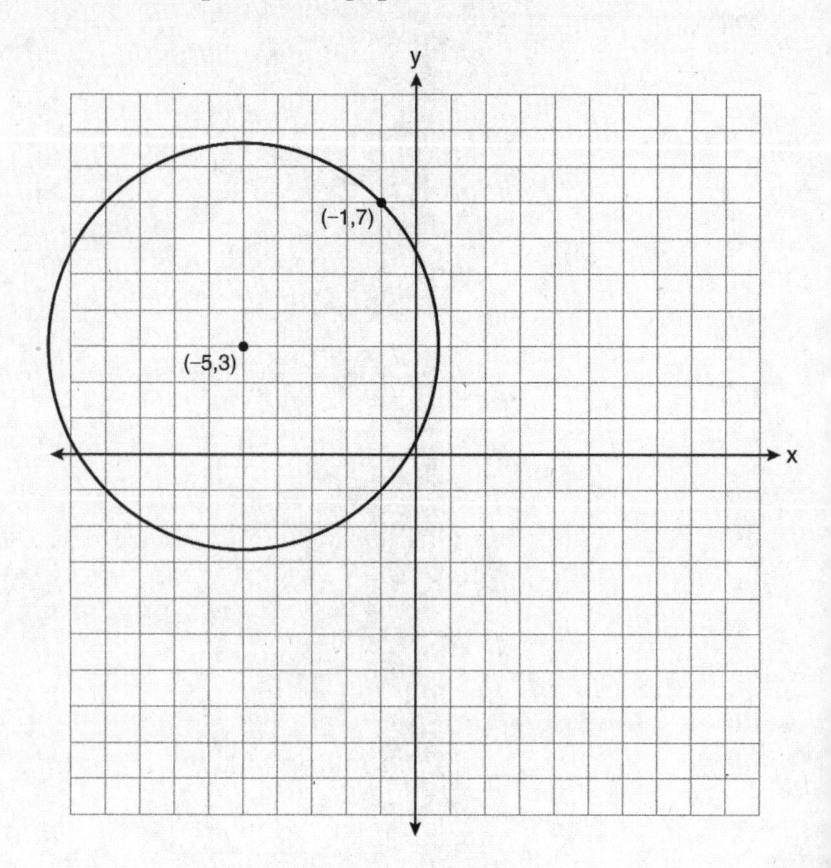

Write an equation that represents the circle.

34 Express $\left(\frac{2}{3}x-1\right)^2$ as a trinomial.

$\left(\frac{2}{3}x-1\right)\left(\frac{2}{3}x-1\right)$

$\frac{4}{9}x^2-\frac{2}{3}x-\frac{2}{3}x+1$

$\frac{4}{9}x^2-\frac{4}{3}+1$

35 Find the total number of different twelve-letter arrangements that can be formed using the letters in the word *PENNSYLVANIA*.

$\frac{12!}{3!\,2!}$

PART III

Answer all 3 questions in this part. Each correct answer will receive 4 credits. Clearly indicate the necessary steps, including appropriate formula substitutions, diagrams, graphs, charts, etc. For all questions in this part, a correct numerical answer with no work shown will receive only 1 credit. [12 credits]

36 Solve algebraically for x: $\frac{1}{x+3} - \frac{2}{3-x} = \frac{4}{x^2-9}$

37 If $\tan A = \frac{2}{3}$ and $\sin B = \frac{5}{\sqrt{41}}$ and angles A and B are in Quadrant I, find the value of $\tan (A + B)$.

38 A study shows that 35% of the fish caught in a local lake had high levels of mercury. Suppose that 10 fish were caught from this lake. Find, to the *nearest tenth of a percent*, the probability that *at least* 8 of the 10 fish caught did *not* contain high levels of mercury.

PART IV

Answer the question in this part. The correct answer will receive 6 credits. Clearly indicate the necessary steps, including appropriate formula substitutions, diagrams, graphs, charts, etc. A correct numerical answer with no work shown will receive only 1 credit. [6 credits]

39 Solve algebraically for x: $\log_{x+3} \frac{x^3 + x - 2}{x} = 2$

Answers August 2010

Algebra 2/Trigonometry

Answer Key

PART I

1. (4)	**6.** (3)	**11.** (2)	**16.** (4)	**21.** (3)	**26.** (3)
2. (1)	**7.** (3)	**12.** (2)	**17.** (4)	**22.** (1)	**27.** (3)
3. (2)	**8.** (4)	**13.** (3)	**18.** (2)	**23.** (2)	
4. (1)	**9.** (3)	**14.** (1)	**19.** (3)	**24.** (2)	
5. (4)	**10.** (2)	**15.** (2)	**20.** (3)	**25.** (3)	

PART II

28. $a(2x-5)(5x+1)$

29. $\sum_{n=1}^{15} 7n$ or $7\sum_{n=1}^{15} n$ or an equivalent expression.

30. Controlled experiment with an appropriate explanation.

31. $y = (10.596)(1.586)^x$

32. 45° and 225°

33. $(x+5)^2 + (y-3)^2 = 32$ or an equivalent equation.

34. $\frac{4}{9}x^2 - \frac{4}{3}x + 1$

35. 39,916,800

PART III

36. $\frac{1}{3}$

37. $\frac{23}{2}$ or 11.5

38. 26.2%

PART IV

39. $-\frac{1}{3}$ and -1

Answers Explained

PART I

1. Use the distributive property to multiply two binomials.

$$\left(3+\sqrt{5}\right)\left(3-\sqrt{5}\right)=9+3\sqrt{5}-3\sqrt{5}-5$$

$$=4$$

This problem could also have been done on the calculator.
The correct choice is **(4)**.

2. To convert degrees to radians, multiply by $\frac{\pi}{180^\circ}$.

$$\left(-420^\circ\right)\cdot\frac{\pi}{180^\circ}=\frac{-420\pi}{180}$$

$$=-\frac{7\pi}{3}$$

CALC Check

The conversion can also be done on your calculator. Set the calculator to radian mode (the mode you want for the answer). Include the degree sign (found in the ANGLE menu) when you enter the angle. The answer will be a decimal; the choices are all in terms of π. You could evaluate each choice to see which one equals –7.330382858. Alternatively, you could divide the calculator answer by π and convert the result to a fraction. If you do it this way, remember that the correct answer is $-\frac{7\pi}{3}$, not $-\frac{7}{3}$.

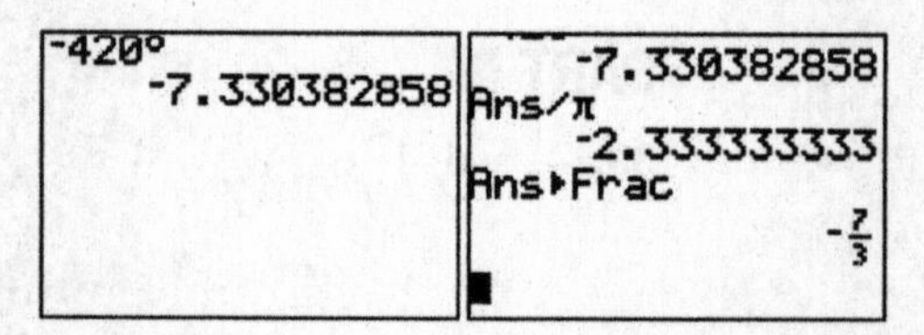

The correct choice is **(1)**.

3. The domain of a function is the set of all possible inputs, x-values, that the function can have. When looking at the graph, you see the smallest x-value is the leftmost point on the graph, where $x = -4$. The solid dot on the graph indicates that $x = -4$ is included in the domain. The arrow on the right end of the graph tells you x has no upper limit. Thus the domain is $\{x \mid x \geq -4\}$. The range is the set of all possible outputs, or y-values that match the graph. The range is $\{y \mid y \geq 2\}$.

The correct choice is **(2)**.

4. You should know that $i^2 = -1$. You can then determine that $i^3 = i^2 \cdot i = -1i = -i$. Substituting those values into the given expression yields $2i^2 + 3i^3 = 2(-1) + 3(-i) = -2 - 3i$.

CALC Check

You can also type the expression into your calculator.

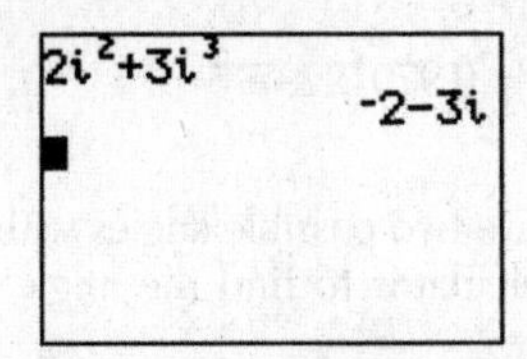

The correct choice is **(1)**.

5. A negative angle in standard position starts at the x-axis and rotates in a clockwise direction. Starting from the positive x-axis and rotating 70° clockwise leaves the terminal side of the angle in Quadrant IV with a reference angle (the acute angle with the x-axis) of 70°.

The correct choice is **(4)**.

6. Draw a rough diagram. Do not worry about scale or accuracy for this. Remember, side a is opposite $\angle A$ and side c is opposite $\angle C$.

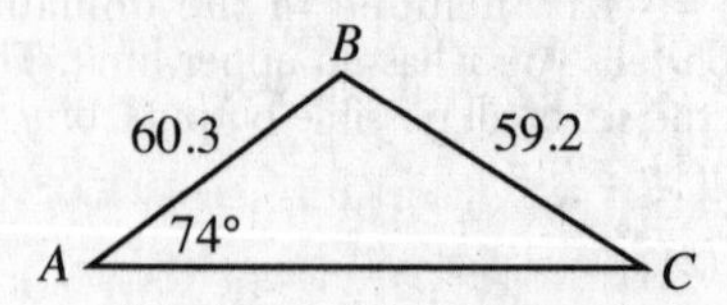

This is the ambiguous case triangle problem. Use the Law of Sines shown on the reference sheet. Set your calculator to degree mode to find possible values for m$\angle C$.

$$\frac{59.2}{\sin 74^\circ} = \frac{60.3}{\sin C}$$

$$59.2 \sin C = 60.3 \sin 74^\circ$$

$$\sin C = \frac{60.3 \sin 74^\circ}{59.2} = 0.9791229775$$

Sine is positive in Quadrants I and II. So there are two possible angles with a sine value of 0.9791229775. Use $\sin^{-1}$ on your calculator to find the angle in Quadrant I:

$$m\angle C = \sin^{-1}(0.9791229775) \approx 78.27^\circ$$

The second angle, in Quadrant II, is 180° minus that answer:

$$m\angle C = 180^\circ - 78.27^\circ = 101.73^\circ$$

When rounded to the nearest tenth, the two possible angles are 78.3° and 101.7°.

Note: When doing an ambiguous case triangle problem, you should usually check that the obtuse angle is part of a possible triangle by finding the value of the third angle of the triangle. In this case, m$\angle B$ = 180° − 74° − 101.7° = 4.3° > 0. So this triangle is possible. Since the problem stated that there were two possible angles, you could skip this last step.

The correct choice is **(3)**.

7. The inverse cosine function $\cos^{-1}\left(-\frac{\sqrt{3}}{2}\right) = A$ can be rewritten as $\cos A = -\frac{\sqrt{3}}{2}$. The reference angle is 30° because $\cos 30° = \frac{\sqrt{3}}{2}$. For negative inputs, the principal value for inverse cosine is an angle in Quadrant II. Subtract 30° from 180° to find the Quadrant II angle, 150°.

CALC Check

The easiest way to do this problem is to type it into your calculator. The calculator gives the principal value for inverse trigonometric functions. Be sure the calculator is in degree mode. Be careful to type the fraction accurately, with the "/2" outside the radical sign.

```
cos-1( -√3/2)
                150
```

The correct choice is **(3)**.

8. Since 9 and 27 are both powers of 3 ($9 = 3^2$ and $27 = 3^3$), the equation can be rewritten with common bases.

$$9^{3x+1} = 27^{x+2}$$

$$(3^2)^{3x+1} = (3^3)^{x+2}$$

$$3^{2(3x+1)} = 3^{3(x+2)}$$

When a power is raised to a power, exponents are multiplied. If two powers of the same base are equal, the exponents must be equal.

$$2(3x+1) = 3(x+2)$$

$$6x + 2 = 3x + 6$$

$$3x = 4$$

$$x = \frac{4}{3}$$

CALC Check

Use TBLSET to set the table to Ask. The table below on the right shows the answer choices listed in order. Fractions were entered, but the calculator displays them in decimal form. Choice (4) has the same value for both expressions.

```
Plot1 Plot2 Plot3
\Y1=9^(3X+1)
\Y2=27^(X+2)
\Y3=
\Y4=
\Y5=
\Y6=
```

X	Y1	Y2
1	6561	19683
.33333	81	2187
.5	243	3788
1.3333	59049	59049

X=

The correct choice is **(4)**.

9. The equation is quadratic. You can try factoring. Since two of the choices have radicals, though, you cannot be sure factoring will work. (If you choose to factor, check your results by substituting them into the original equation.) However, this equation is not factorable. Use the quadratic formula.

$$2x^2 + 7x - 3 = 0 \rightarrow a = 2, b = 7, \text{ and } c = -3$$

$$x = \frac{-b \pm \sqrt{b^2 - 4ac}}{2a}$$

$$= \frac{-7 \pm \sqrt{(7)^2 - 4(2)(-3)}}{2(2)}$$

$$= \frac{-7 \pm \sqrt{73}}{4}$$

The correct choice is **(3)**.

10. To do this problem, you must know that cosecant (csc) is the reciprocal of sine, $\csc A = \frac{1}{\sin A}$. In a right triangle, $\sin A = \frac{\text{opposite}}{\text{hypotenuse}}$. In the triangle shown, $\sin A = \frac{7}{25}$. Take the reciprocal to find csc A, $\csc A = \frac{25}{7}$.

The correct choice is **(2)**.

11. Simplify inside the parentheses first. To divide two powers with the same base, subtract the exponents.

To raise a power to a power, multiply the exponents.

$$\left(\frac{w^{-5}}{w^{-9}}\right)^{\frac{1}{2}} = \left(w^{-5-(-9)}\right)^{\frac{1}{2}} = \left(w^{4}\right)^{\frac{1}{2}}$$

$$= w^{4\left(\frac{1}{2}\right)} = w^{2}$$

CALC Check

One way to check this using the calculator is to enter the original expression and the candidate solution (using X instead of w) into Y_1 and Y_2 and then compare the tables.

Plot1 Plot2 Plot3
\Y1=(X^-5/X^-9)^1/2
\Y2=X^2
\Y3=
\Y4=
\Y5=

X	Y1	Y2
1	1	1
2	4	4
3	9	9
4	16	16
5	25	25
6	36	36
7	49	49

Press + for ΔTbl

The correct choice is **(2)**.

12. When selecting a committee, the order in which students are selected does not matter. When order does not matter, use combinations, ${}_nC_r$. You can find ${}_nC_r$ using the MATH key and the PRB submenu on your calculator. The number of possible committees of 8 that can be selected from 15 total students is ${}_{15}C_8 = 6{,}435$.

The correct choice is **(2)**.

13. Use the normal distribution chart on the reference table. The percentage of scores that will be within one standard deviation of the mean, both above and below, is 34.1% + 34.1% = 68.2%. So 68.2% of 50 games is 0.682(50) = 34.1 or about 34 games.

The correct choice is **(3)**.

14. The terms in the given sequence have a common difference: 12 – 10 = 2, 14 – 12 = 2, 16 – 14 = 2. This is an arithmetic sequence. The formula for the *n*th term of an arithmetic sequence is $a_n = a_1 + d(n-1)$, where a_1 is the first term, 10, and d is the common difference, 2. Thus the formula is $a_n = 10 + 2(n-1)$. Since that is not one of the choices, simplify it:

$$\begin{aligned} a_n &= 10 + 2(n-1) \\ &= 10 + 2n - 2 \\ &= 8 + 2n \end{aligned}$$

CALC Check

If you do not remember the formula, you can simply check all the answers in the table of your calculator. Use X for n, and see which answer choice gives the correct first four terms of the sequence. Remember that the first term corresponds to $n = 1$.

Plot1 Plot2 Plot3
\Y1=8+2X
\Y2=10+2X
\Y3=10(2)^X
\Y4=10(2)^(X-1)
\Y5=
\Y6=

X	Y1	Y2
1	10	12
2	12	14
3	14	16
4	16	18
5	18	20
6	20	22
7	22	24

Press + for ΔTbl

X	Y3	Y4
1	20	10
2	40	20
3	80	40
4	160	80
5	320	160
6	640	320
7	1280	640

Y4=10

The correct choice is **(1)**.

15. This system can be solved algebraically using substitution.

$$\begin{aligned} y &= 3x - 6 \\ y &= x^2 - x - 6 \\ x^2 - x - 6 &= 3x - 6 \\ x^2 - 4x &= 0 \\ x(x-4) &= 0 \\ x = 0 &\text{ or } x = 4 \end{aligned}$$

CALC Check

Alternatively, the system could be solved graphically on your calculator. The intersections appear to be at $x = 0$ and $x = 4$. To be precise, use CALC intersect.

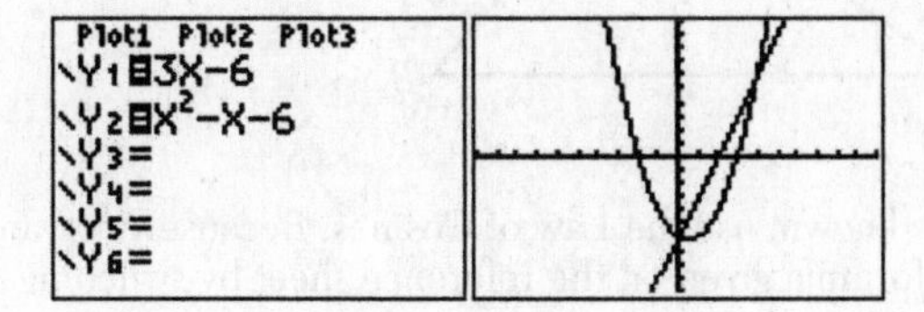

Since this is a multiple-choice problem, an easy way to solve it is simply to check all the candidate answers in the original equations. Use TBLSET to set the table to Ask. Then enter the possible x-values from the choices given. Only $x = 0$ and $x = 4$ give the same value for both functions.

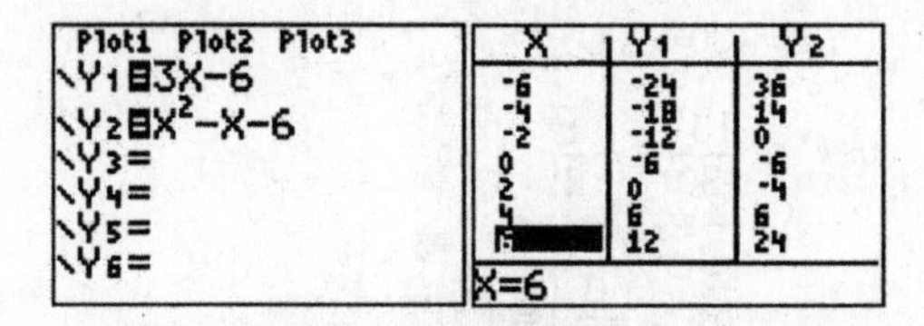

X	Y1	Y2
-6	-24	36
-4	-18	14
-2	-12	0
0	-6	-6
2	0	-4
4	6	6
6	12	24

X=6

The correct choice is **(2)**.

16. You could solve the equation using the quadratic formula. Since all you need to know is the nature of the roots, you can save time by looking at just the discriminant (the radicand in the formula), $b^2 - 4ac$. With $a = 9$, $b = 3$, and $c = -4$, the discriminant is $(3)^2 - 4(9)(-4) = 153$. Since the discriminant is positive, the equation has two distinct (unequal) real answers. Since the discriminant is not a perfect square, the answers will be irrational.

The correct choice is **(4)**.

17. Draw a diagram.

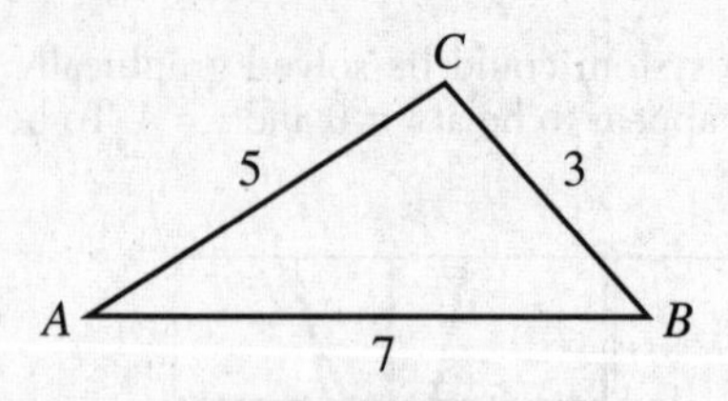

Since all three sides are known, use the Law of Cosines. Because you want to find m∠C, rewrite the formula given on the reference sheet by switching a and c and replacing A with C:

$$c^2 = a^2 + b^2 - 2ab \cos C$$

$$7^2 = 3^2 + 5^2 - 2(3)(5) \cos C$$

$$49 = 34 - 30 \cos C$$

$$15 = -30 \cos C$$

$$\cos C = \frac{15}{-30} = -\frac{1}{2}$$

$$C = \cos^{-1}\left(-\frac{1}{2}\right) = 120°$$

Be careful to use the correct order of operations. Remember that $49 = 34 - 30 \cos C$ is not $49 = 4 \cos C$!

If you remember the geometry of triangles, recall that C must be the largest angle of the triangle since it is across from the longest side. The largest angle of a triangle can never be 22° or 38°. It can be 60° only if the triangle is equilateral, which this one is not. The only possible answer is 120°.

The correct choice is **(4)**.

18. Remember, a negative exponent means a reciprocal, $\frac{x^{-1}-1}{x-1}=\frac{\frac{1}{x}-1}{x-1}$.

This complex fraction is easily simplified by multiplying the numerator and denominator of the fraction by the least common denominator, x:

$$\frac{\frac{1}{x}-1}{x-1}=\frac{\left(\frac{1}{x}-1\right)x}{(x-1)x}$$

$$=\frac{1-x}{x(x-1)}$$

$$=-\frac{1}{x}$$

Note that $1-x$ and $x-1$ are opposites; they simplify to -1. Note also that the remaining x is in the denominator. The answer is not $-x$.

Some students prefer to simplify complex fractions by rewriting the fractions with a common denominator and then dividing and reducing:

$$\frac{\frac{1}{x}-1}{x-1}=\frac{\frac{1}{x}-1\cdot\frac{x}{x}}{x-1}$$

$$=\frac{\frac{1-x}{x}}{\frac{x-1}{1}}$$

$$=\frac{\overset{-1}{\cancel{1-x}}}{x}\cdot\frac{1}{\cancel{x-1}}$$

$$=-\frac{1}{x}$$

CALC Check

You can check this answer by entering the original expression and the candidate solution into your calculator. Then see that they give the same results. Note that the expressions agree for all values of x for which they are both defined. Neither expression is defined for $x = 0$, and the original expression is not defined for $x = 1$.

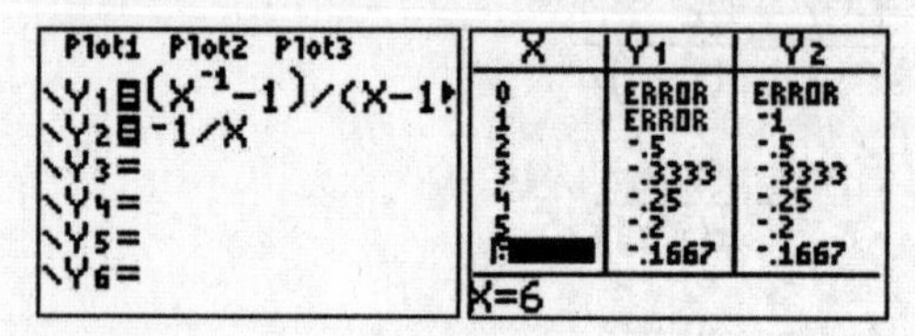

The correct choice is **(2)**.

19. Rationalize the denominator and simplify:

$$\frac{3}{\sqrt{3a^2b}} = \frac{3}{\sqrt{3a^2b}} \cdot \frac{\sqrt{3a^2b}}{\sqrt{3a^2b}}$$

$$= \frac{3\sqrt{3a^2b}}{3a^2b}$$

$$= \frac{3\sqrt{a^2}\sqrt{3b}}{3a^2b}$$

$$= \frac{3a\sqrt{3b}}{3a^2b}$$

$$= \frac{\sqrt{3b}}{ab}$$

The correct choice is **(3)**.

20. In a function, each input (x-value) gives only one output (y-value). For a graph, use the vertical line test. No vertical line drawn on the graph can intersect a function in more than one point. This eliminates choice (2), which is not a function. For a function to be one-to-one, each output (y-value) must come from only one input (x-value). A graph that is one-to-one must pass the horizontal line test. No horizontal line drawn on the graph can intersect the function in more than one point. Choices (1) and (4) are not one-to-one because a horizontal line can intersect each graph more than once. For choice (4), one horizontal line intersects the graph in an infinite number of points. The only choice that passes both the vertical and horizontal line tests is choice (3).

The correct choice is **(3)**.

21. Draw a diagram.

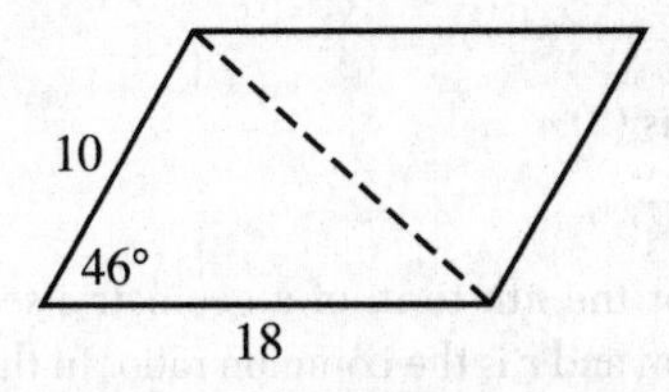

Find the area of the parallelogram by recognizing that it is twice the area of one of the triangles created by drawing the diagonal as shown in the diagram. The formula for the area of a triangle is given on the reference sheet: $K = \frac{1}{2}ab \sin C$. The actual labeling of the sides and the angle are not important as long as the angle in the formula is the angle included between the two sides. Set your calculator to degree mode to calculate the area of the parallelogram:

$$\text{Area} = 2\left(\frac{1}{2}ab\sin C\right)$$

$$= 2\left(\frac{1}{2}\right)(18)(10)\sin 46°$$

$$= 129.48$$

To the nearest square centimeter, the area of the parallelogram is 129 cm^2.

The correct choice is **(3)**.

22. The transformation f(x) + 5 is a vertical translation of the graph of f(x) up 5 units. In a vertical translation, the x-values will not change. The y-values will all increase by 5. The point (–1, –3) will be translated to (–1, –3 + 5), which is (–1, 2).

The correct choice is **(1)**.

23. The roots of the equation $x^3 - 4x^2 + x + 6 = 0$ are the x-intercepts of the graph of $y = x^3 - 4x^2 + x + 6$. From the graph, the equation has roots at $x = -1$, $x = 2$, and $x = 3$. The product of the roots is $(-1)(2)(3) = -6$.

The correct choice is **(2)**.

24. The conjugate of the complex number $a + bi$ is $a - bi$. Keep the sign of the real part and change the sign of the imaginary part. Thus, the conjugate of $-2 + 3i$ is $-2 - 3i$.

The correct choice is **(2)**.

25. The formula for the nth term of a geometric sequence is $a_n = a_1r^{n-1}$, where a_1 is the first term and r is the common ratio. In this problem, $a_1 = 27$. So $a_n = 27r^{n-1}$. The fourth term, 64, corresponds to $n = 4$:

$$64 = 27r^{4-1}$$

$$r^3 = \frac{64}{27}$$

$$r = \sqrt[3]{\frac{64}{27}}$$

$$= \frac{4}{3}$$

If you do not remember the formula for the nth term of a geometric sequence, you can use reasoning if you understand the idea of a geometric sequence. Each new term is r times the preceding term. So $a^2 = a_1r$, $a_3 = a_2r = (a_1r)r = a_1r^2$, and $a_4 = a_3r = (a_1r^2)r = a_1r^3$. Then substitute in the given values of a_1 and a_4 and solve as above.

CALC Check

Alternatively, you can check the choices on the calculator. The common ratio is the multiplier. Since 27 is the first term, multiply by the ratio three times to find the fourth term of 64.

```
27*(4/3)
                36
Ans*(4/3)
                48
Ans*(4/3)
                64
```

The correct choice is **(3)**.

26. The formula for period of a sine graph, $y = a \sin bx$, is period $= \dfrac{2\pi}{b}$. In this problem, $b = 3\pi$, so the period is $\dfrac{2\pi}{b} = \dfrac{2\pi}{3\pi} = \dfrac{2}{3}$. As seen in the graphs, the first positive root of $y = a \sin bx$ is only half a period from the origin. In this problem, the first root should be at $\dfrac{1}{2}\left(\dfrac{2}{3}\right) = \dfrac{1}{3}$. This is shown in choice (3).

CALC Check

Alternatively, you could graph the function on your calculator. Make sure you are in radian mode. Then find the value of the first positive root. In the graph below, the window used was $-1 \leq x \leq 1$ and $-1 \leq y \leq 1$.

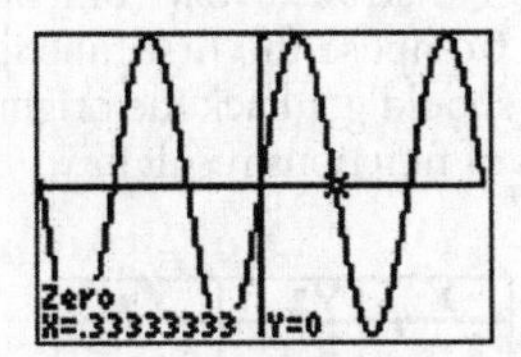

The correct choice is **(3)**.

27. The best way to do this problem is to remember that a log function is the inverse of an exponential function with the same base. The base of the natural log function, ln x, is e. So e^x is the inverse of ln x as shown in choice (3).

A more time consuming way to do the problem is to compare the graphs of each pair of functions. If two functions are inverses of each other, their graphs will be reflections of each other over the line $y = x$. If you do this, make sure your calculator is in radian mode and you use the ZDecimal or ZSquare window. In the graphs below, the second function is graphed with a thick line; all the graphs include the diagonal line $y = x$. Only choice (3) shows a pair of functions that are reflections of each other over $y = x$.

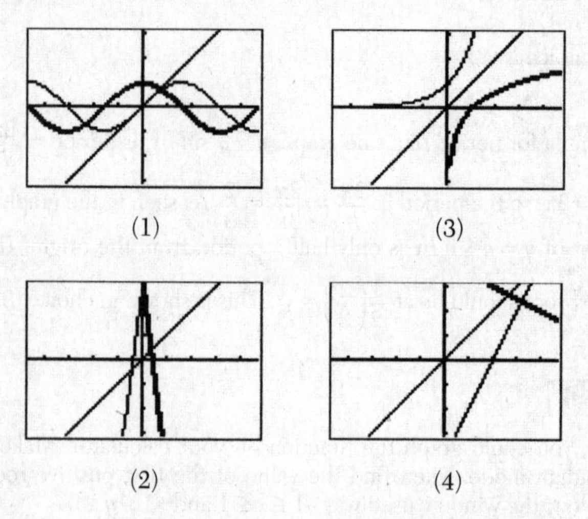

A third way to do the problem is to remember that an inverse function should undo whatever the original function does. Compose the first function with the second function. If they are inverses, you should get back the original value of x. Only in choice (3) does the composition of functions result in x.

```
Plot1 Plot2 Plot3
\Y1=sin(cos(X))
\Y2=3+8(3-8X)
\Y3=e^(ln(X))
\Y4=2(-1/2X+4)-4
\Y5=
\Y6=
```

X	Y1	Y2
1	.5144	-37
2	-.4042	-101
3	-.836	-165
4	-.6081	-229
5	.27987	-293
6	.81929	-357
7	.68449	-421

X=7

X	Y3	Y4
1	1	3
2	2	2
3	3	1
4	4	0
5	5	-1
6	6	-2
7	7	-3

Y4=-3

The correct choice is **(3)**.

PART II

28. First factor out the greatest common factor, which is a. Then factor the remaining trinomial:

$$\begin{aligned} 10ax^2 - 23ax - 5a &= a(10x^2 - 23x - 5) \\ &= a(5x + 1)(2x - 5) \end{aligned}$$

CALC Check

You can check your factoring of the trinomial $10x^2 - 23x - 5$ using your calculator.

Plot1 Plot2 Plot3
\Y1=10X²-23X-5
\Y2=(5X+1)(2X-5)
\Y3=
\Y4=
\Y5=
\Y6=

X	Y1	Y2
0	-5	-5
1	-18	-18
2	-11	-11
3	16	16
4	63	63
5	130	130
6	217	217

Press + for ΔTbl

$10ax^2 - 23ax - 5a$ factors to $a(5x + 1)(2x - 5)$.

29. Each term in the series is a multiple of 7. So each term can be represented by the expression $7n$ for an integer n. For the first term, 7, $n = 1$. For the last term, 105, $n = \frac{105}{7} = 15$. Thus the index n goes from 1 to 15. The series can be written as $\sum_{n=1}^{15} 7n$. Alternatively, you can factor out 7 from each term in the sum, $7 + 14 + 21 + \ldots + 105 = 7(1 + 2 + 3 + \ldots + 15) = 7\sum_{n=1}^{15} n$.

$$7 + 14 + 21 + \ldots + 105 = \sum_{n=1}^{15} 7n \text{ or } 7\sum_{n=1}^{15} n$$

30. A controlled experiment compares the results from two groups; the experimental condition is changed for only one of the groups. This describes what Howard did. One tank of fish was left uncovered near the pond to approximate actual conditions in the pond. This was the control group. The other tank was covered to block out sunlight. This was the experimental group.

In an observation, Howard would have observed the fish eggs without changing the conditions of any of them. In other words, he would not have placed them into tanks or covered them.

In a survey, Howard would have asked the eggs questions to get their opinions on the effect of sunlight on egg hatching. Doing this might indicate that Howard has spent too much time in the sun.

Howard's investigation was a controlled experiment.

31. The problem says to let $x = 1$ represent 1986. This means the given data will have $x = 1, 2, 3, \ldots, 9$. If you did not get the correct regression equation, check your x-values first. Enter these x-values into list L1 on your graphing calculator and the corresponding y-values into list L2. Double check your data; a single typing error will result in the wrong regression equation. Use the exponential regression function on your calculator.

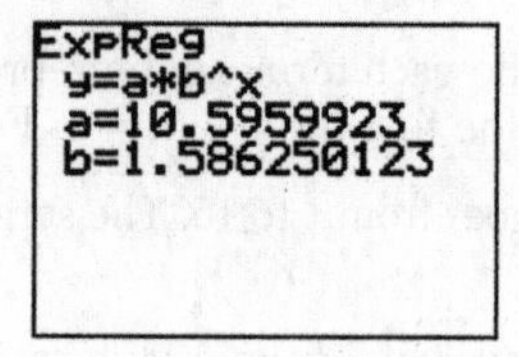

Write your final answer in the correct form, $y = ab^x$. Round the coefficients to the nearest thousandth, $y = 10.596(1.586)^x$.

CALC Caution

If you do not follow the directions and forget to change the x-values so that 1986 is $x = 1$, you will have x-values that are too large for the exponential regression in the calculator. Your calculator screen will look like the one below. This is a hint from your calculator to read the problem more closely.

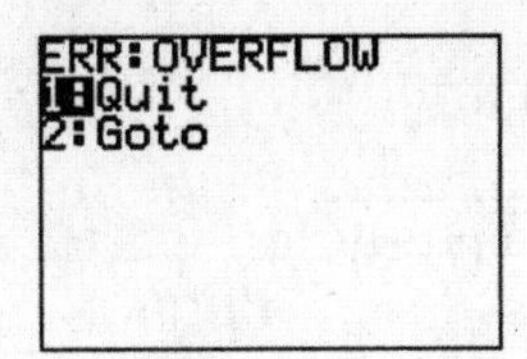

To the nearest thousandth, the exponential regression equation is $y = 10.596(1.586)^x$.

32. Solve for the trigonometric function, tan C:

$$2 \tan C - 3 = 3 \tan C - 4$$
$$-3 = \tan C - 4$$
$$1 = \tan C$$

Tangent is positive in Quadrants I and III. In Quadrant I, $C = \tan^{-1}(1) = 45°$. In Quadrant III, $C = 180° + 45° = 225°$.

CALC Check

To check this on your calculator, set your calculator to degree mode and your window to $0 \le x \le 360$ and $-10 \le y \le 10$. Use CALC Intersect or TABLE Ask.

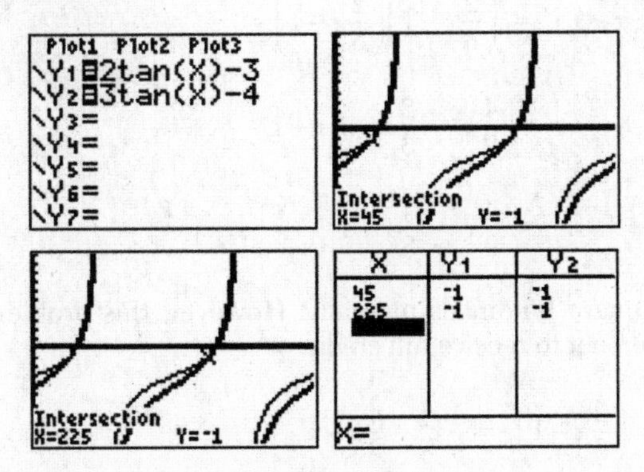

The solutions are 45° and 225°.

33. To write the equation of a circle, you need to know the radius and the coordinates of the center. The center is marked on the graph, (–5, 3). To find the radius, use the distance formula to find the length between the center and the known point on the circle, (–1, 7). Do not try to estimate the radius simply by counting boxes; you will not be precise. Note that you can count boxes to find Δx and Δy in the distance formula.

$$r = \sqrt{(\Delta x)^2 + (\Delta y)^2}$$

$$= \sqrt{(-1-(-5))^2 + (7-3)^2}$$

$$= \sqrt{4^2 + 4^2}$$

$$= \sqrt{32}$$

The equation of a circle is $(x-h)^2 + (y-k)^2 = r^2$, where (h, k) are the coordinates of the center and r is the radius.

The equation of the given circle is $(x-(-5))^2 + (y-3)^2 = (\sqrt{32})^2$. This simplifies to $(x+5)^2 + (y-3)^2 = 32$.

The equation of the circle is $(x+5)^2 + (y-3)^2 = 32$.

34. Write the square as the product of two binomials, and use the distributive property. This is sometimes taught as FOIL for binomials.

$$\left(\frac{2}{3}x-1\right)^2 = \left(\frac{2}{3}x-1\right)\left(\frac{2}{3}x-1\right)$$

$$= \frac{4}{9}x^2 - \frac{2}{3}x - \frac{2}{3}x + 1$$

$$= \frac{4}{9}x^2 - \frac{4}{3}x + 1$$

Some students can square binomials mentally. However, this problem requires work to be written out to receive full credit.

$$\left(\frac{2}{3}x-1\right)^2 = \frac{4}{9}x^2 - \frac{4}{3}x + 1$$

35. An arrangement of objects where order matters is a permutation. *PENNSYLVANIA* has 12 letters. However, it includes two repeated letters, 3 *N*s and 2 *A*s. When letters are repeated, the number of different permutations is found by dividing the total number of permutations by the number of possible permutations of each of the repeated letters:

$$\frac{12!}{(3!)(2!)} = 39{,}916{,}800$$

CALC Caution

If you are using a TI-83 or an older TI-84 that has not been updated, be careful to use parentheses in the denominator or you will get the wrong answer as shown in the screen on the left. The new MathPrint feature on the TI-84 makes it easy to calculate accurately, as shown in the screen on the right.

```
12!/3!2!
         159667200
```

```
12!
3!2!
          39916800
```

PART III

36. The simplest solution requires that you recognize that $3 - x$ is the opposite of $x - 3$: $3 - x = -(x - 3)$. Then rewrite the equation:

$$\frac{1}{x+3} - \frac{2}{3-x} = \frac{4}{x^2 - 9}$$

$$\frac{1}{x+3} - \frac{2}{-(x-3)} = \frac{4}{x^2 - 9}$$

$$\frac{1}{x+3} + \frac{2}{x-3} = \frac{4}{x^2 - 9}$$

To solve a rational equation, factor the denominators.

$$\frac{1}{x+3} + \frac{2}{x-3} = \frac{4}{(x+3)(x-3)}$$

Next get common denominators for all the terms in the equation. Remember that x cannot equal either 3 or -3 because the denominator cannot equal zero.

$$\frac{(x-3)}{(x-3)}\frac{1}{x+3} + \frac{(x+3)}{(x+3)}\frac{2}{x-3} = \frac{4}{(x+3)(x-3)}$$

Set the numerators equal to each other.

$$1(x-3) + 2(x+3) = 4$$

Instead of finding a common denominator, you can multiply the equation by the least common denominator of all the terms. Careful simplification clears all the fractions. Remember that x cannot equal either 3 or -3.

$$(x+3)(x-3)\left(\frac{1}{x+3} + \frac{2}{x-3} = \frac{4}{(x+3)(x-3)}\right)$$

$$\cancel{(x+3)}(x-3)\frac{1}{\cancel{(x+3)}} + (x+3)\cancel{(x-3)}\frac{2}{\cancel{(x-3)}} = \cancel{(x+3)}\cancel{(x-3)}\left(\frac{4}{\cancel{(x+3)}\cancel{(x-3)}}\right)$$

Both methods yield the same equation to solve.

$$1(x-3)+2(x+3)=4$$
$$x-3+2x+6=4$$
$$3x+3=4$$
$$3x=1$$
$$x=\frac{1}{3}$$

Finally, use your calculator to check the candidate solution in the original equation.

$$\frac{1}{\left(\frac{1}{3}\right)+3}-\frac{2}{3-\left(\frac{1}{3}\right)}=\frac{4}{\left(\frac{1}{3}\right)^2-9}$$

$$-0.45=-0.45 \checkmark$$

The solution is $x=\frac{1}{3}$.

If you do not simplify the equation first, solving it is more complicated but still possible. First factor the denominators.

$$\frac{1}{x+3}-\frac{2}{3-x}=\frac{4}{x^2-9}$$

$$\frac{1}{x+3}-\frac{2}{3-x}=\frac{4}{(x+3)(x-3)}$$

Get common denominators for all terms in the equation.

$$\frac{(3-x)(x-3)}{(3-x)(x-3)}\frac{1}{x+3}-\frac{(x+3)(x-3)}{(x+3)(x-3)}\frac{2}{3-x}=\frac{(3-x)}{(3-x)}\frac{4}{(x+3)(x-3)}$$

Set the numerators equal to each other.

$$(3-x)(x-3)-2(x+3)(x-3)=4(3-x)$$

Instead of finding a common denominator, you can multiply the original equation by the least common denominator of all the terms.

$$(x+3)(3-x)(x-3)\left(\frac{1}{x+3}-\frac{2}{3-x}=\frac{4}{(x+3)(x-3)}\right)$$

$$\cancel{(x+3)}(3-x)(x-3)\frac{1}{\cancel{(x+3)}}-(x+3)\cancel{(3-x)}(x-3)\frac{2}{\cancel{(3-x)}}$$

$$=\cancel{(x+3)}(3-x)\cancel{(x-3)}\left(\frac{4}{\cancel{(x+3)}\cancel{(x-3)}}\right)$$

Again, both methods yield the same equation.

$$(3-x)(x-3)-2(x+3)(x-3)=4(3-x)$$

$$3x-9-x^2+3x-2(x^2-9)=12-4x$$

$$-3x^2+6x+9=12-4x$$

Put the equation in standard form.

$$3x^2-10x+3=0$$

Factor and solve.

$$(3x-1)(x-3)=0$$

$$x=\frac{1}{3} \text{ or } x=3$$

Use your calculator to check $x=\frac{1}{3}$ in the original equation.

$$\frac{1}{\left(\frac{1}{3}\right)+3}-\frac{2}{3-\left(\frac{1}{3}\right)}=\frac{4}{\left(\frac{1}{3}\right)^2-9}$$

$$-0.45=-0.45\ \checkmark$$

Use your calculator to check $x = 3$ in the original equation.

$$\frac{1}{3+3} - \frac{2}{3-(3)} = \frac{4}{(3)^2 - 9}$$

$$\text{Undefined} = \text{Undefined} \quad \text{NO}$$

Only $x = \dfrac{1}{3}$ checks.

The solution is $x = \dfrac{1}{3}$.

CALC Check

In addition to using a calculator to check your answers, you could use a calculator to solve this equation graphically. A graphical solution was worth partial, but not full, credit.

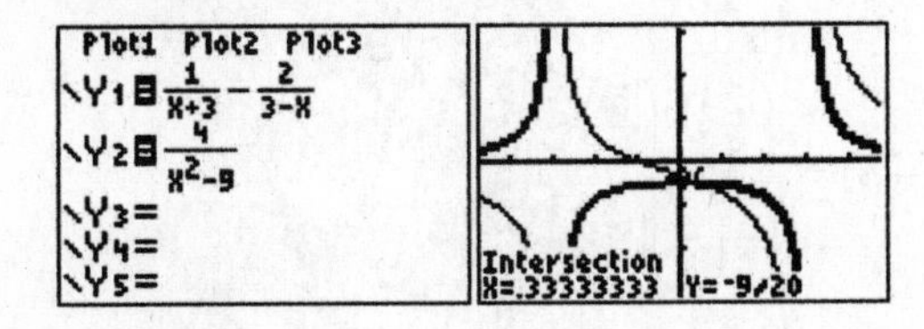

The solution is $x = \dfrac{1}{3}$.

37. The problem uses the formula for $\tan(A + B)$ shown on the reference sheet. To use the formula, you need to know $\tan A = \frac{2}{3}$, which was given, and $\tan B$, which was not given. To find $\tan B$ from $\sin B$, draw a diagram of a right triangle with acute angle B. Label the opposite and hypotenuse with the values given for $\sin B$. Then use the Pythagorean theorem to find the adjacent side. Angle B is in Quadrant I, so $\tan B = \frac{5}{4}$.

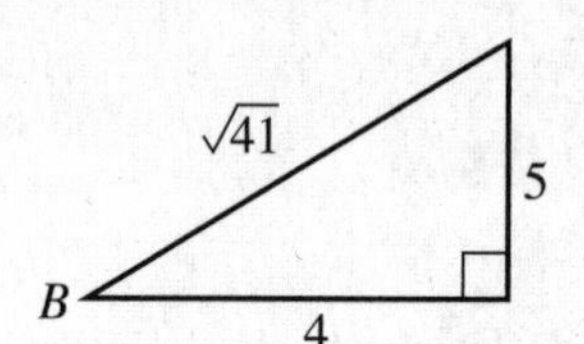

The reference sheet shows $\tan(A+B) = \dfrac{\tan A + \tan B}{1 - \tan A \tan B}$. Substitute for $\tan A$ and $\tan B$:

$$\tan(A+B) = \frac{\left(\frac{2}{3}\right) + \left(\frac{5}{4}\right)}{1 - \left(\frac{2}{3}\right)\left(\frac{5}{4}\right)}$$

$$= 11.5$$

$$= \frac{23}{2}$$

Alternatively, you can do this problem on your calculator. First find angles A and B using inverse trigonometric functions. Then find $\tan(A + B)$. For this problem, it does not matter which mode you choose. Many students prefer using degrees, as shown below.

```
tan-1(2/3)
      33.69006753
Ans→A
      33.69006753
```

```
sin-1(5/√41)
      51.34019175
Ans→B
      51.34019175
```

```
tan(A+B)
              11.5
```

Be sure to write out the steps shown on the calculator on your paper to receive full credit.

$\tan(A+B) = 11.5$ or $\frac{23}{2}$

38. This is an example of the binomial distribution. To find the probability that at least 8 of the 10 fish did not contain high levels of mercury, you must add the individual probabilities that 8, 9, or 10 fish did not contain high levels of mercury. Use the formula

$${}_{\text{number of trials}}C_{\text{number of successes}}(\text{probability of success})^{\text{number of successes}}(\text{probability of failure})^{\text{number of failures}}.$$

In the given problem, there are 10 trials (fish caught). There are 8, 9, or 10 successes (did not contain high levels of mercury) and 2, 1, or 0 failures (did contain high levels of mercury). The probability of failure, having high levels of mercury, was given as 35% = 0.35. The probability of success is 1 – 0.35 = 0.65. Let A stand for at least 8 of the fish did not contain high levels of mercury.

$$P(A) = {}_{10}C_8(0.65)^8(0.35)^2 + {}_{10}C_9(0.65)^9(0.35)^1 + {}_{10}C_{10}(0.65)^{10}(0.35)^0 = 0.2616$$

The question asked for the answer as a percent rounded to the nearest tenth. So 0.2616 = 26.16% = 26.2% to the nearest tenth.

To the nearest tenth of a percent, the probability that at least 8 of the fish did not contain high levels of mercury is 26.2%.

PART IV

39. The key to this problem is to remember that the equation $\log_b a = n$ is equivalent to the equation $b^n = a$. In words, $n = \log_b a$ says that n is the exponent that goes on base b to give the answer a.

$$\log_{x+3}\frac{x^3+x-2}{x} = 2$$

$$(x+3)^2 = \frac{x^3+x-2}{x}$$

Multiply both sides by x. $\quad x(x+3)^2 = x^3 + x - 2$

Square $x + 3$. $\quad x(x^2+6x+9) = x^3 + x - 2$

Distribute x on the left side. $\quad x^3 + 6x^2 + 9x = x^3 + x - 2$

Write in standard form. $\quad 6x^2 + 8x + 2 = 0$

Divide by the common factor of 2. $\quad 3x^2 + 4x + 1 = 0$

Factor and set each factor equal to 0. $\quad (3x+1)(x+1) = 0$

Solve for x. $\quad x = -\frac{1}{3}$ or $x = -1$

Because this is a log equation, check these answers to ensure that neither involves taking a log of a nonpositive number, which would be imaginary. A log function should also have a positive base.

For $x = -\frac{1}{3}$, the base is $-\frac{1}{3} + 3 = \frac{8}{3} > 0$. The argument is

$$\frac{\left(-\frac{1}{3}\right)^3 + \left(-\frac{1}{3}\right) - 2}{-\frac{1}{3}} = \frac{64}{9} > 0.$$

$$\log_{\frac{8}{3}}\left(\frac{64}{9}\right) = 2$$

$$\left(\frac{8}{3}\right)^2 = \frac{64}{9}$$

$$\frac{64}{9} = \frac{64}{9} \checkmark$$

For $x = -1$, the base is $-1 + 3 = 2 > 0$. The argument is $\dfrac{(-1)^3 + (-1) - 2}{-1} = 4 > 0$.

$$\log_2(4) = 2$$

$$2^2 = 4$$

$$4 = 4 \checkmark$$

CALC Check

In addition to using a calculator to check your answers, you could use a calculator to solve this equation graphically. A graphical solution was worth partial, but not full, credit.

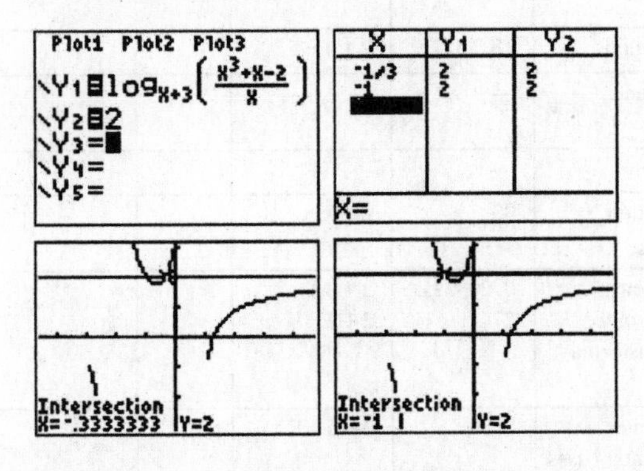

The solution set is $\left\{-\frac{1}{3}, -1\right\}$.

Topic	Question Numbers	Number of Points	Your Points	Your Percentage
1. Exponents and Radicals: operations, equivalent expressions, simplifying, rationalizing	11, 19	$2+2=4$		
2. Complex Numbers: operations, powers of i, rationalizing	1, 4, 24	$2+2+2=6$		
3. Quadratics and Higher-Order Polynomials: operations, factoring, binomial expansion, formula, quadratic equations and inequalities, nature of the roots, sum and product of the roots, quadratic-linear systems, completing the square, higher-order polynomials	9, 15, 16, 23, 28, 34	$2+2+2+$ $2+2+2=12$		
4. Rationals: operations, equations	18, 36	$2+4=6$		
5. Absolute Value: equations and inequalities	—	—		
6. Direct and Inverse Variation	—	—		
7. Circles: center-radius equation, completing the square	33	2		
8. Functions: relations, functions, domain, range, one-to-one, onto, inverses, compositions, transformations of functions	3, 20, 22, 27	$2+2+$ $2+2=8$		
9. Exponential and Logarithmic Functions: equations, common bases, logarithm rules, e, ln	8, 39	$2+6=8$		
10. Trigonometric Functions: radian measure, arc length, cofunctions, unit circle, inverses	2, 5, 7, 10	$2+2+$ $2+2=8$		
11. Trigonometric Graphs: graphs of basic functions and inverse functions, domain, range, restricted domain, transformations of graphs, amplitude, frequency, period, phase shift	26	2		
12. Trigonometric Identities, Formulas, and Equations	32, 37	$2+4=6$		
13. Trigonometry Laws and Applications: area of a triangle, sine and cosine laws, ambiguous cases	6, 17, 21	$2+2+2=6$		
14. Sequences and Series: sigma notation, arithmetic, geometric	14, 25, 29	$2+2+2=6$		

Topic	Question Numbers	Number of Points	Your Points	Your Percentage
15. Statistics: studies, central tendency, dispersion including standard deviation, normal distribution	13, 30	2 + 2 = 4		
16. Regressions: linear, exponential, logarithmic, power, correlation coefficient	31	2		
17. Probability: permutation, combination, geometric, binomial	12, 35, 38	2 + 2 + 4 = 8		

HOW TO CONVERT YOUR RAW SCORE TO YOUR ALGEBRA 2/TRIGONOMETRY REGENTS EXAMINATION SCORE

Below is the conversion chart that can be used to determine your final score on the August 2010 Regents Examination. To estimate your final exam score, locate in the column labeled "Raw Score" the total number of points you scored out of a possible 88 points. Since partial credit is allowed in Parts II, III, and IV of the test, you may need to approximate the credit you would receive for a solution that is not completely correct. Then locate in the adjacent column to the right the scaled score that corresponds to your raw score. The scaled score is your final Regents Examination score.

Raw Score	Scaled Score	Raw Score	Scaled Score	Raw Score	Scaled Score
88	**100**	58	**78**	28	**42**
87	**99**	57	**77**	27	**40**
86	**99**	56	**76**	26	**39**
85	**98**	55	**75**	25	**37**
84	**98**	54	**74**	24	**36**
83	**97**	53	**74**	23	**34**
82	**97**	52	**73**	22	**33**
81	**96**	51	**71**	21	**32**
80	**95**	50	**70**	20	**30**
79	**94**	49	**69**	19	**29**
78	**94**	48	**68**	18	**27**
77	**93**	47	**67**	17	**26**
76	**92**	46	**66**	16	**25**
75	**91**	45	**65**	15	**23**
74	**91**	44	**63**	14	**22**
73	**90**	43	**62**	13	**20**
72	**89**	42	**61**	12	**19**
71	**88**	41	**60**	11	**17**
70	**88**	40	**58**	10	**16**
69	**87**	39	**57**	9	**15**
68	**86**	38	**56**	8	**13**
67	**86**	37	**54**	7	**12**
66	**85**	36	**53**	6	**10**
65	**84**	35	**51**	5	**9**
64	**83**	34	**50**	4	**7**
63	**82**	33	**49**	3	**5**
62	**81**	32	**47**	2	**4**
61	**81**	31	**46**	1	**2**
60	**80**	30	**44**	0	**0**
59	**79**	29	**43**		